化学专业英语

(第 4 版)

马永祥　孙晓君　编

图书在版编目（CIP）数据

化学专业英语 / 马永祥，孙晓君编. -- 兰州：兰州大学出版社，2019.8（2023.7重印）
ISBN 978-7-311-05648-3

Ⅰ.①化… Ⅱ.①马… ②孙… Ⅲ.①化学－英语－高等学校－教材 Ⅳ.①O6

中国版本图书馆CIP数据核字(2019)第161671号

责任编辑	张爱民　王曦莹　熊　芳
数字编辑	段安骏
封面设计	王曦莹

书　　名	化学专业英语（第4版）
作　　者	马永祥　孙晓君　编
出版发行	兰州大学出版社　（地址：兰州市天水南路222号　730000）
电　　话	0931-8912613(总编办公室)　0931-8617156(营销中心)
网　　址	http://press.lzu.edu.cn
电子信箱	press@lzu.edu.cn
印　　刷	甘肃日报报业集团有限责任公司印务分公司
开　　本	787 mm×1092 mm　1/16
印　　张	17.5(插页2)
字　　数	565千
版　　次	2019年8月第4版
印　　次	2023年7月第5次印刷
书　　号	ISBN 978-7-311-05648-3
定　　价	45.00元

（图书若有破损、缺页、掉页，可随时与本社联系）

前 言

在科技英语词汇中，涉及化学和化工领域的词汇数量最大，从事这一专业工作的技术人员在掌握专业英语时，面对的词汇量数以万计，因而使该领域人员在学习专业英语时遇到的困难最大、周期最长，为使化学、化工专业学生缩短掌握专业英语的周期和在学习中少走弯路，收到事半功倍的效果，我们于1983年选编了此教材，并经过数年使用后，于1987年由兰州大学出版社正式出版发行，发行后陆续被国内许多高等院校化学及其相关专业选作高年级教学用书，一些从事化学工作的中青年教师、出国人员、科研和工程技术人员也作为自学教材进行了学习。

本书已修订多次，累计印书达十几万册。在第二版的修订中，删减了部分无机化学、有机化学的课文，增加了一些物理化学、结构化学的内容。

第三版修订又增加了三课新内容（组合化、离子液、杂环化学）和常用的三个附录（常用有机化合物英文缩写、常用分析测试仪器中英文名对照、美国《化学文摘》中常用词缩写），修订了一些印刷错误和改进了编辑中的缺陷。

此次修订，首先对本书的装帧做了全新的改变；其次，我们融合新媒体平台，首次采用二维码链接数字资源的形式，读者扫码即可免费获取配套学习资源，即全文朗读、参考译文、课后练习与答案，以及延伸阅读文章；再者订正了前版的印刷错误和修订了编辑中的不准确内容。经过此次修订，使本教材内容更加充实、更具时代性、印刷质量更高，也更符合当下数字阅读的需求，必将使广大读者更为满意。

本书共40课，内容涉及无机化学（元素及周期表、命名、配合物、酸碱概念等)，有机化学(命名、有机合成、金属有机、聚合物等)，物理化学(热力学、动力学、结构化学等)，化工基础(结晶、蒸馏、物料衡算)，化学文献等，并附有化学实验室常用仪器、杂环化合物的命名等5篇附录，最后附有常用词组、词头和词尾的索引，以及总词汇表。

本书内容丰富，取材新颖，领域广泛，文体各异，句型繁多，词汇量大，并且均有注音和一些构词规律，对一些语法现象也进行了解释和概括。熟读本书不仅可以熟练地阅读本专业英语资料，且能用英语撰写稿件和进行国际学术交流。

在本次修订过程中，兰州大学外籍教师 Nielsen James Walter 承担了全部课文的朗读工作，兰州大学化学化工学院李敏、范兆麟两位同学参与了部分校对工作，在此对他们的辛勤付出一并表示衷心的感谢。

由于我们水平有限，虽经多次修订和完善，仍难免有不足之处和错误，盼望读者提出宝贵意见。

编者

2019年6月

目 录

001	**Lesson 1**	The Elements And The Periodic Table 元素和周期表
009	**Lesson 2**	The Nonmetal Elements 非金属元素
014	**Lesson 3**	Groups Ⅰb And Ⅱb Elements Ⅰb和Ⅱb族元素
020	**Lesson 4**	Group Ⅲb—Ⅷb Elements Ⅲb—Ⅷb族元素
026	**Lesson 5**	Interhalogen And Noble Gas Compounds 杂卤素和惰性气体化合物
030	**Lesson 6**	The Classification Of Inorganic Compounds 无机化合物分类
036	**Lesson 7**	The Nomenclature Of Inorganic Compounds 无机化合物的命名
043	**Lesson 8**	BröNsted'S And Lewis' Acid-base Concepts 布朗斯特和路易氏酸碱概念
048	**Lesson 9**	The Coordination Complex 配位化合物
053	**Lesson 10**	Alkanes 烷烃
059	**Lesson 11**	Unsaturated Compounds 不饱和化合物
064	**Lesson 12**	The Nomenclature Of Cyclic Hydrocarbons 环烃的命名
069	**Lesson 13**	Substitutive Nomenclature 取代基命名法
075	**Lesson 14**	The Compounds Containing Oxygen 含氧化合物
081	**Lesson 15**	Preparation Of A Carboxylic Acid By The Grignard Method 格氏法制备羧酸

Lesson 16	The Structures Of Covalent Compounds 共价化合物的结构	086
Lesson 17	Oxidation And Reduction In Organic Chemistry 有机化学中的氧化和还原	092
Lesson 18	Synthesis Of Alcohols And Design Of Organic Synthesis 醇类的合成与有机合成设计	096
Lesson 19	Organometallics—metal π Complexes 金属有机化合物——金属π—配合物	100
Lesson 20	The Role Of Protective Groups In Organic Synthesis 有机合成中保护基的使用	106
Lesson 21	Heterocyclic Chemistry 杂环化学	111
Lesson 22	Polymers 聚合物	116
Lesson 23	Ionic Liquid 离子液体	123
Lesson 24	Volumetric Analysis 容量分析	129
Lesson 25	Combinatorial Chemistry 组合化学	133
Lesson 26	Vapor-phase Chromatography 气相色谱法	138
Lesson 27	Infrared Spectroscopy 红外光谱	143
Lesson 28	Nuclear Magnetic Resonance(Ⅰ) 核磁共振(Ⅰ)	147
Lesson 29	Nuclear Magnetic Resonance(Ⅱ) 核磁共振(Ⅱ)	152

157	**Lesson 30**	A Map Of Physical Chemistry 物理化学概貌
161	**Lesson 31**	The Chemical Thermodynamics 化学热力学
166	**Lesson 32**	Chemical Equilibrium And Kinetics 化学平衡和动力学
171	**Lesson 33**	The Rates Of Chemical Reactions 化学反应速度
175	**Lesson 34**	Nature Of The Colloidal State 胶态性质
179	**Lesson 35**	Electrochemical Cells 化学电池
184	**Lesson 36**	Boiling Points And Distillation 沸点和蒸馏
189	**Lesson 37**	Extractive And Azeotropic Distillation 萃取蒸馏和共沸蒸馏
194	**Lesson 38**	Crystallization 结晶
198	**Lesson 39**	Material Accounting——the Law Of Conservation Of Mass Really Works 物料衡算
202	**Lesson 40**	The Literature Matrix Of Chemistry 化学文献溯源
208	**APPENDIX** I	Common Laboratory Equipment 化学实验室常用仪器
214	**Appendix** II	Specialist Heterocyclic Nomenclature 杂环化合物的命名
222	**Appendix** III	Abbreviations For Common Organic Compounds 常用有机化合物英文缩写
226	**Appendix** IV	Analytic Apparatus In Common Use 常用分析测试仪器中英文名对照

Appendix V   Abbreviations And Acronyms Used In Cas Publication   228
美国《化学文摘》中常用词缩写
Appendix VI   Word Group   236
词组
Appendix VII   Prefix And Suffix   241
词头和词尾
Vocabulary

Lesson 1
THE ELEMENTS AND THE PERIODIC TABLE

The number of protons in the nucleus of an atom is referred to as the atomic number, or proton number, Z. The number of electrons in an electrically neutral atom is also equal to the atomic number, Z. The total mass of an atom is determined very nearly by the total number of protons and neutrons in its nucleus. This total is called the mass number, A. The number of neutrons in an atom, the neutron number, is given by the quantity A-Z.

The term element refers to a pure substance with atoms all of a single kind. To the chemist the "kind" of atom is specified by its atomic number, since this is the property that determines its chemical behavior. At present all the atoms from Z=1 to Z=107 are known: there are 107 chemical elements. Each chemical element has been given a name and a distinctive symbol. For most elements the symbol is simply the abbreviated form of the English name consisting of one or two letters, for example:

oxygen=O nitrogen=N neon=Ne magnesium=Mg

Some elements, which have been known for a long time, have symbols based on their Latin names, for example:

iron=Fe(ferrum)copper=Cu(cuprum)lead=Pb(plumbum)

A complete listing of the elements may be found in Table 1.

Beginning in the late seventeenth century with the work of Robert Boyle, who proposed the presently accepted concept of an element, numerous investigations produced a considerable knowledge of the properties of elements and their compounds1. In 1869, D. Mendeleev and L. Meyer, working independently, proposed the periodic law. In modern form, the law states that the properties of the elements are periodic functions of their atomic numbers. In other words, when the elements are listed in order of increasing atomic number, elements having closely similar properties will fall at definite intervals along the list. Thus it is possible to arrange the list of elements in tabular form with elements having similar properties placed in vertical columns[2]. Such an arrangement is called a periodic table.

Each horizontal row of elements constitutes a period. It should be noted that the lengths of the periods vary. There is a very short period containing only 2 elements, followed by two short periods of 8 elements each, and then two long periods of 18 elements each3. The next period includes 32 elements, and the last period is apparently incomplete. With this arrangement, elements in the same vertical column have similar characteristics. These columns constitute the chemical families or groups. The groups headed by the members of the two 8-element periods are designated as main group elements, and the members of the other groups are called transition or inner transition elements.

In the periodic table, a heavy stepped line divides the elements into metals and nonmetals. Elements to the left of this line (with the exception of hydrogen) are metals, while those to the right are nonmetals. This division is for convenience only; elements bordering the line - - the metalloids - have properties characteristic of both metals and nonmetals. It may be seen that most of the elements, including all the transition and inner transition elements, are metals.

Except for hydrogen, a gas, the elements of group ⅠA make up the alkali metal family. They are very reactive metals, and they are never found in the elemental state in nature. However, their compounds are widespread. All the members of the alkali metal family form ions having a charge of 1^+ only. In contrast, the elements of group IB – copper, silver, and gold – are comparatively inert. They are similar to the alkali metals in that they exist as 1^+ ions in many of their compounds. However, as is characteristic of most transition elements, they form ions having other charges as well4.

The elements of group ⅡA are known as the alkaline earth metals. Their characteristic ionic charge is 2^+. These metals, particularly the last two members of the group, are almost as reactive as the alkali metals. The group ⅡB elements – zinc, cadmium, and mercury are less reactive than are those of group ⅡA5, but are more reactive than the neighboring elements of group ⅠB. The characteristic charge on their ions is also 2^+.

With the exception of boron, group ⅢA elements are also fairly reactive metals. Aluminum appears to be inert toward reaction with air, but this behavior stems from the fact that the metal forms a thin, invisible film of aluminum oxide on the surface, which protects the bulk of the metal from further oxidation. The metals of group ⅢA form ions of 3^+ charge. Group ⅢB consists of the metals scandium, yttrium, lanthanum, and actinium.

Group ⅣA consists of a nonmetal, carbon, two metalloids, silicon and germanium, and two metals, tin and lead. Each of these elements forms some compounds with formulas which indicate that four other atoms are present per group ⅣA atom, as, for example, carbon tetrachloride, CCl4. The group ⅣB metals-titanium, zirconium, and hafnium – also form compounds in which each group ⅣB atom is combined with four other atoms; these compounds are nonelectrolytes when pure.

The elements of group ⅤA include three nonmetals – nitrogen, phosphorus, and arsenic – and two metals – antimony and bismuth. Although compounds with the formulas N_2O_5, PCl_5, and $AsCl_5$ exist, none of them is ionic. These elements do form compounds-nitrides, phosphides, and arsenides-in which ions having charges of minus three occur. The elements of group ⅤB are all metals. These elements form such a variety of different compounds that their characteristics are not easily generalized.

With the exception of polonium, the elements of group ⅥA are typical nonmetals. They are sometimes known as the chalcogens, from the Greek word meaning "ash formers". In their binary compounds with metals they exist as ions having a charge of 2^-. The elements of group ⅦA are all nonmetals and are known as the halogens, from the Greek term meaning "salt formers". They are the most reactive nonmetals and are capable of reacting with practically all the metals and with most nonmetals, including each other.

The elements of groups ⅥB, ⅦB, and ⅧB are all metals. They form such a wide variety of compounds that it is not practical at this point to present any examples as being typical of the behavior of the respective groups[6].

The periodicity of chemical behavior is illustrated by the fact that, excluding the first period, each period begins with a very reactive metal. Successive elements along the period show decreasing metallic character, eventually becoming nonmetals, and finally, in group ⅦA, a very reactive nonmetal is found. Each period ends with a member of the noble gas family.

New words and Expressions

element['elimənt] n. 元素
proton['prəutɔn] n. 质子
nucleus['nju:kləs] 复 nuclei
 ['nju:kliai] n. 核
atomic[ə'tɔmik] a. 原子的
atomic number 原子序数
neutral['nju:trəl] a. 中性的
mass[mæs] n. 质量
mass number 质量数
atom['ætəm] n. 原子
neutron['nju:trɔn] n. 中子
single kind 同一类
chemical['kemikəl] a. 化学的
　　　　　　　　　n. 化学品
symbol['simbəl] n. 符号
accept[ək'sept] vt. 接受,承认
concept[kɔnsept] n. 概念
compound[kəm'paund] n. 化合物
property['prɔpəti] n. 性质,特性
periodic[piəri'ɔdik] a. 周期的
　 periodic table　周期表
　 periodic law　周期律
state[steit] vt. 说明,认为
function['fʌŋkʃən] n. 官能;函数
tabular['tæbjulə] a. 表的
vertical['və:tikəl] a. 竖的,垂直的

column['kɔləm] n. 柱,塔;纵列
horizontal[hɔri'zɔntl] a. 水平的;横式的
row[rou] n. 排,横列
period['piəriəd] n. 周期
family['fæmili] n. (周期表的)族
group[gru:p] n. 族,基,团
　　　　　　　 vt. 把……分成组
transition[træn'ziʃən] n. 过渡,转变
metalloid['metəlɔid] n. 准金属
characteristic[kæriktə'ristik] n.a.
　　　　　　　　　　　　 特性,特点
alkali['ælkəlai] n. 碱
alkali metal　碱金属
reactive[ri'æktiv] a. 活泼的,反应的
inert[i'nə:t] a. 惰性的,不活泼的
coinage['kɔinidʒ] n. 造币,货币
ionic[ai'ɔnik] a. 离子的
alkaline['ælkəlain] a. 碱的
　 alkaline earth metal　碱土金属
thin[θin] a. 薄的,稀薄的; n. 薄层
invisible[in'vizəbl] a. 肉眼看不见的
film[film] n. 膜,胶片
aluminum oxide　氧化铝
surface['sə:fis] n.a. 表面
formula['fɔ:mjulə] n. 分子式,公式
oxidation[ɔksi'deiʃən] n. 氧化

carbon tetrachloride 四氯化碳
combine [kəm'bain] v. 化合,结合(with)
non-electrolyte ['nɔn-i'lektrəlait]
　　　　　　　　n. 非电解质
nitride ['naitraid] n. 氮化物
phosphide ['fɔsfaid] n. 磷化物
arsenide ['ɑ:sinaid] n. 砷化物
chalcogen ['kælkədʒən] n. 硫属,硫族
ash former [æʃ'fɔ:mə] 灰源体,成灰(者)
binary ['bainəri] a.n. 二元(的),双(的)
halogen ['hælədʒən] n. 卤素
salt former [sɔ:lt'fɔ:mə] 盐源体,成盐(者)
periodicity [piəriə'disiti] n. 周期性

前　缀

in- [in-] (il-, im-, ir-) 不,无,非
　　incomplete, inorganic, invisible, impure, irregular
non- [nɔn-] 非,不,无
　　nonmetal, nonelectrolyte, noninflammable (不燃的), nontoxic(无毒的)
di- [dai-] 二,双,联,重,偶
　　dipositive, dioxide; dimolecular, diatomic, diphenyl(联苯), dibenzoyl(联苯甲酰), dichromate(重铬酸盐), dipole(偶极)

后　缀

-on [-ɔn, -ən] (名词词尾)
　①组成原子微粒
　　proton, neutron, ion, electron
　②非金属元素词尾
　　argon, boron, carbon, silicon
-ic [-ik] (形容词词尾)
　　ionic, periodic, atomic, catalytic
-ide [-aid] …化物(名词词尾)
　　oxide, chloride, hydride, halide, nitride

词　组

referred to as… 称为……,被认为是……
equal to… 等于,与……相等
refer to… 涉及,指的是
(be) abbreviated form… 是……的缩写,是……之略
consist of… 由……组成
(be) based (up)on… 根据……,以……为准
begin with… 从……开始
in modern form 按近代方式
a function of 随……而变,……的函数
in other words 换句话说
in order of… 按……(排列)
followed by… 接着,继之有,后面是
main group element 主族元素
for convenience 为方便起见
divide A into B and C 把A分成B和C
with the exception of… 除……之外
except for… 除……之外
make up… 形成,组成
in contrast 相反,与此对比
similar to… 类似于
exist as… 以[……形式]存在
…as well [同样]也
known as 就是通常说的……,以……著称
protect A from B 保护A不受B…[影响];使A免于B
composed of… 由……组成
to some extent 在某种程度上
a variety of… 各种各样的
capable of (+ing) 能够……,有……可能

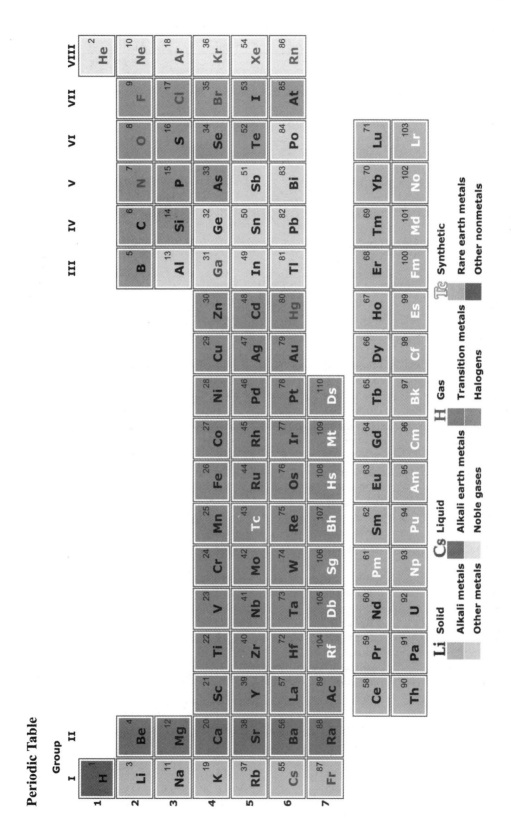

Table 1　IUPAC* Names and Symbols of the Elements

Name	Pronunciation	Symbol	At.No.	Chinese
Actinium	[ækˈtiniəm]	Ac	89	锕
Aluminum	[əˈljuːminəm]	Al	13	铝
Americium	[æməˈrisiəm]	Am	95	镅
Antimony	[ˈæntiməni]	Sb	51	锑
Argon	[ˈɑːgɔn]	Ar	18	氩
Arsenic	[ˈɑːsnik]	As	33	砷
Astatine	[ˈæstətin]	At	85	砹
Barium	[ˈbɛəriəm]	Ba	56	钡
Berkelium	[bəˈkliəm]	Bk	97	锫
Beryllium	[beˈriljəm]	Be	4	铍
Bismuth	[ˈbizməθ]	Bi	83	铋
Boron	[ˈbɔːrɔn]	B	5	硼
Bromine	[ˈbroumiːn]	Br	35	溴
Cadmiun	[ˈkædmiəm]	Cd	48	镉
Calcium	[ˈkælsiəm]	Ca	20	钙
Californium	[ˈkæliˈfɔːniəm]	Cf	98	锎
Carbon	[ˈkɑːbən]	C	6	碳
Cerium	[ˈsiəriəm]	Ce	58	铈
Cesium	[ˈsiːziəm]	Cs	55	铯
Chlorine	[ˈklɔːrin]	Cl	17	氯
Chromium	[ˈkroumiəm]	Cr	24	铬
Cobalt	[kəˈbɔːlt]	Co	27	钴
Copper	[ˈkɔpə]	Cu	29	铜
Curium	[ˈkjuəriəm]	Cm	96	锔
Dysprosium	[disˈprousiəm]	Dy	66	镝
Einsteinium	[ainˈstainiəm]	Es	99	锿
Erbium	[ˈəːbiəm]	Er	68	铒
Europium	[juəˈroupiəm]	Eu	63	铕
Fermium	[ˈfɛəmiəm]	Fm	100	镄
Fluorine	[ˈflu(ː)ərin]	F	9	氟
Francium	[ˈfrænsiəm]	Fr	87	钫
Gadolinium	[ˌgædəˈliniəm]	Gd	64	钆
Gallium	[ˈgæliəm]	Ga	31	镓
Germanium	[dʒəːˈmeiniəm]	Ge	32	锗
Gold	[ˈgould]	Au	79	金
Hafnium	[ˈhæfniəm]	Hf	72	铪
Helium	[ˈhiːljəm]	He	2	氦
Holmium	[ˈhɔlmiəm]	Ho	67	钬
Hydrogen	[ˈhaidrədʒən]	H	1	氢
Indium	[ˈindiəm]	In	49	铟
Iodine	[ˈaiədiːn]	I	53	碘
Iridium	[aiˈridiəm]	Ir	77	铱
Iron	[ˈaiən]	Fe	26	铁
Krypton	[ˈkriptɔn]	Kr	36	氪
Lanthanum	[ˈlænθənəm]	La	57	镧
Lawrencium	[lɔːˈrensiəm]	Lr	103	铹
Lead	[led]	Pb	82	铅
Lithium	[ˈliθiəm]	Li	3	锂
Lutetium	[ljuːˈtiːʃiəm]	Lu	71	镥
Magnesium	[mægˈniːziəm]	Mg	12	镁
Manganese	[ˈmæŋgəniːz]	Mn	25	锰
Mendelevium	[ˌmendəˈliviəm]	Md	101	钔

续表

Name	Pronunciation	Symbol	At.No.	Chinese
Mercury	[ˈməːkjuri]	Hg	80	汞
Molybdenum	[məˈlibdiːnəm]	Mo	42	钼
Neodymium	[ˌni(ː)əˈdimiəm]	Nd	60	钕
Neon	[ˌniːɔn]	Ne	10	氖
Neptunium	[nepˈtjuniəm]	Np	93	镎
Nickel	[ˈnikəl]	Ni	28	镍
Niobium	[naiˈoubiəm]	Nb	41	铌
Nitrogen	[ˈnaitridʒən]	N	7	氮
Nobelium	[nouˈbeliəm]	No	102	锘
Osmium	[ˈɔzmiəm]	Os	76	锇
Oxygen	[ˈɔksidʒən]	O	8	氧
Palladium	[pəˈleidiəm]	Pd	46	钯
Phosphorus	[ˈfɔsfərəs]	P	15	磷
Platinum	[ˈplætinəm]	Pt	78	铂
Plutonium	[pluːˈtounjəm]	Pu	94	钚
Polonium	[pəˈlouniəm]	Po	84	钋
Potassium	[pəˈtæsjəm]	K	19	钾
Praseodymium	[ˌpreiziouˈdimiəm]	Pr	59	镨
Promethium	[prəˈmiːθiəm]	Pm	61	钷
Protactinium	[ˌproutækˈtiniəm]	Pa	91	镤
Radium	[ˈreidiəm]	Ra	88	镭
Radon	[ˈreidɔn]	Rn	86	氡
Rhenium	[ˈriːniəm]	Re	75	铼
Rhodium	[ˈroudiəm]	Rh	45	铑
Rubidium	[ruːˈbidiəm]	Rb	37	铷
Ruthenium	[ruːˈθiniəm]	Ru	44	钌
Samarium	[səˈmɛəriəm]	Sm	62	钐
Scandium	[ˈskændiəm]	Sc	21	钪
Selenium	[siˈliːnjəm]	Se	34	硒
Silicon	[ˈsilikən]	Si	14	硅
Silver	[ˈsilvə]	Ag	47	银
Sodium	[ˈsoudjəm]	Na	11	钠
Strontium	[ˈstrɔnʃiəm]	Sr	38	锶
Sulfur	[ˈsʌlfə]	S	16	硫
Tantalum	[ˈtæntələm]	Ta	73	钽
Technetium	[tekˈniːʃiəm]	Tc	43	锝
Tellurium	[teˈljuəriəm]	Te	52	碲
Terbium	[ˈtəːbiəm]	Tb	65	铽
Thallium	[ˈθæliəm]	Tl	81	铊
Thorium	[ˈθɔːriəm]	Th	90	钍
Thulium	[ˈθjuːliəm]	Tm	69	铥
Tin	[tin]	Sn	50	锡
Titanium	[tiˈteiniəm]	Ti	22	钛
Tungsten	[ˈtʌŋsten]	W	74	钨
Uranium	[juəˈreinjəm]	U	92	铀
Vanadium	[vəˈneidiəm]	V	23	钒
Xenon	[ˈzenɔn]	Xe	54	氙
Ytterbium	[iˈtəːbjəm]	Yb	70	镱
Yttrium	[ˈitriəm]	Y	39	钇
Zinc	[ˈziŋk]	Zn	30	锌
Zirconium	[zəːˈkouniəm]	Zr	40	锆

*IUPAC 是 International Union of Pure and Applied Chemistry，即："国际纯粹化学和应用化学联合会"的缩写。

Notes

1. beginning in the late seventeenth century with the work of Robert Boyle, who proposed the presently accepted concept of an element, numerous investigations produced a considerable knowledge of the properties of elements and their compounds. 其中 "numerous investigations…compounds." 为主句，Beginning 引导的分词短语是表示时间的状语，"who…an element" 是修饰 Robert Boyle 的定语从句。**译文：**"早在17世纪末期，罗伯特·波义耳就开始了这项工作，他提出了现在公认的元素概念，大量的研究使我们对元素及其化合物的性质有了相当的了解。"

2. Thus it is possible to arrange the list of elements in tabular form with elements having similar properties placed in vertical columns. 该句中 it 是形式主语，代替 "to arrange…columns." 不定式短语，此短语中 with elements…columns 是独立主格结构作方式状语，修饰 arrange。**译文：** "于是，将具有类似性质的元素排成纵列，从而把元素排成表格形式是可能的。"

3. each：在此处作副词，一般在句子末尾，意思是 "每个"，"各（个地）"，也用作形容词和代词，如 "each horizontal row" 和 "each of these elements"。

4. However, as is characteristic of most transition elements, they form ions having other charges as well. 这里 "they form…as well" 是主句，"as is…elements" 是关系代词 as 引导的定语从句，修饰整个句子，as 本身在从句中作主语。**译文：** "然而，像许多过渡元素所具有的特点一样，它们也形成具有其它电荷的离子。"

5. than are those of group IIA：Than 后的从句，有时主、谓语颠倒。

6. They form such a wide variety of compounds that it is not practical at this point to present any examples as being typical of the behavior of the respective groups. "They form…compounds" 是主句，"that it is…respective groups." 是 that 引起的结果状语从句，从句中 as+现在分词短语 "being…groups" 是 examples 的定语。**译文：** "它们形成了各种不同的化合物，在这一点上我们甚至不能举出任何能表现各族元素典型变化的例子。"

Exercises

1. Fill in the blanks with proper words or phrases.

    Group IA is made up by _____, _____, _____, _____, rubidium, cesium and francium, while Group VIIA is composed of _____, _____, _____, _____ and astatine from the top of the list.

    A. lithium   B. fluorine   C. hydrogen   D. iodine   E. chlorine

    F. potassium   G. bromine   H. sodium

2. Match the following explanation with the proper word.

    _____ horizontal, _____ reactive, _____ inert, _____ invisible, _____ periodicity

    A. having only a limited ability to react chemically

    B. parallel to or in the plane of the horizon or a base line

    C. impossible or nearly impossible to see; imperceptible by the eye

    D. the property of circulating at regular intervals

    E. participating readily in reactions

Lesson 2
THE NONMETAL ELEMENTS

We noted earlier that nonmetals exhibit properties that are greatly different from those of the metals. As a rule, the nonmetals are poor conductors of electricity (graphitic carbon is an exception) and heat; they are brittle, are often intensely colored, and show an unusually wide range of melting and boiling points. Their molecular structures, usually involving ordinary covalent bonds, vary from the simple diatomic molecules of H_2, Cl_2, I_2, and N_2 to the giant molecules of diamond, silicon and boron.

The nonmetals that are gases at room temperature are the low-molecular weight diatomic molecules and the noble gases that exert very small intermolecular forces. As the molecular weight increases, we encounter a liquid (Br_2) and a solid (I_2) whose vapor pressures also indicate small intermolecular forces. Certain properties of a few nonmetals are listed in Table 2.

Table 2. Molecular Weights and Melting Points of Certain Nonmetals

Diatomic Molecules	Molecular Weight	Melting Point ℃	Color
H_2	2	−259.1[1]	None
N_2	28	−210	None
F_2	38	−223	Pale yellow
O_2	32	−218	Pale blue
Cl_2	71	−102	Yellow-green
Br_2	160	−7.3	Red-brown
I_2	254	113	Gray-black

Simple diatomic molecules are not formed by the heavier members of Groups Ⅴ and Ⅵ at ordinary conditions. This is in direct contrast to the first members of these groups, N_2 and O_2. The difference arises because of the lower stability of π bonds formed from p orbitals of the third and higher main energy levels as opposed to the second main energy level[2]. The larger atomic radii and more dense electron clouds of elements of the third period and higher do not allow good parallel overlap of p orbitals necessary for a strong π bond. This is a general phenomenon - strong π bonds are formed only between elements of the second period. Thus, elemental nitrogen and oxygen form stable molecules with both σ and π bonds, but other members of their groups form more stable structures based on σ bonds only at ordinary conditions. Note[3] that Group Ⅶ elements form diatomic molecules, but π bonds are not required for saturation of valence.

Sulfur exhibits allotropic forms. Solid sulfur exists in two crystalline forms and in an amorphous form. Rhombic sulfur is obtained by crystallization from a suitable solution, such as CS_2, and it melts at 112℃. Monoclinic sulfur is formed by cooling melted sulfur and it melts at 119℃. Both forms of crystalline sulfur melt into S-gamma, which is composed of S_8 molecules. The S_8 molecules are puckered rings and survive heating to about 160℃. Above 160℃, the S_8

rings break open, and some of these fragments combine with each other to form a highly viscous mixture of irregularly shaped coils. At a range of higher temperatures the liquid sulfur becomes so viscous that it will not pour from its container. The color also changes from straw yellow at sulfur's melting point to a deep reddish-brown as it becomes more viscous.

As[4] the boiling point of 444℃ is approached, the large-coiled molecules of sulfur are partially degraded and the liquid sulfur decreases in viscosity. If the hot liquid sulfur is quenched by pouring it into cold water, the amorphous form of sulfur is produced. The structure of amorphous sulfur consists of large-coiled helices with eight sulfur atoms to each turn of the helix; the overall nature of amorphous sulfur is described as[5] rubbery because it stretches much like ordinary rubber. In a few hours the amorphous sulfur reverts to small rhombic crystals and its rubbery property disappears.

Sulfur, an important raw material in industrial chemistry, occurs as[5] the free element, as[5] SO_2 in volcanic regions, as H_2S in mineral waters, and in a variety of sulfide ores such as iron pyrite FeS_2, zinc blende ZnS, galena PbS and such, and in common formations of gypsum $CaSO_4 \cdot 2H_2O$, anhydrite $CaSO_4$, and barytes $BaSO_4 \cdot 2H_2O$. Sulfur, in one form or another, is used in large quantities for making sulfuric acid, fertilizers, insecticides, and paper.

Sulfur in the form of SO_2 obtained in the roasting of sulfide ores is recovered and converted to sulfuric acid, although in previous years much of this SO_2 was discarded through exceptionally tall smoke stacks. Fortunately, it is now economically favorable to recover these gases, thus greatly reducing this type of atmospheric pollution. A typical roasting reaction involves the change:

$$2\ ZnS + 3\ O_2 \rightarrow 2\ ZnO + 2\ SO_2$$

Phosphorus, below 800℃, consists of tetratomic molecules, P_4. Its molecular structure provides for a covalence of three, as may be expected from the three unpaired p electrons in its atomic structure, and each atom is attached to three others[6]. Instead of a strictly orthogonal orientation, with the three bonds 90° to each other, the bond angles are only 60°. This supposedly strained structure is stabilized by the mutual interaction of the four atoms (each atom is bonded to the other three), but it is chemically the most active form of phosphorus. This form of phosphorus, the white modification, is spontaneously combustible in air. When heated to 260℃ it changes to red phosphorus, whose structure is obscure. Red phosphorus is stable in air but, like all forms of phosphorus, it should be handled carefully because of its tendency to migrate to the bones when ingested, resulting in serious physiological damage.

Elemental carbon exists in one of two crystalline structures - diamond and graphite. The diamond structure, based on tetrahedral bonding of hybridized sp^3 orbitals, is encountered among Group Ⅳ elements. We may expect that as the bond length increases, the hardness of the diamond-type crystal decreases. Although the tetrahedral structure persists among the elements in this group-carbon, silicon, germanium, and gray tin-the interatomic distances increase from 1.54 Å for carbon to 2.80Å for gray tin. Consequently, the bond strengths among the four elements range from very strong to quite weak. In fact, gray tin is so soft that it exists in the form of microcrystals or merely as a powder. Typical of the Group Ⅳ diamond-type crystalline elements, it is a nonconductor and shows other nonmetallic properties[7].

New words and Expressions

exhibit[ig'zibit] vt.n. 显示,表示,呈现
graphitic[græ'fitik] a. 石墨的
graphite['græfait] n. 石墨
brittle['britl] a. 易碎的
intensely[in'tensli] ad. 强烈地
diatomic molecule 双原子分子
melting point['meltiŋ'pɔint] 熔点
boiling point['bɔiliŋ'pɔint] 沸点
molecule['mɔlikju:l] n. 分子,克分子
molecular[məu'lekjulə] a. 分子的
　　~ weight 分子量
structure['strʌktʃə] n. 结构
valent['veilənt] a. 价的
covalent[kəu'veilənt] a. 共价的
valence['veiləns] n.(原子)价
bond[bɔnd] n. 键; vi. 结合,相接
diamond['daiəmənd] n. 金刚石
exert[ig'zə:t] vt. 产生,采用
liquid['likwid] n.a. 液体
solid['sɔlid] n.a. 固体
vapor['veipə] n. 蒸汽
pressure['preʃə] n. 压力
intermolecular[ˌintə(:)məu'lekjulə]
　　　　　　　　　　a. 分子间的
arise[ə'raiz] vi. 出现,发生
stability[stə'biliti] n. 稳定性,安定性
stable['steibl] a. 稳定的,安定的
orbital['ɔ:bitl] n.a. 轨道
energy level['enədʒi'levl] 能级
radius['reidiəs](复 radii['reidiai])n. 半径
electron[i'lektrɔn] n. 电子
　　~ cloud 电子云
parallel['pærəlel] a. 平行的,同一方向的
overlap[ˌəuvə'læp] n.v. 重叠
phenomenon[fi'nɔminən] n. 现象
allotropic[ˌælə'trɔpik] a. 同素异形的
crystalline['kristəlain] a.n. 结晶
crystal['kristl] n. 结晶
amorphous[ə'mɔ:fəs] a. 无定形的

rhombic['rɔmbik] a. 斜方(形)的
pale yellow 浅黄色
pale blue 浅蓝色
yellow-green 黄绿色
red-brown 红棕色
reddish-brown 微红棕色
gray-black 灰黑色
crystallization[ˌkristəlai'zeiʃən] n. 结晶
solution[sə'lu:ʃən] n. 溶液
melt[melt] v. 熔化
monoclinic['mɔnə'klinik] a. 单斜晶系的
pucker['pʌkə] v. 折叠,缩拢
ring[riŋ] n. 环
survive[sə'vaiv] v. 还活着,仍存在
fragment['frægmənt] n. 碎片,断片
viscous['viskəs] a. 粘(稠)的,胶粘的
viscosity[vis'kɔsiti] n. 粘度
mixture['mikstʃə] n. 混合物
coil[kɔil] n. 卷线,螺旋圈; v. 卷,盘绕
pour[pɔ:] v. 注,流出
container[kən'teinə] n. 容器
straw yellow 稻草黄色
degrade[di'greid] v. 降解,降低
quench[kwentʃ] vt. 骤冷,淬火
helix['hi:liks](复 helices['helisi:z])
　　　　　　　　　　n. 螺旋线
rubbery['rʌbəri] a. 橡胶状的
rubber['rʌbə] n. 橡胶
stretch[stretʃ] v. 伸长,拉伸
raw material[rɔ:mə'tiəriəl] 原料
volcanic[vɔl'kænik] a. 火山的
mineral water 矿泉水,矿物水
iron pyrite['aiənpai'rait] 黄铁矿
zinc blende['ziŋk'blend] 闪锌矿
galena[gə'li:nə] n. 方铅矿
common formation 共生(矿)
gypsum['dʒipsəm] n. 石膏
anhydrite[æn'haidrait] n. 无水石膏,硬石膏
barytes[bə'raitiz] n. 重晶石

sulfuric[sʌl'fjuərik] a. 硫的,含硫的
~ acid 硫酸
fertilizer['fə:tilaizə] n. 肥料
insecticide[in'sektisaid] n. 杀虫剂
roast[roust] vt. 焙烧
sulfide['sʌlfaid] n. 硫化物
discard[dis'kɑ:d] v. 排放,丢弃
smoke stack['smoukstæk] 烟囱
atmospheric[ætməs'ferik] a. 大气的
pollution[pə'lu:ʃn] n. 污染
orthogonal[ɔ:'θɔgənl] a. 直角的,正交的
orientation[ɔ:rien'teiʃən] n. 方向
mutual['mju:tjuəl] a. 相互的,共同的
modification[mɔdifi'keiʃən] n. 修正,变体
combustible[kəm'bʌstəbl] a. 易燃的
obscure[əb'skjuə] a. 不清楚的,暗的
handle['hændl] vt. 处理,掌握
migrate[mai'greit] vi. 移动,迁移
ingest[in'dʒest] vt. 咽下
physiological[fiziə'lɔdʒikəl] a. 生理的
damage['dæmidʒ] n. 损害,损伤
tetrahedral[ˌtetrə'hedrəl] a. 四面体的
hybridize['haibridaiz] v. 杂化

前 缀

un-[ʌn-]不,未,非,无,脱,去
  unpaired, unassociated(未缔合的),
  unknown(未知物),
  unedible(非食用的),
  uncolored(无色的),
  ungrease(脱脂的), unhair(去毛)
co-[kəu-]共,一起,同,联
  covalent, copolymer, co-worker,
  coproduct(联产物)
inter-[intə(:)-]相互,中间,之间
  interaction, intermolecular, interface
  (界面), interconversion(互变),
  intercondenser(中间冷却器)
micro-[ˌmaikrou-]微,显微
  microcrystal(微晶), microanalysis(微量
  分析), microscope, microburette(微量滴
  定管), microwave

后 缀

-cide[-said]除(杀,防)……剂(名词词尾)
  insecticide, herbicide bactericide,
  fungicide(防霉剂)
-hedral[-hedrəl]…面体的(形容词词尾)
  tetrahedral, hexahedral, octahedral,
  polyhedral
-able, -ible[-əbl, -ibl]可(易,适于)……的
(形容词词尾)
  combustible, invisible, compressible,
  variable, exchangeable, countable.
-ure[-ə]表示行为及结果(名词词尾)
  temperature, pressure, measure, exposure

词 组

different from…　与……不同
at…temperature　在……温度下
list in table…　列于(示于)表……中
at ordinary condition　在通常条件下
in contrast to（with）　与……相反
form from…　由……形成
as opposed to…　与……相反
exist in …form　以……形式存在
change from A to B　从A转变为B
combine with…　与……化合(结合)
decrease(vi.)in…　……减少,……降低
one…or another　各种各样的,某种
in(large)quantities　大量地
use…for(+ing)　用……来(……)
in the form of…　以……形式
attached to…　接于……,与……相接
instead of …　代替,不……(而)
bonded to …　键合(结合)于……,
　　　　　　与……键合(结合)
tendency to…　倾向于,有……趋势
result in …　引起,导致
range from…to…　从……到……,
　　　　　　在……与……间变化
subjected to…　受到,易受
convert A (in)to B　把A转化为B
accompanied by…　伴有……,随着
such A as B　像B这样一类的A

Notes

1. －259.1℃可有如下三种读法

   （1）minus two hundred and fifty nine point one degrees Centigrade［ˈsentigreid］

   （或 Celsius［ˈselsjəs］）

   （2）two hundred and fifty nine point one degrees Centigrade below zero

   （3）minus two five nine point one degrees Centigrade

2. "The difference arises because of the lower stability of π-bonds formed from p orbitals of the third and higher main energy levels as opposed to the second main energy level."系主从复合句，主句是"The difference…higher main energy levels"，"as + 过去分词…"是 as 引出的非限制性定语从句，修饰 level，从句中省去了 is，而 as 代表整个主句。**译文：**"差别的出现是由于与第二主能级相反，第三和更高主能级的 p-轨道形成的 π-键稳定性较低"。

3. Note：常引出祈使句，它的前面省去了 please，即 please note（请注意……），note 后一般带有 that 引导的宾语从句。

4. as：此处作连接词，引导时间状语从句。

5. as：在这些地方引出主语补语。

6. Its molecular structure provides for a convalence of three, as may be expected from the three unpaired p electrtons in its atomic structure, and each atom is attached to three others. 该句为并列复合句，第一分句为"Its molecular … atomic structure."，句中"as may … structure"是 as 引导的非限制性定语从句，as 为关系代词，代表"a covalence of three"，第二个分句是"and each atom … others."**译文：**"它的分子结构提供了三个共价，可预期三个共价是由该原子结构中的三个未成对的 p-电子形成的，且每个原子都与另外三个相接。"

7. Typical of the Group Ⅳ diamond-type crystalline elements, it is a nonconductor and shows other nonmetallic properties. 本句是具有两个谓语的简单句，"typical of … elements"是主语 it 的同位语。**译文：**"碳作为典型的第Ⅳ族金刚石型晶体的元素，它是不良导体，且显示其它非金属性质。"

Exercises

1. Fill in the blanks with proper words or phrases.

   Sulfur is a ____ element with symbol S and ____ 16. It is ____, multivalent, and nonmetallic. Under normal conditions, sulfur atoms form cyclic octatomic molecules with a chemical S8. Elemental sulfur is a ____ yellow, crystalline solid at room ____ . Sulfur is the tenth common element by mass in the universe, and the ____ most common on Earth.

   A. atomic number   B. fifth   C. chemical   D. formula

   E. temperature   F. abundant   G. most   H. bright

2. Choose a proper prefix to make the following words to its antonyms.

   ____ organic, ____ paired, ____ metal, _ toxic, ____ pure, ____ colored

   A. un-   B. in-   C. non-

Lesson 3
GROUPS ⅠB AND ⅡB ELEMENTS

Physical properties of Group ⅠB and ⅡB

These elements have a greater bulk use as metals than in compounds, and their physical properties vary widely.

Gold is the most malleable and ductile of the metals. It can be hammered into sheets of 0.00001 inch in thickness; one gram of the metal can be drawn into a wire 1.8 m in length[1]. Copper and silver are also metals that are mechanically easy to work. Zinc is a little brittle at ordinary temperatures, but may be rolled into sheets at between 120° to 150℃; it becomes brittle again about 200℃. The low-melting temperatures of zinc contribute to the preparation of zinc-coated iron, galvanized iron; clean iron sheet may be dipped into vats of liquid zinc in its preparation. A different procedure is to sprinkle or air blast zinc dust onto hot iron sheeting for a zinc melt and then coating.

Cadmium has specific uses because of its low-melting temperature in a number of alloys. Cadmium rods are used in nuclear reactors because the metal is a good neutron absorber.

Mercury vapor and its salts are poisonous, though the free metal may be taken internally under certain conditions. Because of its relatively low boiling point and hence volatile nature, free mercury should never be allowed to stand in an open container in the laboratory. Evidence shows that inhalation of its vapors is injurious.

The metal alloys readily with most of the metals (except iron and platinum) to form amalgams, the name given to any alloy of mercury.

Copper sulfate, or blue vitriol ($CuSO_4 \cdot 5H_2O$) is the most important and widely used salt of copper. On heating, the salt slowly loses water to form first the trihydrate ($CuSO_4 \cdot 3H_2O$), then the monohydrate ($CuSO_4 \cdot H_2O$), and finally the white anhydrous salt. The anhydrous salt is often used to test for the presence of water in organic liquids. For example, some of the anhydrous copper salt added to alcohol (which contains water) will turn blue because of the hydration of the salt.

Copper sulfate is used in electroplating. Fishermen dip their nets in copper sulfate solution to inhibit the growth of organisms that would rot the fabric. Paints specifically formulated for use on the bottoms of marine craft contain copper compounds to inhibit the growth of barnacles and other organisms.

When dilute ammonium hydroxide is added[2] to a solution of copper(Ⅱ) ions, a greenish precipitate of Cu(OH)$_2$ or a basic copper(Ⅱ) salt is formed. This dissolves as more ammonium hydroxide is added. The excess ammonia forms an ammoniated complex with the copper(Ⅱ) ion of the composition, $Cu(NH_3)_4^{2+}$. This ion is only slightly dissociated; hence in an ammoniacal solution very few copper (Ⅱ) ions are present. Insoluble copper compounds, except copper sulfide, are dissolved by ammonium hydroxids. The formation of the copper (Ⅱ) ammonia ion is often used as a test for Cu^{2+} because of its deep, intense blue color.

Copper (Ⅱ) ferrocyanide $[Cu_2Fe(CN)_6]$ is obtained as a reddish-brown precipitate on the addition of a soluble ferrocyanide to a solution of copper (Ⅱ) ions. The formation of this salt is also used as a test for the presence of copper (Ⅱ) ions.

Compounds of Silver and Gold

Silver nitrate, sometimes called lunar caustic, is the most important salt of silver. It melts readily and may be cast into sticks for use in cauterizing wounds. The salt is prepared by dissolving silver in nitric acid and evaporating the solution.

$$3Ag+4HNO_3 \rightarrow 3AgNO_3+NO+2H_2O$$

The salt is the starting material for most of the compounds of silver, including the halides used in photography. It is readily reduced by organic reducing agents, with the formation of a black deposit of finely divided silver; this action is responsible for black spots left on the fingers from the handling of the salt. Indelible marking inks and pencils take advantage of this property of silver nitrate.

The halides of silver, except the fluoride, are very insoluble compounds and may be precipitated by the addition of a solution of silver salt to a solution containing chloride, bromide, or iodide ions.

The addition of a strong base to a solution of a silver salt precipitates brown silver oxide (Ag_2O). One might expect the hydroxide of silver to precipitate, but it seems likely that silver hydroxide is very unstable and breaks down into the oxide and water—if, indeed, it is ever formed at all[3]. However, since a solution of silver oxide is definitely basic, there must be hydroxide ions present in solution.

$$Ag_2O+H_2O \rightleftharpoons 2Ag^++2OH^-$$

Because of its inactivity, gold forms relatively few compounds. Two series of compounds are known—monovalent and trivalent. Monovalent (aurous) compounds resemble silver compounds (aurous chloride is water insoluble and light sensitive), while the higher valence (auric) compounds tend to form complexes. Gold is resistant to the action of most chemicals—air, oxygen, and water have no effect. The common acids do not attack the metal, but a mixture of hydrochloric and nitric acids (aqua regia) dissolves it to form gold (Ⅲ) chloride or chloroauric acid. The action is probably due to free chlorine present in the aqua regia.

$$3HCl+HNO_3 \longrightarrow NOCl+Cl_2+2H_2O$$
$$2Au+3Cl_2 \longrightarrow 2AuCl_3$$
$$AuCl_3+HCl \longrightarrow HAuCl_4$$

chloroauric acid ($HAuCl_4 \cdot H_2O$ crystallizes from solution).

Compounds of Zinc

Zinc is fairly high in the activity series. It reacts readily with acids to produce hydrogen and displaces less active metals from their salts. The action of acids on impure zinc is much more rapid than on pure zinc, since bubbles of hydrogen gas collect on the surface of pure zinc and slow down the action. If another metal is present as an impurity, the hydrogen is liberated from the surface of the contaminating metal rather than from the zinc. An electric couple to facilitate the action is probably set up between the two metals.

$$Zn + 2H^+ \longrightarrow Zn^{2+} + H_2$$

Zinc oxide (ZnO), the most widely used zinc compound, is a white powder at ordinary temperature, but changes to yellow on heating. When cooled, it again becomes white. Zinc oxide is obtained by burning zinc in air, by heating the basic carbonate, or by roasting the sulfide. The principal use of[4] ZnO is as a filler in rubber manufacture, particularly in automobile tires. As a body for paints it has the advantage over white lead of not darkening on exposure to an atmosphere containing hydrogen sulfide. Its covering power, however, is inferior to that of white lead.

Notes

malleable['mæliəbl] a. 展性的
ductile['dʌktail] a. 延性的
sheet[ʃi:t] n. 箔,纸张,薄板; vt. 复盖,铺展
roll[roul] vt. 辗薄,滚压平; n. 卷状物
zinc-coated iron['ziŋk'koutid 'aiən] 镀锌铁,白铁
galvanize['gælvənaiz] vt. 镀锌,电镀
galvanized iron 镀锌铁
dip[dip] v. 浸渍
vat[væt] n. 大盆,桶,瓮
procedure[prə'si:dʒə] n. 步骤,方法
sprinkle['spriŋkl] vt. 喷雾,洒
blast[blɑ:st] n. 鼓风
dust[dʌst] n. 粉尘,灰尘
alloy[ə'lɔi] n.v. 合金,熔合,搀合
nuclear['nju:kliə] a. 核的
reactor[ri'æktə] n. 反应堆,反应器

rod[rɔd] n. 细长棒
absorber[əb'sɔ:bə] n. 吸收器,吸收剂
poisonous['pɔiznəs] a. 有毒的
volatile['vɔlətail] a. 易挥发的,挥发性的
inhalation[ˌinhə'leiʃən] n. 吸入
injurious[in'dʒuəriəs] a. 有害的
amalgam[ə'mælgəm] n. 汞齐
sulfate=sulphate['sʌlfeit] n. 硫酸盐(酯)
blue vitriol[blu:'vitriəl] 胆矾,硫酸铜
hydrate['haidreit] n. 水合物; v. 水合,水化
anhydrous[æn'haidrəs] a. 无水的
organic[ɔ:'gænik] a. 有机的
alcohol['ælkəhɔl] n. 醇,乙醇
hydration[hai'dreiʃən] n. 水合,水化
electroplate[i'lektrəpleit] vt. 电镀
net[net] n. 网,网状物
inhibit[in'hibit] vt. 防止,阻止
organism['ɔ:gənizəm] n. 有机体,生物体
greenish['gri:niʃ] a. 微带绿色的

rot[rɔt] vt. 完全破坏,腐烂
fabric[ˈfæbrik] n. 织物,纤维品
paint[peint] n. 油漆,涂料
formulate[ˈfɔːmjuleit] vt. 用公式表示
formulated[ˈfɔːmjuleitid] a. 公式化了的,一般化了的,配制的
barnacle[ˈbɑːnəkl] n. 藤壶
dilute[daiˈljuːt] a. 稀的
ammonia[əˈməunjə] n. 氨
ammonium[əˈməunjəm] n. 铵
hydroxide[haiˈdrɔksaid] n. 氢氧化物
basic[ˈbeisik] a. 碱性的,碱式的
dissolve[diˈzɔlv] vt. 溶解
ammoniated[əˈməunieitid] a. 与氨化合了的,氨合的
complex[ˈkɔmpleks] n. 络合物,配合物;a. 复杂的
composition[ˌkɔmpəˈziʃən] n. 构成,组成
dissociate[diˈsəuʃieit] vt. 离解
ammoniacal[ˌæməuˈnaiəkəl] a. 氨的
soluble[ˈsɔljubl] a. 可溶的
ferrocyanide[ˌferouˈsaiənaid] n. 亚铁氰化物
nitrate[ˈnaitreit] n. 硝酸盐(酯)
lunar caustic[ˈluːnəˈkɔːstik] 医用硝酸银,含银腐蚀剂
cast[kɑːst] vt. 浇铸
cauterize[ˈkɔːtəraiz] vt. 灼,腐蚀
nitric[ˈnaitrik] acid. 硝酸
evaporate[iˈvæpəreit] vi. 蒸发
halide[ˈhælaid] n. 卤化物
reduce[riˈdjuːs] vt.n. 还原,减小
agent[ˈeidʒənt] n. 剂,试剂
deposit[diˈpɔzit] vt.n. 沉淀
finely[ˈfainli] ad. 细小地
spot[spɔt] n. 斑点,地点
indelible[inˈdelibl] a. 不能拭去的
marking ink[ˈmɑːkiŋˈiŋk] 不褪色墨水
fluoride[ˈfluəraid] n. 氟化物
bromide[ˈbroumaid] n. 溴化物

iodide[ˈaiədaid] n. 碘化物
darken[ˈdɑːkən] vi. 变黑
exposure[iksˈpəuʒə] n. 暴露,暴光
base[beis] n. 碱
strong base 强碱
precipitate[preˈsipitit] n. 沉淀物 [preˈsipiteit] v. 沉淀
inactivity[inækˈtiviti] n. 不活泼性
aurous[ˈɔːrəs] a. 亚金的,一价金的
~ chloride 氯化亚金
resemble[riˈzembl] vt. 类似,像
sensitive[ˈsensitiv] a. 敏感的,灵敏的
auric[ˈɔːrik] a. 金的,三价金的
hydrochloric[ˈhaidrəˈklɔrik] acid. 盐酸
aqua regia[ˈækwəˈriːdʒiə] 王水
chloroauric[ˈklɔːrəˈɔːrik] acid. 氯金酸
activity[ækˈtiviti] n. 活动性、活泼,活度
~ series 活动顺序
displace[disˈpleis] vt. 代替,置换
bubble[ˈbʌbl] n. 泡;v. 起泡,鼓泡
impurity[imˈpjuəriti] n. 杂质
contaminate[kənˈtæmineit] vt. 污染
electric couple 电对,电偶
facilitate[fəˈsiliteit] v. 促进
basic carbornate[ˈkɑːbənit] 碱式碳酸盐,
filler[ˈfilə] n. 填料,填充剂
tire[ˈtaiə] n. 轮胎
body[ˈbɔdi] n. 底质,底料,坯(体)
atmosphere[ˈætməsfiə] n. 大气,大气压
white lead 铅白,碱式碳酸锌
zincate[ˈziŋkeit] n. 锌酸盐

前 缀

re-[ri(ː)-] (1)再,重新
　　recycle, redissolve, recrystallize
　　(2)互,反,还
　　react, recover(回收), replace(取代)
mono-[ˈmɔnə-] (uni-[ˈjuːni-]) 单,一
　　monoclinic, monomer(单体), monoxide

tri-['trai-]三
  trivalent, tripod(三脚架)
tetra-['tetrə-]四
  tetrachloride, tetrahedron
pent(a)-['pent(ə)-]五,戊
  pentahydrate, pentane
hex(a)-['heksə-]六,己
hept(a)['heptə-]七,庚
oct(a)-['ɔkt(ə)-]八,辛
nona-['nɔnə-]九,壬
deca-['dekə-]十,癸

-en[-ən]变得(动词词尾)
  darken, whiten, lessen(减少,减轻)

词 组

hammer into… 锤成
contribute to… 有助于
added to… 被加入到……中
the addition of A to B 把A加到B中
set up… 形成,产生
responsible for… 负……责任,
      造成……结果
break down(into) 分解为……,
      打碎(成)……
resistant to 抗……的(作用),
     防……的(作用)
the action of A on B A对B的作用
slow down(up)… 使……慢下来
have the advantage over(of)… 优于
inferior to… 不如……,在……之下

后 缀

-er[-ə](名词词尾)
 ①人:worker, teacher, driver
 ②工具:container, recorder
 ③物料:filler, drier, foamer(发泡剂)
-ish[-iʃ]略带……的,微……的(形容词词
 尾)greenish, reddish

Notes

1. 本句中"in thickness"和"in length"表示厚度和长度,一般放在单位的后面,也可以不用介词短语,直接用形容词,如 The box is 8 inches long 6 inches broad and 5 inches high. 这只盒子8英寸长,6英寸宽,5英寸高。

2. add:主要用法如下
 (1)动词:(be)added(to)
 Copper hydroxide dissolves as more ammonium hydroxide is added.
 Dilute ammonium hydroxide is added to a solution of copper(Ⅱ) ions.
 (2)名词:"the addition of…to…";可作主语用,或在它的前面加by作方式状语,加on作时间状语。
 The addition of a strong base to a solution of a silver salt precipitates brown silver oxide.
 Copper(Ⅱ) ferrocyanide is obtained on the addition of a soluble ferrocyanide to a solution of Cu^{2+}.

3. One might expect the hydroxide of silver to precipitate, but it seems likely that silver hydroxide is very unstable and breaks down into the oxide and water – if, indeed, it is ever formed at all. 本句中"One might … precipitate"和"but it seems … water"是并列复合句,前者的谓语 might expect是虚拟语气,表示与实际情况相反。"if,… at all"为一插入语从句,而indeed又是这个

从句中的插入语(插入语可以是从句,也可以是短语或单个的词),这类插入语从句与状语很接近,它们和句子其它部分关系不很密切,去掉之后句子仍然成立,它不修饰句中的谓语或某个词,而是说明全句。**译文**:"你可能想到氢氧化银会沉淀,但它看来不稳定,可能分解成氧化物和水——如果说氢氧化银确实生成过。"

4. use是化学中常用的词,主要用法如下:

(1)名词:"(the)… use of…"

The principal use of ZnO is as a filler in rubber manufacture.

(2)动词:"be used in(+n)…";"be used as…"

The halides of silver are used in photography.

The formation of this salt is also used as a test for Cu^{2+}.

(3)形容词:useful

The reaction is useful only for the synthesis of primary nitroalkanes.

Exercises

1. Fill in the blanks with proper words or phrases.

Mercury is a heavy, silvery-white ____ metal. Compared to other metals, it is a poor ____ of heat, but a good conductor of electricity.

Copper sulfate is ____ industrially by treating copper metal with hot concentrated ____ or its oxides with ____ sulfuric acid.

Silver nitrate can be prepared by reacting silver with nitric acid, ____ silver nitrate, water, and oxides of nitrogen. This is ____ under a fume hood because of toxic nitrogen oxides evolved ____ the reaction.

  A. produced   B. during   C. sulfuric acid   D. liquid,

  E. conductor   F. dilute   G. performed   H. resulting in

2. Match the following explanation with the proper word.

  ____ galvanize, ____ volatile, ____ anhydrous, ____ dissociate, ____ evaporate.

  A. without water; especially without water of crystallization

  B. cover with zinc

  C. lose or cause to lose liquid by vaporization leaving a more concentrated residue

  D. evaporating readily at normal temperatures and pressures

  E. to undergo a reversible or temporary breakdown of a molecule into simpler molecules or atoms

Lesson 4
GROUP ⅢB—ⅧB ELEMENTS

Group ⅢB includes the elements scandium, yttrium, lanthanum, and actinium[1], and the two rare-earth series of fourteen elements each[2]-the lanthanide and actinide series. The principal source of these elements is the high gravity river and beach sands built up by a water-sorting process during long periods of geologic time. Monazite sand, which contains a mixture of rare earth phosphates, and an yttrium silicate in a heavy sand are now commercial sources of a number of these scarce elements.

Separation of the elements is a difficult chemical operation. The solubilities of their compounds are so nearly alike that a separation by fractional crystallization is laborious and time-consuming. In recent years, ion exchange resins in high columns have proved effective. When certain acids are allowed to flow down slowly through a column containing a resin to which ions of Group ⅢB metals are adsorbed, ions are successively released from the resin[3]. The resulting solution is removed from the bottom of the column or tower in bands or sections. Successive sections will contain specific ions in the order of release by the resin. For example, lanthanum ion (La^{3+}) is most tightly held to the resin and is the last to be extracted, lutetium ion (Lu^{3+}) is less tightly held and appears in one of the first sections removed. If the solutions are recycled and the acid concentrations carefully controlled, very effective separations can be accomplished. Quantities of all the lanthanide series (except promethium, Pm, which does not exist in nature as a stable isotope) are produced for the chemical market.

The predominant group oxidation number of the lanthanide series is +3, but some of the elements exhibit variable oxidation states. Cerium forms cerium (Ⅲ) and cerium (Ⅳ) sulfates, $Ce_2(SO_4)_3$ and $Ce(SO_4)_2$, which are employed in certain oxidation-reduction titrations. Many rare earth compounds are colored and are paramagnetic, presumably as a result of unpaired electrons in the $4f$ orbitals.

All actinide elements have unstable nuclei and exhibit radioactivity. Those with higher atomic numbers have been obtained only in trace amounts. Actinium ($_{89}Ac$), like lanthanum, is a regular Group ⅢB element.

Group ⅣB Elements

In chemical properties these elements resemble silicon, but they become increasingly more metallic from titanium to hafnium. The predominant oxidation state is +4 and, as with silica (SiO_2), the oxides of these elements occur naturally in small amounts. The formulas and mineral names of the oxides are TiO_2, rutile; ZrO_2, zirconia; HfO_2, hafnia.

Titanium is more abundant than is usually realized. It comprises about 0.44%[4] of the earth's crust. It is over 5.0% in average composition of first analyzed moon rock. Zirconium and titanium oxides occur in small percentages in beach sands.

Titanium and zirconium metals are prepared by heating their chlorides with magnesium metal. Both are particularly resistant to corrosion and have high melting points.

Pure TiO_2 is a very white substance which is taking the place of white lead in many paints. Three-fourths of the TiO_2 is used in white paints, varnishes, and lacquers. It has the highest index of refraction (2.76) and the greatest hiding power of all the common white paint materials. TiO_2 also is used in the paper, rubber, linoleum, leather, and textile industries.

Group ⅤB Elements: Vanadium, Niobium, and Tantalum

These are transition elements of Group ⅤB, with a predominant oxidation number of +5. Their occurrence is comparatively rare.

These metals combine directly with oxygen, chlorine, and nitrogen to form oxides, chlorides, and nitrides, respectively. A small percentage of vanadium alloyed with steel gives a high tensile strength product which is very tough and resistant to shock and vibration. For this reason vanadium alloy steels are used in the manufacture of high-speed tools and heavy machinery. Vanadium oxide is employed as a catalyst in the contact process of manufacturing sulfuric acid. Niobium is a very rare element, with limited use as an alloying element in stainless steel. Tantalum has a very high melting point (2850℃) and is resistant to corrosion by most acids and alkalies.

Groups ⅥB and ⅦB Elements

Chromium, molybdenum, and tungsten are Group ⅥB elements. Manganese is the only chemically important element of Group ⅦB. All these elements exhibit several oxidation states, acting as metallic elements in lower oxidation states and as nonmetallic elements in higher oxidation states. Both chromium and manganese are widely used in alloys, particularly in alloy steels.

Group ⅧB Metals

Group ⅧB contains the three triads of elements. These triads appear at the middle of long periods of elements in the periodic table, and are members of the transition series. The elements of any given horizontal triad have many similar properties, but there are marked differences between the properties of the triads, particularly between the first triad and the other two. Iron, cobalt, and nickel are much more active than members of the other two triads, and are also much more abundant in the earth's crust. Metals of the second and third triads, with many common properties, are usually grouped together and called the platinum metals.

These elements all exhibit variable oxidation states and form numerous coordination compounds.

Corrosion

Iron exposed to the action of moist air rusts rapidly, with the formation of a loose, crumbly deposit of the oxide. The oxide does not adhere to the surface of the metal, as does aluminum oxide and certain other metal oxides, but peels off, exposing a fresh surface of iron to the action of the air. As a result, a piece of iron will rust away completely in a relatively short time unless steps are taken to prevent the corrosion. The chemical steps in rusting are rather obscure, but it has been established that the rust is a hydrated oxide of iron, formed by the action of both oxygen and moisture, and is markedly speeded up by the presence of minute amounts of carbon dioxide[5].

Corrosion of iron is inhibited by coating it with numerous substances, such as paint, an aluminum powder gilt, tin, or organic tarry substances or by galvanizing iron with zinc. Alloying iron with metals such as nickel or chromium yields a less corrosive steel. "Cathodic protection" of iron for lessened corrosion is also practiced. For some pipelines and standpipes zinc or magnesium rods in the ground with a wire connecting them to an iron object have the following effect: with soil moisture acting as an electrolyte for an Fe-Zn couple the Fe is lessened in its tendency to become Fe^{2+}. It acts as a cathode rather than an anode[6].

New words and Expressions

lanthanide['lænθənaid] n. 镧系
actinide['æktinaid] n. 锕系
rare earth['rɛə ə:θ] 稀土
rare earth element 稀土元素
beach['bi:tʃ] n. (海、河、湖的)滩头
sort['sɔ:t] vt. 分级,拣选
water-sorting 水(力)拣选,水选
monazite['mɔnəzait] n. 独居石
phosphate['fɔsfeit] n. 磷酸盐(酯)
silicate['silikit] a. 硅酸盐(酯)
scarce[skɛəs] a. 稀有的,不足的
~ element 稀有元素
solubility[ˌsɔlju'biliti] n. 溶解度
fractional['frækʃnl] a. 分馏的,分级的
~ crystallization 分级结晶
laborious[lə'bɔ:riəs] a. 费劲的,吃力的
time-consuming[kən'sju:miŋ] 费时,浪费时间
resin['rezin] n. 树脂
ion exchange resin 离子交换树脂

tower['tauə] n. 塔
section['sekʃn] n. 断面,部分
extract[iks'trækt] vt. 抽提,萃取;
['ekstrækt] n. 萃出物
recycle[ri:'saikl] vt.n 再循环
concentration[ˌkɔnsen'treiʃən] n. 浓度,浓缩
isotope['aisətəup] n. 同位素
reduction[ri'dʌkʃən] n. 还原
titrate['titreit] v. 滴定
titration[ti'treiʃn] n. 滴定
paramagnetic[ˌpærəmæg'netik] a. 顺磁的
unpaired[ʌn'pɛəd] a. 未配对的
rare-rare earth 稀有稀土
radioactivity['reidiəuæk'tiviti] n. 放射性
trace[treis] n. 微量,痕量;vt. 探索,考察
regular['regjulə] a. 正规的,正式的
silica['silikə] n. 硅石,二氧化硅
rutile['ru:tail] n. 金红石,氧化钛
zirconia[zə:'kəuniə] a. 氧化锆

hafnia[ˈhæfniə] n. 氧化铪

corrosion[kəˈrəuʒən] n. 腐蚀

hiding[ˈhaidiŋ] n. 掩蔽, 躲藏

varnish[ˈvɑːniʃ] n. 清漆, 罩光漆

lacquer[ˈlækə] n. 大漆, 中国漆

index of refraction[ˈindeks əvriˈfrækʃən] 折光指数

linoleum[liˈnəuljəm] n. 漆布, 油毡

leather[ˈleðə] n. 皮革

refractory[riˈfræktəri] n. 耐火材料

occurrence[əˈkʌrens] n. 埋藏(量)

tensile strenth[ˈtensail streŋθ] 抗张强度

tough[tʌf] a. 不易磨损的, 耐磨的

shock[ʃɔk] n. 冲击

vibration[vaiˈbreiʃən] n. 振动

catalyst[ˈkætəlist] n. 催化剂

contact process[ˈkɔntektˈprəuses] 接触法

stainless[ˈsteinlis] a. 不锈的

　~ steel 不锈钢

acid[ˈæsid] a.n. 酸

manganous[ˈmæŋgənəs] a. 亚锰的, 二价锰的

manganic[mæŋˈgænik] a. 锰的, 三价锰的

manganate[ˈmæŋgəneit] n. 锰酸盐

dioxide[daiˈɔksaid] n. 二氧化物

permanganate[pəˈmæŋgənit] n. 高锰酸盐

analytic(al)[ˌænəˈlitik(əl)] a. 分析的

analytical chemistry　分析化学

standard solution[ˈstændəd səˈljuːʃən] 标准溶液

endpoint[ˈendˈpɔint] n. 终点

triad[ˈtraiəd] n. 三个一组, 三素组, 三价元素

coordination[kəuˌɔːdiˈneiʃən] n. 配位

　~ compound 配位化合物

moist[mɔist] a. 潮湿的

rust[rʌst] v. 生锈; n. 铁锈

loose[luːs] a. 松散的

crumbly[ˈkrʌmbli] a. 易碎的

deposit[diˈpɔzit] vt.n. 沉淀(物); 附着(物)

adhere[ədˈhiə] vi. 粘附

peel[piːl] vi. 脱皮, 剥落(off)

moisture[ˈmɔistʃə] n. 湿气

minute[maiˈnjuːt] a. 微小的

coat[kəut] vt. 涂; n. 涂层

gilt[gilt] n. 炫目的外表, 涂层

tarry[ˈtɑːri] a. 焦油状的, 柏油状的

cathodic protection[kæˈθɔdik prəˈtekʃən] 阴极保护

pipe[paip] n. 管道, 输送管

pipeline[ˈpaipˈlain] n. 管线

standpipe[ˈstændˈpaip] n. 竖管, 筒形塔

electrolyte[iˈlektrəulait] n. 电解质

cathode[ˈkæθəud] n. 阴极

anode[ˈænəud] n. 阳极

mineral[ˈminərəl] a. 矿物的, 无机的; n. 矿物

前　缀

iso-[ˈaisə-] 同, 等, 异

　①同; isotope, isotropic(各向同性的)

　②等: isotactic(等规立构的) isothermal(等温的)

　③异: isomer, isobutane, isonicotine(异烟碱)

后　缀

-a[-ə] 取掉某些元素的-ium, -um 或-on 词尾, 然后加-a, 便构成该元素的氧化物, 如 zirconia, hafnia, alumina, silica

-ad[-əd, -æd] 表示集合数字(名词词尾)

　triad 三个(一组), 三素组, 三价元素

　monad 一个, 一价物, 一价元素

　dyad 二个(一组), 二素组, 二价元素

　tetrad 四个(一组), 四素组, 四价元素

-less[-lis] 不, 无(形容词词尾)

　stainless, colorless

-ness[-nis]性质,状态(名词词尾)
　　toughness,hardness
-age[-idʒ]总和,状态,结果(名词词尾)
　　average,percentage,linkage 键(合)

词　组

build up… 造成
in recent years　近些年来
a number of…　许多,一些
as a result of…　由于……
give A to B　把A给B
(as) listed below (above)…　列于下(上)面的
as with…　正如
in small amounts　少量地
take the place of…　代替……
employ…as…　用……作……;
　　　　　　　　　把……用作……
exposed to…　容易受到……
expose A to B　使A受到B的(作用,影响)
as a result　结果,从而
speed up　加速
act as…　作为……,起……的作用
connect…to…　把……接到……上

Notes

1. 元素的一般构词规律
   (1) 金属元素词尾:一般为-lum,-um(个别非金属元素,如helium属例外),
   　有相当一部分为习惯命名,如zinc,antimony等。
   (2) 非金属元素词尾:(a)-on 如boron,carbon及氡以外的稀有气体;(b)-gen,如hydrogen,oxygen;(c)-ine,如astatine及卤素;(d)俗名:如sulfur,phosphorus

2. "…of…each":当多个系列中每个系列有同等数目的成员时,可用此结构表示,如two short periods of 8 elements each

3. When certain acids are allowed to flow down slowly through a column containing a resin to which ions of Group ⅢB metals are adsorbed, ions are successively released from the resin. 该句系复合句,主句是"ions are…the resin","when certain … are adsorbed"为时间状语从句,从句中带有一个修饰a resin的定语从句,即"to which … adsorbed"。

4. 0.44%应读作"zero point forty-four percent",100%读作"a hundred percent."

5. The chemical steps in rusting are rather obscure, but it has been established that the rust is a hydrated oxide of iron, formed by the action of both oxygen and moisture, and is markedly speeded up by the presence of minute amounts of carbon dioxide. 该句是由"The chemical … obscure"和"but it has … carbon dioxide"组成的并列复合句,后一个分句中it是形式主语,真正的主语是that引导的从句,从句中有两个并列的谓语,一个是以系动词is+表语a hydrated oxide of iron构成的,后一个是被动态,formed引起的分词短语是表示原因的状语。
   译文:"生锈的化学过程不十分清楚,但已确认铁锈是氧和湿气作用形成的水合氧化铁,并且由于微量二氧化碳的存在而显著加速。"

6. 请注意区别英语中阴极和阳极,阴离子和阳离子的表达,并牢记:
   cathode[ˈkæθəud] n. 阴极　　　　　　anode[ˈænəud] n. 阳极
   cathodic[kæˈθɔdik] a. 阴极的　　　　anodic[æˈnɔdik] a. 阳极的

cation[ˈkætaiən] n. 阳离子　　　　　　　anion[ˈænaiən] n. 阴离子
cationic[ˈkætaiˈɔnik] a. 阳离子的　　　　anionic[ˈænaiˈɔnik] a. 阴离子的

Exercises

1. Fill in the blanks with proper words or phrases.

   Lanthanoid series of chemical elements ____ the 15 metallic chemical elements with atomic numbers 57 ____ 71. These elements, ____ the chemically similar elements scandium and yttrium, are often collectively ____ the rare earth elements.

   ____ the metals that form inter oxide layers, iron oxides ____ more volume than other metal and thus flake off, ____ fresh surfaces for ____.

   A. unlike　　B. corrosion　　C. comprises　　D. along with,

   E. known as　　F. exposing　　G. occupy　　H. through

2. Match the following explanation with the proper word.

   ____ laborious, ____ extract, ____ corrosion, ____ vibration, ____ moisture

   A. the gradual destruction of metal by the effect of water or chemicals

   B. taking a lot of time and effort

   C. a continuous slight shaking movement

   D. small amounts of water that are present in the air, in a substance, or on a surface

   E. to carefully remove a substance from something which contains it, using a machine, chemical process

Lesson 5
INTERHALOGEN AND NOBLE GAS COMPOUNDS

Interhalogen and noble gas compounds comprise a relatively limited family of highly reactive and unstable molecules whose primary importance is their role in testing chemical bonding theory[1]. At first it may seem rather strange to treat the chemistry of the halogens and the noble gases, two groups that represent the extremes of chemical activity and inertia, in the same section. The superficial differences between the halogens and the noble gases are much reduced, however, if we focus our attention on the comparison of halide ions (particularly F^-) with the isoelectronic noble gas atoms and the noble gas compounds with halogen atoms or the halogens in their higher positive-oxidation states.

Noble gases are exceptional in their reluctance to either gain or lose an electron. Halide ions-because of their excess negative charge, relative to the isoelectronic noble gas atoms-have both a lower ionization energy and a lesser electron affinity. On the other hand, noble gas cations have greater electron affinities and greater ionization energies than do isoelectronic halogen atoms. From such considerations, it is obvious that inert gases should be less reactive than are halide ions, but their compounds should show even higher reactivity than the halogens. The big question remaining is: Are there any chemically significant conditions under which noble gases can be persuaded to yield electrons sufficiently to produce stable compounds? The answer is definitely, yes! (The same question can be asked of halogen atoms, which have ionization energies comparable to those of the inert gases.)

Another obvious point of similarity between halogen and noble gas compounds is the characteristically large number of electrons that must be accommodated in the valence shell. For a noble gas atom bonded to any number of other atoms, the octet rule must be exceeded; for a halogen atom to be bonded to more than one other atom, the same must be true. It is a curious historical fact that the mythical inertia of a closed shell did much to diminish the energy expended in the search for noble gas compounds, long after numerous examples of superoctet valence shells were known, particularly among interhalogen compounds.

We may roughly classify the interhalogen compounds into two categories: those in oxidation state zero (the binary analogs of the elementary diatomics[2]) and those in which one of the halogens is in a formally positive oxidation state. Heterodiatomic halogens are generally formed readily on mixing the required pair of halogens in a 1∶1 ratio. The bond energies are always higher in the heteropolar molecules than are the average bond energies of the two constituents and in some cases higher than either. It is this factor that drives the reactions. All heterodiatomics[2] are more or less stable under ambient conditions except for BrF, which spontaneously disproportionates to BrF_3 and Br_2. The bonding in the halogen diatomics can be attributed to a single σ bond, formed by overlap of p orbitals. In the heterodiatomics, the principal new features are the poorer orbital overlaps that are possible between atoms of widely different principal quantum number(n), the polarity arising from the difference in electronegativity, the contribution of ionic terms to increase bond energy, and the relief in interelectronic repulsion in the fluorides, relative to difluorine.

Dihalogens (except for F_2) usually react by dissociation into atoms or by heterolytic dissociation under the influence of an attacking reagent. Thus, reaction of Cl_2 with hydroxide may be viewed as displacement of Cl^- from Cl_2 by OH^-:

$$Cl_2 + OH^- \longrightarrow HOCl + Cl^-$$

The tendency to undergo heterolytic fission increases on descending the group, and the I_2 molecule can actually be cleaved to two stable species:

$$I_2 + 2C_5H_5N + AgNO_3 \longrightarrow [(C_5H_5N)_2I]NO_3 + AgI$$

The increased homolytic bond energies of the heterodiatomic halogens decrease the tendency toward homolytic reactions, but the increased polarity increases the tendency toward heterolytic reactions. Thus, ICl is a much better electrophilic iodinating agent than is I_2 and unlike I_2 even iodinates aromatic compounds.

In the interhalogen compounds, one of the halogen atoms may be assigned a positive oxidation state. As may be expected, the general trends reflect the increasing difficulty of withdrawing electrons from the central atom on ascending the group and with increasing oxidation state. As might be anticipated, all known stable compounds are fully electron-paired, and the series IF, IF_3, IF_5, and IF_7 give us a homogeneous sequence of molecules exemplifying all possible odd-coordination numbers and their associated geometries (there is no known nine-coordinate neutral binary molecule, although many nine-coordinate complexes are known). We may complement this series with some of the fluorides discussed earlier and with the xenon fluorides to be discussed below, thereby completing the primary family of molecular structures and electron configurations for main-group elements. In each of the cases shown, the experimental evidence indicates that the molecule adopts the structure predicted by VSEPR theory. A valence-bond description of the molecules with more than eight-valence electrons requires the inclusion of d orbitals in the hybridization scheme. An MO scheme without the participation of d orbitals requires location of electrons in antibonding orbitals, and therefore bond order of less than one, for molecules with more than eight-valence electrons. The mixing of empty d orbitals into the scheme can lower the energy of the antibonding electrons (make them less antibonding) and thereby increase the bond strength.

New words and Expressions

interhalogen[ˌintə(:)'hælədʒen] *n.* 杂卤素
cleave[kli:v] *vt.* 劈开
noble['nəubl] *a.* 贵重的;(气体)惰性的
~ gas 惰性气体
inertia[i'nəʃiə] *n.* 惰性
superficial[ˌsju:pə'fiʃəl] *a.* 表面的

isoelectronic [ˌaisəuilek'trɔnik] a. 等电子的
oxidation state [ˌɔksi'deiʃən steit] n. 氧化态
reluctance [ri'lʌkəns] n. (to; at) 抵抗
ionization [ˌaiənai'zeiʃən] n. 电离(作用)
affinity [ə'finit] n. 亲和力
cation ['kætaiən] n. 阳离子
accommodate [ə'kɔmədeit] vt. 容纳
valence shell ['veiləns ʃel] n. 价电子层
octet [ɔk'tet] n. 八偶
mythical ['miθikəl] a. 设想的，虚构的
category ['kætigəri] n. 类型
analog ['ænəlɔg] n. 类似物
diatomic [daiə'tɔmik] a. 双原子的
elementary [ˌeli'mentəri] a. 元素的，基本的，单元的
heterodiatomic [ˌhetərədaiə'tɔmik] a. 杂双原子的
heteropolar [ˌhetərə'pəulə] a. 异极的
constituent [kən'stitjuənt] n.a. 成分，组分
ambient ['æmbiənt] a. 周围的
~ temperature 室温
spontaneous [spɔn'teinjəs] a. 自发的
disproportionate [ˌdisprə'pɔːʃənit] a. 不均衡的; vi. [ˌdis'prə'pɔːʃəneit] 歧化
principal quantum number 主量子数
polarity [pəu'læriti] n. 极性
electronegativity [iˌlektrənegə'tiviti] n. 电负性
relief [ri'liːf] n. 减轻，释放，解除
repulsion [ri'pʌlʃən] n. 推斥
dissociation [diˌsəusi'eiʃən] n. 离解(作用)
heterolytic [hetərə'litik] a. 异(性)的
influence ['influəns] n. 影响
displacement [dis'pleismənt] n. 置换，取代
fission ['fiʃən] n. 裂开
species ['spiːʃiz] n. 物种
descend [di'send] vi. 下降
ascend [ə'send] vi. 上升
homolytic [hɔmə'litik] a. 同性的，均一的
electrophilic [iˌlektrəu'filik] a. 亲电的
iodinate ['aiədineit] vt. 碘化

aromatic [ˌærəu'mætik] a. 芳香族的
withdraw [wið'drɔː] vt. 吸引，收回
electron-paired a. 电子成对的
series ['siəriːz] n. 系列，组
configuration [kənˌfigju'reiʃən] n. 构型
VSEPR theory (=valence-shell electron pair repulsion theory) 价层电子对互斥理论
scheme [skiːm] n. 图式，方式
location [ləu'keiʃən] n. 定域
antibonding ['ænti'bɔndiŋ] a. 反键的
bond order   键级

前缀

super- ['sjuːpə-] ① 上 superficial, supernatant (上层清液) ② 超 supercentrifuge, superconductor ③ 过 superheated, supersaturated
heter(o)- ['hetər(ə)-] 其它，异，杂 heterolytic, heteropolar, heterocyclic, heterophase (多相)
homo- ['hɔ(əu)mə-] (heter(o)-的反义词词头) 同，相同，均匀 homolytic, homolog, homophase, homogeneous

词组

seem rather strange to do…   做…似乎很不习惯
focus one's attention to …   集中某人的注意于……上
be attributed to…   被归结于……
arise from (out of)…   产生于…，由…引起的
classify…into…   将…分类为…
displacement of A from B by C   用C取代B中之A
on descending (ascending) the group   沿该族元素下行(上行)
in the search for …   在寻找…中
comparison (of…) with…   将…与…比较
have both…and …   既具有…又具有…
more or less   或多或少(地)，程度不同(地)

under ambient condition 在通常条件下	the tendency toward… 向…的趋势
under the influence of… 在…影响之下	be assigned… 被指认为…
react by …into… 经由…过程反应变为……	as may be expected (anticipated) 如所预期的
be cleaved to… 被分裂为…	complement A with B 用B补充A

Notes

1. "Interhalogen and noble gas compounds comprise a relatively limited family of highly reactive and unstable molecules whose primary importance is their role in testing chemical bonding theory"。该句系复合,主句是"Interhalogen and……molecules"。whose引导一非限制性定语从句,修饰"Interhalogen and noble gas compounds"。**译文**:"杂卤素和惰性气体化合物包含了高度活泼和不稳定分子的比较有限的一族,这些化合物头等重要性是它们在考察化学键理论中的作用"。

2. 某些形容词+词尾s,可构成名词,如:

diatomic 双原子的	diatomics 双原子化合物
heterodiatomic 杂双原子	heterodiatomics 杂双原子化合物
organometallic 金属有机的	organometallics 金属有机物(化学)
thermodynamic 热力(学)的	thermodynamics 热力学
kinetic 运动的,动力(学)的	kinetics 动力学
chemical 化学的	chemicals 化学品

Exercises

1. Fill in the blacks with proper words or phrases.

    An interhalogen ____ is a molecule which ____ two or more different halogen atoms and no atoms of elements from ____ group.
    Most interhalogen compounds known are ____. Their formulae are ____ XYn, where n = 1, 3, 5 or 7, and X is the ____ electronegative of the two halogens. They are all prone to ____, and ionize to give rise to polyhalogen ions.

    A. contains B. binary C. less D. hydrolysis,
    E. compound F. any other G. generally

2. Match the following explanation with the proper word.
    ____ superficial, ____ elementary, ____ spontaneous, ____ homolytic, ____ configuration

    A. happening or arising without apparent external cause
    B. existing in or relating to the top layer of something
    C. of or being the essential or basic part
    D. an arrangement of parts or elements
    E. breakdown to equal pieces

Lesson 6
THE CLASSIFICATION OF INORGANIC COMPOUNDS

The Classes of Compounds

Thousands and tens of thousands of compounds are known to the chemist today. It would be impossible to learn properties and behavior of even a fraction of this number if it had to be done on the basis of individual compounds[1]. Fortunately, most chemical compounds can be grouped together in a few classes. Then, if we can properly classify a compound, we are at once aware of the general properties of the compound from knowledge of the properties of that class or group of compounds. For example, HCl is classed as an acid, and by becoming familiar with the behavior of acids as a distinct class, we are at once aware of the general properties of the compound[2]. A great many of the compounds we are to study may be classified as acids, bases, salts, metallic oxides, or nonmetallic oxides. Of these five classes of compounds, the first three-acids, bases, and salts-are by far the most important[3].

When an acid, base, or salt is dissolved in water the resulting solution is a conductor of the electric current and is termed an electrolyte. If no conduction of current occurs, the compound is known as a nonelectrolyte.

Classification of Common Compounds

By looking at the chemical formulas we may classify many common compounds in the following way.

1. Acids, in the conventional sense, may be recognized by noting that the H is written first in the formula and that the rest of the compound is generally nonmetallic. Ex., HCl, H_2SO_4, HClO.

2. Conventional bases have OH radicals written last in the formula. The first part of the formula is usually a metal. Ex., NaOH, $Ca(OH)_2$, $Fe(OH)_3$.

3. A salt consists of a metal, written first, combined with a non-metal or radical written last in a formula. Ex., NaCl, $Fe_2(SO_4)_3$, $Ca(ClO)_2$.

4. Oxides are compounds containing oxygen and only one other element.

If the element other than oxygen is a nonmetal, the oxide is classed as a nonmetal oxide or an acidic anhydride. The latter name comes about because water added to nonmetal oxides under certain conditions produces acids. Likewise, if water is removed from an acid containing oxygen, the acid anhydride (without water) results.

The other class of oxides, metallic oxides or basic anhydrides, consist of oxygen combined with a metal. When water is added under proper conditions to basic anhydrides, bases result and vice versa.

Acids

All acids in the conventional sense contain hydrogen*, which may be replaced by metals. The negative portion of the acid molecule is composed of a nonmetal or a radical (negative valence group). These negative valence groups (except oxide and hydroxide) are often referred to as acid radicals. All acids are covalent compounds in which the atoms are held together by a sharing of electrons. When an acid is dissolved in water, ions are formed as a result of the transfer of a hydrogen ion (proton) from the acid molecule to the water molecule—for example,

$$H:\ddot{C}l:+H:\ddot{O}:H \longrightarrow [H:\overset{H}{\underset{}{\ddot{O}}}:H]^+ + [:\ddot{C}l:]^-$$

This is a case of coordinate valence, in which an unused pair of electrons from the water molecule combines with a hydrogen ion to form a hydronium ion. The hydronium ion is a hydrated hydrogen ion or proton ($H^+ \cdot H_2O$) and, while the ionization of acids in aqueous solution depends on its formation, we shall ordinarily use the simple H^+ in writing equations. Such equations are thereby simplified and easier to balance.

The chief characteristic of an acid is its ability to furnish hydrogen ions (protons); therefore, an acid is usually defined as a substance which may furnish protons.

Properties of Acids. In general, aqueous solutions of acids are characterized by the following properties:[①]

1. They have a sour taste. Lemons, oranges, and other citrus fruits owe their sour taste to[4] the presence of citric acid; the taste of sour milk is due to the presence of lactic acid.

2. They turn blue litmus paper red. Litmus is a dye which has a red color in acid solution and a blue color in basic solution; paper which has been soaked in litmus is referred to as litmus paper. Substances of this type, which enable us to determine whether a given solution is acid or basic, are called indicators. Methyl orange and phenolphthalein are other indicators frequently used by chemists.

3. They react with certain metals to produce hydrogen. Reactions of this type were studied in connection with the preparation of hydrogen.

4. They react with bases to produce salts and water.

Common strong acids are H_2SO_4, HNO_3, HCl, HBr, and HI. Most other acids are generally only partially ionized and consequently only moderately strong or weak.

Bases

All metallic hydroxides are classed as conventional bases. Of the common bases only NaOH,

---

* This concept of acids will be extended later.

KOH, $Ca(OH)_2$ and $Ba(OH)_2$ are appreciably soluble in water. If these compounds are dissolved in water, the OH^- is common to all of their solutions.

An aqueous solution of NH_3 is also classed as a base, since OH^- ions are present in the solution.

In each of these compounds we find a combination of a metal (or NH_4) with the hydroxide group. Just as the characteristic part of an acid is hydrogen ion, so the characteristic part of a base in water solution is the hydroxide ion, OH^-. Later the concept of a base will be extended to include substances which do not furnish hydroxide ions in solution.

Properties of Bases. In general, water solutions of metallic hydroxides (bases) exhibit the following properties:

1. Bitter taste.
2. Soapy or slippery feeling.
3. Turn red litmus paper blue.
4. React with acids to form salts and water.
5. Most metallic hydroxides are insoluble in water. Of the common ones, only NaOH, KOH, $Ca(OH)_2$, $Ba(OH)_2$, and NH_3 are soluble.

The common strong bases are NaOH, KOH, $Ca(OH)_2$, and $Ba(OH)_2$.

Salts

An acid reacts with a base to produce a salt and water. Hydrogen from the acid combines with hydroxide from the base to form water molecules.

The reaction of an acid with a base is called neutralization. If all the water is removed by evaporation from the solution after the reaction, the positive ions from the base and the negative ions from the acid form a crystal lattice of solid salt.

It was shown that the compound sodium chloride, a salt, is an electrovalent compound and is ionized in the solid or crystalline state. The crystal is made up of positive sodium ions and negative chloride ions oriented in a definite pattern. In general, most salts in the crystalline state are electrovalent and are composed of ions oriented in a definite way.

New words and Expressions

chemist ['kemist] n. 化学家
become [bi'kʌm] vt. 适合, 与……相称
radical ['rædikəl] a.n. 基, 根, 原子团
acidic [ə'sidik] a. 酸的, 酸性的
anhydride [æn'haidraid] n. 酐
basic anhydride 碱酐
negative ['negətiv] a. 负的, 阴的

acid radical 酸根
share [ʃɛə] v. 共享; 分享
transfer ['trænsfəː] n. 转移
[træns'fəː] v. 转移
coordinate [kəu'ɔːdinit] a.n. 配位
[kəu'ɔːdineit] v. 配位
~ valence 配价

hydronium[hai'drəuniəm]ion
　水合氢离子

aqueous[ˈeikwiəs] a.水的

equation[iˈkweiʃən] n.方程式

balance[ˈbæləns] n.天平,平衡;v.平衡

substance[ˈsʌbstəns] n.物质

sour[ˈsauə] a.酸的,酸味的;v.变酸

taste[teist] n.味道,气味,味觉

lemon[ˈlemən] n.柠檬

orange[ˈɔrindʒ] n.柑,桔;橙色

citric[ˈsitrik] n.柠檬的
　~ acid　柠檬酸

citrus[ˈsitrəs] n.柑桔属

lactic[ˈlæktik] n.乳的
　~ acid　乳酸

litmus[ˈlitməs] n.石蕊
　~ paper　石蕊试纸

dye[ˈdai] n.染料

soak[səuk] v.n.浸泡,浸渍

enable[iˈneibl] vt.使……能够

indicator[ˈindikeitə] n.指示剂

methyl[ˈmeθil] n.甲基
　~ orange　甲基橙

phenolphthalein[ˌfiːnɔlˈfθæliin] n.酚酞

react[ri(ː)ˈækt] vi.反应(with,on)

reaction[ri(ː)ˈækʃən] n.反应

strong acid　强酸

ionize[ˈaiənaiz] vt.离子化,电离

bitter[ˈbitə] a.n.苦味

soapy[ˈsəupi] a.肥皂般的,滑腻的

slippery[ˈslipəri] a.滑的

neutralization[ˌnjutrəlaiˈzeiʃən] n.中和

evaporation[iˈvæpəˈreiʃən] n.蒸发

positive[ˈpɔzətiv] a.正的,阳的
　~ ion　正离子

negative ion　负离子

lattice[ˈlætis] n.格子,点阵

crystal lattice　晶格

electrovalent[iˌlektrouˈveilənt] a.电价的

orient[ˈɔːriənt] vt.定向,取向

pattern[ˈpætən] n.模型,型式

前　缀

chem(o)-[kem(ə)-]化学(的)
　chemist,chemistry,chemical,
　chemosynthesis,chemosphere（臭氧层）

anhydr(o)-[ænˈhaidr(ai,ə)-]脱水,无水,酐
　anhydride,anhydrous,anhydrite,
　anhydroglucose

后　缀

-or[-ə]剂,器,体,人(名词词尾),由词尾为-t(e)的动词加-or形成,如:
　剂　indicator,initiator,inhibitor
　器　reactor,extractor,desiccator
　体　conductor,acceptor,insulator
　人　inventor,bailor(委托人)

-ar[-ə](形容词词尾)
　molecular,nuclear,linear

词　组

thousands and tens of thousands　成千上万

be known to…　被……所知

a fraction of…　一小部分

on the basis of…　根据……,在……基础上

aware of…　知道,意识到……

familiar with…　熟悉,通晓

a great(good)many of…　很多,大量

look at…　考察,注视

vice versa[ˈvaisiˈvəːsə]　反之亦然

in the conventional sence　按传统的观念（常识）

come about　发生,出现

other than…　与……不同的,除……以外的

hold together　结合在一起

define as　定义为……

enable…to(+ing)　使……能够(……)

in connection(connexion) with……
与……有关,在……方面

just as…,so… 正如……那样……也

Notes

1. It would be impossiple to learn the properties and behavior of even a fraction of this number if it had to be done on the basis of individual compounds. 该句是虚拟语气主从复合句,主句是"It would be…of this number","if it…compounds。"是虚拟条件状语从句。本句译文"如果根据个别化合物来了解这么多化合物的性质和行为,即使其中的一小部分也是不可能的。"

2. For example, HCl is classed as an acid, and by becoming familiar with the behavior of acids as a distinct class, we are at once aware of the general properties of the compound。该句是并列复合句,第一分句是"HCl is…acid,"其中as an acid是主语补足语,另一个分句是"by becoming…the compound."其中介词by+动名词短语作状语。译文:"例如,盐酸归类为酸,由于已熟悉作为不同类别的酸的性质,我们就会立即知道这一化合物的一般性质。"

3. by far the most important(最重要的):注意"by far + 形容词最高级(或比较级)"意思为"最……的"

4. to: due to;in virtue of

5. 化学反应的表述方法

(1)通用型动词:表达化学反应常用的一般性动词有react,convert.form,prepare,produce,change,yield,give,obtain,synthesize等,其用法举例如下:

Acids react with certain metals to produce hydrogen.

Sulfur in the form of SO_2 obtained in the roasting of sulfide ores is converted to sulfuric acid.

A mixture of nitrogen and hydrogen forms ammonia in the reaction.

这些动词中,react和synthesize应用得最广泛,而且也常用其名词,如reaction.

The reaction of ethylene with chlorine gives an addition product.

(2)专用型动词:除使用上述通用型动词外,对不同类型的反应可使用特定的动词,其中有combine(化合),break down(分解),decompose(分解),neutralize(中和),replace(取代),oxidize(氧化),reduce(还原),add(加成),dissociate(离解),hydrolyze(水解),hydrate(水合),hydrogenate(加氢),dehydrogenate(脱氢),nitrate(硝化),chlorate(氯化)等,如:

Hydrogen from the acid combines with hydroxide from the base to from water molecules.

Silver hydroxde is very unstable and breaks down into the oxide and water.

Sodium hydroxide neutralizes HCl to form sodium chloride and water.

The two hydrogen atoms in H_2SO_4 may be replaced with sodium and potassium to yield the mixed salt, $NaKSO_4$.

Tertiary amines are oxidized by either hydrogen peroxide or peracids to produce amine oxides.

Lithium borohydride will reduce esters more rapidly

HCN (hydrogen cyanide) adds to the carbonyl group of aldehydes and ketones to form cyanohydrins.

Exercises

1. Fill in the blacks with proper words or phrases.

   Any acid with a pK$_a$ ___ which is less than about −2 is ___ as a strong acid. This results from the very high buffer ___ of solutions with a pH value of 1 or ___.

   A weak acid is a substance that ___ dissociates when it is dissolved in a solvent. In solution there is an ___ between the acid.

   The ___ of the strong bases appear in the first and second groups of the ___ table (alkali and earth alkali metals).

      A. less   B. value   C. capacity   D. partially

      E. periodic   F. equilibrium   G. classed   H. cations

2. Match the following explanation with the proper word.

   ___ radical, ___ transfer, ___ aqueous, ___ indicator, ___ neutralization

   A. similar to or containing or dissolved in water

   B. the act of moving something from one location to another

   C. a chemical reaction in which an acid and a base interact with the formation of a salt

   D. something that can be regarded as a sign of something else

   E. two or more atoms bound together as a single unit and forming part of a molecule

Lesson 7
THE NOMENCLATURE OF INORGANIC COMPOUNDS

With the discovery of thousands of new inorganic compounds it has become necessary to revise the traditional rules of nomenclature. An international committee has recommended a set of rules for naming compounds, and these are now being adopted throughout the world. Many of the older names are still used, however, and our ensuing discussion will include in many cases both the old and new, with emphasis on the latter. One of the principal changes is that proposed by Albert Stock and now known as the Stock system for the naming of compounds of metals (oxides, hydroxides, and salts) in which the metal may exhibit more than one oxidation state. In these cases the oxidation state of the metal is shown by a Roman numeral in parentheses immediately following the English name of the metal which corresponds to its oxidation number[1]. If the metal has only one common oxidation number, no Roman numeral is used. Another important change is in the naming of complex ions and coordination compounds. We will defer the nomenclature of the latter until these compounds are discussed.

Naming Metal Oxides, Bases, and Salts

The student should have a good start in learning nomenclature if he has learned the Valence Table 3 which gives both charges on ions and names for the more common ones[2]. A compound is a combination of positive and negative ions in the proper ratio to give a balanced charge and the name of the compound follows from names of the ions, for example, NaCl, is sodium chloride; $Al(OH)_3$ is aluminum hydroxide; $FeBr_2$ is iron(II) bromide or ferrous bromide; $Ca(C_2H_3O_2)_2$ is calcium acetate; $Cr_2(SO_4)_3$ is chromium(III) sulfate or chromic sulfate, and so on. Table 4 gives some additional examples of the naming of metal compounds. Of the two common systems used, the Stock system is preferred. Note that even in this system, however, the name of the negative ion will need to be obtained from Valence Table 4.

Negative ions, anions, may be monatomic or polyatomic. All monatomic anions have names ending with ide. Two polyatomic anions which also have names ending with ide are the hydroxide ion, OH^-, and the cyanide ion, CN^-.

Many polyatomic anions contain oxygen in addition to another element. The number of oxygen atoms in such oxyanions is denoted by the use of the sulffixes ite and ate, meaning fewer and more oxygen atoms, respectively. In cases where it is necessary to denote more than two oxyanions of the same element, the prefixes hypo and per, meaning still fewer and still more oxygen atoms, respectively, may be used. A series of oxyanions is named in Table 5.

Table 3. Some Common Ions

1+		2+		2+		3+	
ammonium	NH_4^+	barium	Ba^{2+}	magnesium	Mg^{2+}	aluminum	Al^{3+}
copper(I)	Cu^+	calcium	Ca^{2+}	manganese(II)	Mn^{2+}	chromium(III)	Cr^{3+}
hydrogen	H^+	chromium(II)	Cr^{2+}	mercury(II)	Hg^{2+}	iron(III)	Fe^{3+}
potassium	K^+	copper(II)	Cu^{2+}	mercury(I)	Hg_2^{2+}		
silver	Ag^+	iron(II)	Fe^{2+}	tin(II)	Sn^{2+}		
sodium	Na^+	lead(II)	Pb^{2+}	strontium	Sr^{2+}		
				zinc	Zn^{2+}		

3−		2−		1−		1−	
arsenate	AsO_4^{3-}	carbonate	CO_3^{2-}	acetate	$C_2H_3O_2^-$	hydrogen sulfite	HSO_3^-
arsenite	AsO_3^{3-}	chromate	CrO_4^{2-}	bromide	Br^-	hydride	H^-
phosphate	PO_4^{3-}	dichromate	$Cr_2O_7^{2-}$	chlorate	ClO_3^-	hydroxide	OH^-
phosphite	PO_3^{3-}	oxalate	$C_2O_4^{2-}$	chloride	Cl^-	hypochlorite	ClO^-
		oxide	O^{2-}	chlorite	ClO_2^-	iodate	IO_3^-
		sulfide	S^{2-}	cyanide	CN^-	nitrate	NO_3^-
		sulfate	SO_4^{2-}	fluoride	F^-	iodide	I^-
		sulfite	SO_3^{2-}	hydrogen carbonate (bicarbonate)	HCO_3^-	nitrite	NO_2^-
				hydrogen sulfate	HSO_4^-	perchlorate	ClO_4^-
						permanganate	MnO_4^-

Table 4. Names of some metal oxides, bases, and salts

FeO	iron(II) oxide	ferrous oxide
Fe_2O_3	iron(III) oxide	ferric oxide
$Sn(OH)_2$	tin(II) hydroxide	stannous hydroxide
$Sn(OH)_4$	tin(IV) hydroxide	stannic hydroxide
Hg_2SO_4	mercury(I) sulfate	mercurous sulfate
$HgSO_4$	mercury(II) sulfate	mercuric sulfate
NaClO	sodium hypochlorite	sodium hypochlorite
$K_2Cr_2O_7$	potassium dichromate	potassium dichromate
$Cu_3(AsO_4)_2$	copper(II) arsenate	cupric arsenate
$Cr(C_2H_3O_2)_3$	chromium(III) acetate	chromic acetate

Table 5. Names of oxyanions

Fewest Oxygen Atoms hypo—ite		Fewer Oxygen Atoms —ite		More Oxygen Atoms —ate		Most Oxygen Atoms per —ate	
ClO^-	hypochlorite	ClO_2^-	chlorite	ClO_3^-	chlorate	ClO_4^-	perchlorate
BrO^-	hypobromite	BrO_2^-	bromite	BrO_3^-	bromate	BrO_4^-	perbromate
IO^-	hypoiodite	IO_2^-	iodite	IO_3^-	iodate	IO_4^-	periodate
PO_2^{3-}	hypophosphite	PO_3^{3-}	phosphite	PO_4^{3-}	phosphate		
		NO_2^-	nitrite	NO_3^-	nitrate		
		SO_3^{2-}	sulfite	SO_4^{2-}	sulfate		
				CO_3^{2-}	carbonate		

Naming Nonmetal oxides

The older system of naming and one still widely used employs Greek prefixes for both the number of oxygen atoms and that of the other element in the compound[3]. The prefixes used are (1)mono-, sometimes reduced to mon-, (2)di-, (3)tri-, (4)tetra-, (5)penta-, (6)hexa-, (7)hepata-, (8)octa-, (9)nona- and (10)deca-. Generally the letter a is omitted from the prefix(from tetra on)when naming a nonmetal oxide and often mono-is omitted from the name altogether.

The Stock system is also used with[4] nonmetal oxides. Here the Roman numeral refers to the oxidation state of the element other than oxygen.

In either system, the element other than oxygen is named first, the full name being used followed by oxide. Table 6 shows some examples.

Table 6. Names of some nonmetal oxides

Formula		
CO	carbon(Ⅱ) oxide	carbon monoxide
CO_2	carbon(Ⅳ) oxide	carbon dioxide
SO_3	sulfur(Ⅵ) oxide	sulfur trioxide
N_2O_3	nitrogen(Ⅲ) oxide	dinitrogen trioxide
P_2O_5	phosphorus(Ⅴ) oxide	diphosphorus pentoxide
Cl_2O_7	chlorine(Ⅶ) oxide	dichlorine heptoxide

Naming Acids

Acid names may be obtained directly from a knowledge of Valence Table 3 by changing the name of the acid ion(negative ion)in the table as follows:

Ion in Table 3	Corresponding Acid
_____ate	_____ic
_____ite	_____ous
_____ide	hydro_____ic

Table 7 shows examples of this relationship.

Table 7. Names of some acids

Formula of Acid	Acid Ion in Table 3	Name of Acid
$HC_2H_3O_2$	$C_2H_3O_2^-$ acetate	acetic acid
H_2CO_3	CO_3^{2-} carbonate	carbonic acid
$HClO_2$	ClO_2^- chlorite	chlorous acid
$HClO_4$	ClO_4^- perchlorate	perchloric acid
HCN	CN^- cyanide	hydrocyanic acid
HBr	Br^- bromide	hydrobromic acid
H_4SiO_4	SiO_4^{4-} silicate	silicic acid
H_3AsO_4	AsO_4^{3-} arsenate	arsenic acid
$HMnO_4$	MnO_4^- permanganate	permanganic acid

There are a few cases where name of the acid is changed slightly from that of the acid radical; for example H_2SO_4 is sulfuric acid rather than sulfic. Similarly, H_3PO_4, is phosphoric acid rather than phosphic.

The less common negative ions are not included in the Valence Table 3. For example, BO_3^{3-} is the borate ion and H_3BO_3 is boric acid; TeO_4^{2-} is the tellurate ion and H_2TeO_4 is telluric acid, and so on.

Acid and Basic Salts

It is conceivable that in the neutralization of an acid by a base, only a part of the hydrogen might be neutralized; thus

$$NaOH + H_2SO_4 \longrightarrow NaHSO_4 + H_2O$$

The compound $NaHSO_4$ has acid properties, since it contains hydrogen, and is also a salt, since it contains both a metal and an acid radical. Such a salt containing acidic hydrogen is termed an acid salt. Phosphoric acid (H_3PO_4) might be progressively neutralized to form the salts, NaH_2PO_4, Na_2HPO_4, and Na_3PO_4. The first two are acid salts, since they contain replaceable hydrogen. A way of naming these salts is to call Na_2HPO_4 disodium hydrogen phosphate and NaH_2PO_4 sodium di-hydrogen phosphate. These acid phosphates are important in controlling the alkalinity of the blood. The third compound, sodium phosphate Na_3PO_4, which contains no replaceable hydrogen, is often referred to as normal sodium phosphate, or trisodium phosphate to differentiate it from the two acid salts.

Historically, the prefix bi- has been used in naming some acid salts; in industry, for example, $NaHCO_3$ is called sodium bicarbonate and $Ca(HSO_3)_2$ calcium bisulfite. Since the bi- is somewhat misleading, the system of naming discussed above is preferable.

If the hydroxyl radicals of a base are progressively neutralized by an acid, basic salts may be formed:

$$Ca(OH)_2 + HCl \longrightarrow Ca(OH)Cl + H_2O$$

Basic salts have properties of a base and will react with acids to form a normal salt and water. The OH group in a basic salt is called an hydroxyl group. The name of $Bi(OH)_2NO_3$ would be bismuth dihydroxynitrate.

Mixed Salts

If the hydrogen atoms in an acid are replaced by two or more different metals, a mixed salt results. Thus the two hydrogen atoms in H_2SO_4 may be replaced with sodium and potassium to yield the mixed salt $NaKSO_4$, sodium potassium sulfate. $NaNH_4HPO_4$ is a mixed acid salt that may be crystallized from urine.

================= **New words and Expressions** =================

rule[ruːl] n. 规律

nomenclature[nəuˈmenklətʃə] n. 命名,术语

name[neim] vt. 命名,给……取名

sodium chloride[ˈsəudjəm klɔːraid] 氯化钠

aluminum hydroxide[əˈljuminəm haiˈdrɔksaid] 氢氧化铝
iron(Ⅱ) bromide[ˈaiənˈbroumaid] 溴化亚铁
ferrous[ˈferəs] a. 亚铁的，二价铁的
~ bromide 溴化亚铁
acetate[ˈæsitit] n. 醋酸盐(酯)
calcium acetate 醋酸钙
chromium(Ⅲ)sulfate[ˈkrəumiəm ˈsʌlfeit] 硫酸铬
chromic[ˈkrəumik] a. 铬的，三价铬的
arsenate[ˈɑːsinit] n. 砷酸盐
arsenite[ˈɑːsinait] n. 亚砷酸盐
phosphite[ˈfɔsfait] n. 亚磷酸盐
chromate[ˈkrəumit] n. 铬酸盐
oxalate[ˈɔksəleit] n. 草酸盐(酯)
dichromate[daiˈkrəumit] n. 重铬酸盐
sulfite[ˈsʌlfait](= sulphite) n. 亚硫酸盐
chlorate[ˈklɔːrit] n. 氯酸盐
chlorite[ˈklɔːrait] n. 亚氯酸盐
cyanide[ˈsaiənaid] n. 氰化物
hydrogen carbonate(=bicarbonate) n. 酸式碳酸盐
hydrogen sulfate(=bisulfate) n. 酸式硫酸盐
hydrogen sulfite 酸式亚硫酸盐
hydride[ˈhaidraid] n. 氢化物
hypochlorite[ˌhaipəˈklɔːrait] n. 次氯酸盐
iodate[ˈaiədeit] n. 碘酸盐
nitrite[ˈnaitrait] n. 亚硝酸盐(酯)
perchlorate[pəˈklɔːreit] n. 过氯酸盐
ferric[ˈferik] a. 铁的，三价铁的
stannous[ˈstænəs] a. 亚锡的，二价锡的
stannic[ˈstænik] a. 锡的，四价锡的
mercurous[ˈməkjurəs] a. 亚汞的，一价汞的
mercuric[məˈkjuərik] a. 汞的，二价汞的
cupric[ˈkjuːprik] a. 铜的，二价铜的
monatomic[ˌmɔnəˈtɔmik] a. 单原子的
polyatomic[ˌpɔliəˈtɔmik] a. 多原子的，多元的

anion[ˈænaiən] n. 阴离子
oxyanion[ˈɔksiˈænaiən] n. 含氧阴离子
hypobromite[ˌhaipəˈbrəumait] n. 次溴酸盐
hypoiodite[ˌhaipəˈaiədait] n. 次碘酸盐
hypophosphite[ˈhaipəuˈfɔsfait] n. 次磷酸盐
bromite[ˈbrəumait] n. 亚溴酸盐
iodite[ˈaiədait] n. 亚碘酸盐
bromate[ˈbrəumeit] n. 溴酸盐
perbromate[pəˈbrəumeit] n. 过溴酸盐
periodate[pəːˈraiədeit] n. 高碘酸盐
acid ion 酸根离子
acetic acid[əˈsiːtik ˈæsid] 醋酸，乙酸
carbonic acid[kɑːˌbɔnikˈæsid] 碳酸
chlorous acid[ˈklɔːrəsˈæsid] 亚氯酸
perchloric acid[pəˈklɔːrikˈæsid] 高(过)氯酸
hydrocyanic acid[ˈhaidrəusaiˈænikˈæsid] 氢氰酸
hydrobromic acid[ˈhaidrəˈbrəumikˈæsid] 氢溴酸
silicic[siˈlisik] a.(含)硅的
~ acid 硅酸
arsenic[ɑːˈsenik] a.(含)砷的
~ acid 砷酸
permanganic[ˌpəːmæŋˈɡænik] a.
~ acid 高锰酸
phosphoric[fɔsˈfɔrik] a.(含)磷的
~ acid 磷酸
borate[ˈbɔːreit] n. 硼酸盐(酯)
boric[ˈbɔːrik] a.(含)硼的
~ acid 硼酸
tellurate[ˈteljuəreit] n. 碲酸盐(酯)
telluric[teˈljuərik] a.(含)碲的
~ acid 碲酸
neutralize[ˈnjuːtrəlaiz] vt. 中和
acid salt 酸式盐
alkalinity[ˌælkəˈliniti] n. 碱度，碱性
normal[ˈnɔːməl] a. 正(常)的，当量的
bisulfite[baiˈsʌlfait] n. 酸式亚硫酸盐，

亚硫酸氢盐
hydroxyl[haiˈdrɔksil] n. 羟基
hydyoxyl radical 氢氧根
hydroxy group 羟基(基团)
basic salt[ˈbeisikˈsɔːlt] 碱式盐
normal salt 中性盐,正盐
bismuth dihydroxynitrate[ˈbizməθ
　　ˈdaihaiˌdrɔksiˈnaitreit] 碱式硝酸铋
sodium potassium sulfate 硫酸钠钾
urine[ˈjuərin] n. 尿

前　缀

hypo-[haipə-] 次,低,过少,连
　hypochlorite, hypophosphate;
　hypoosmotic(低渗的),
　hypoglycemia(低血糖);
　hyposulfuric acid(连二硫酸)
oxy-[ɔksi-] 含氧, oxyanion, oxy-acid;
　氧化, oxydenate, oxydehydrogenation;
　氧(代), oxychlorination, oxycarbide;
　羟基, oxyamine, oxy-mercuration
per-[pə-] ①高, permanganate, perchloric acid; ②过, peroxide, percarbonate; ③全,
perhalogenation, perhydroanthracene
bi-[bai-] 二,双,联,酸式
　binary　二元的;
　bimolecular　双分子的, bicyclic;
　biphenyl[baiˈfenil] 联苯;
　bicarbonate　酸式碳酸盐

后　缀

-ate[-eit](动词词尾)evaporate
　[-it]…酸盐(酯)(名词词尾)
　nitrate, sulfate, carbornate
-ite[-ait] 亚……酸盐(酯)(名词词尾)
　nitrite, sulfite
-ic[-ik](与-ous对应,形容词词尾)
　nitric, sulfuric
-ous[-əs] 亚……的(与-ic对应,原子价较低的,形容词词尾)nitrous, sulfurous

词　组

follow from…　根据……得出
from…on　从……开始,从……起
differentiate A from B　把A和B分开
correspond to…　相当于

Notes

1. In these cases the oxidation state of the metal is shown by a Roman numeral in parentheses immediately following the English names of the metal which corresponds to its oxidation number. 该句中"In these…of the metal"是主句,其中"following…the metal"现在分词短语作状语,"which…number"是定语从句,修饰Roman numeral. 本句译文:"在这种情况下,金属氧化态是紧接着金属的英语名称之后在圆括号中用罗马数字表示,该数字与金属的氧化数一致。"

2. The student should have a good start in learning nomenclature if he has learned the Valence Table 3 which gives both charges on ions and names for the more common ones. 本句是主从复合句,主句是"The student…nomenclature","if he…common ones"是条件状语从句,从句中还包括了一个定语从句"which gives …common ones",它修饰the Valence Table 3. 定语从句中有两个宾语,即"charges on ions"和names, "for the more common ones"介词短语作names的定语,其中ones代表ions. 本句译文:"一个学生如果掌握了给出离子电荷和较常见离子名称的价键表3,他在掌握命名方面就必定有了一个好的开端。"

3. The older system of naming and one still widely used employs Greek prefixes for both the number of oxygen atoms and that of the other element in the compound. 该句是一个简单句,主语是 system 和 one,employs 是谓语动词,one 指前面的 system,used 过去分词作后置定语,修饰 one,that 代替前面的 the number,这种结构较常见,如 There are a few cases where name of the acid is changed slightly from that of the acid radical."有几个从酸根名字而得出的酸名称稍有变化的例子。"这里 that 代表前面的 name.

4. with:for

Exercises

1. Fill in the blanks with proper words or phrases.

The IUPAC nomenclature is a ____ of naming chemical compounds and for describing the science of chemistry ____. It is maintained by the International Union of Pure and ____ Chemistry. The rules for naming organic and inorganic compounds are ____ in two publications, the Blue Book and the Red Book. A third ____, Green Book, contains recommendations for the use of symbols for physical ____, while a fourth, the Gold Book, ____ a large number of technical used in chemistry.

    A. Applied   B. system   C. printed   D. publication,

    E. defines   F. quantities   G. terms   H. in general

2. Choose the right name for the following compounds.

____ 高氯酸钠,____ 次氯酸钙,____ 碳酸氢钠,____ 硫酸铜,____ 氢氧化钠,____ 氧化钙

    A. sodium hydroxide

    B. sodium perchlorate

    C. copper(Ⅱ) sulfate

    D. sodium hydrogen carbonate

    E. calcium oxide

    F. calcium hypochlorite

Lesson 8
BRÖNSTED'S AND LEWIS' ACID-BASE CONCEPTS

Brönsted-Lowry Concept(1923). According to this concept, an acid is a substance that releases protons (a proton donor) and a base is a substance that combines with protons (a proton acceptor). Neutralization reactions involve the transfer of protons. Salts are merely aggregates of ions that are produced in some but not all neutralization reactions.

Upon loss of a proton, an acid forms a base, since by the reverse reaction the substance formed can gain a proton. In the same way a base forms an acid upon gaining a proton. These relationships can be represented by equations, and the sum is an equation for a neutralization reaction.

$$Acid_1 \rightleftharpoons H^+ + Base_1 \tag{1}$$

$$Base_2 + H^+ \rightleftharpoons Acid_2 \tag{2}$$

$$Acid_1 + Base_2 \rightleftharpoons Acid_2 + Base_1 \tag{3}$$

We see, then, that a neutralization reaction is a competition for protons between two bases. The acid and base represented in Equation 1 (and Equation 2) are a conjugate pair; that is, an acid forms its conjugate base upon loss of a proton and a base forms its conjugate acid when it gains proton. Substances with the same subscript in Equation 3 are conjugate pairs. This concept can be illustrated by some examples.

$Acid_1$ $Base_2$		$Acid_2$		$Base_1$
1. $H_3O^+ + OH^-$	$\rightleftharpoons$	H_2O	+	H_2O
2. $HCl + H_2O$	$\rightleftharpoons$	H_3O^+	+	Cl^-
3. $H_2O + NH_3$	$\rightleftharpoons$	NH_4^+	+	OH^-
4. $H_2O + CO_3^{2-}$	$\rightleftharpoons$	HCO_3^-	+	OH^-
5. $H_2O + HCO_3^-$	$\rightleftharpoons$	H_2CO_3	+	OH^-

The first equation is a neutralization reaction according to the Arrhenius concept as well as Brönsted-Lowry but, since Arrhenius recognized only the base OH^-, the other reactions are excluded according to his concept[1]. Note that in the second and third reactions that water acts as a base and an acid, respectively. Water, like many other substances that can both donate and accept protons, is amphoteric according to this concept. It functions as an acid in the presence of bases stronger than itself (NH$_3$, for example) and as a base in the presence of an acid stronger than itself (HCl, for example). The positions of the equilibria given above depend upon the relative proton-donating ability of the two acids in each case (or the relative proton-accepting ability of the two bases).

These relationships allow the establishment of scales of acid and base strength. Relative acid strength of proton donors is measured by the extent to which reactions with a common base proceed at equilibrium[2]. Relative base strength of proton acceptors is established in similar fashion. Table 8. lists several familiar substances arranged in the order of decreasing acid strength. Since the conjugate base of a weak acid is strong, and vice versa, the bases formed in the reactions given are in the reverse order by strength; that is, the strongest base is at the bottom of the table.

Table 8. Acid-Base Reactions in water

Acid				Base
$HClO_4$	+	$H_2O \rightleftharpoons H_3O^+$	+	ClO_4^-
HCl	+	$H_2O \rightleftharpoons H_3O^+$	+	Cl^-
H_2SO_4	+	$H_2O \rightleftharpoons H_3O^+$	+	HSO_4^-
H_3PO_4	+	$H_2O \rightleftharpoons H_3O^+$	+	$H_2PO_4^-$
$HC_2H_3O_2$	+	$H_2O \rightleftharpoons H_3O^+$	+	$C_2H_3O_2^-$
H_2S	+	$H_2O \rightleftharpoons H_3O^+$	+	HS^-
NH_4^+	+	$H_2O \rightleftharpoons H_3O^+$	+	NH_3
H_2O	+	$H_2O \rightleftharpoons H_3O^+$	+	OH^-
NH_3	+	$H_2O \rightleftharpoons H_3O^+$	+	NH_2^-

Strongest acids at top; strongest bases at bottom.

The relative strength of a base, according to the Brönsted-concept, is a measure of its ability to accept a proton. Between any two substances that compete for protons, the one that gains protons over the other is more basic. For example, water may compete with the chloride ion for a proton, but Cl$^-$ is so weak as a base that water easily accepts a proton and forms H_3O^+ in this reaction:

$$HCl + H_2O \longrightarrow H_3O^+ + Cl^-$$

On the other hand, water loses out in the competition for protons when water competes with ammonia for protons. Ammonia takes protons from the water, forcing the water to act as an acid in the reaction

$$NH_3 + H_2O \longrightarrow NH_4^+ + OH^-$$

The relative strength of the competitors for protons can be derived from Table 8. The strongest proton acceptors are the bases at the bottom of the list, while the strongest acids or proton donors are acids at the top of the list.

Lewis Concept (1923). G. N. Lewis proposed another acid-base concept in the same year that the Brönsted-Lowry concept appeared. Lewis defined an acid as a molecule or ion that can accept an electron pair from another molecule or ion, and a base as a substance that can share its electron pair with an acid. Thus, an acid is an eletron-pair acceptor and a base is an electron-pair donor. When a Lewis acid reacts with a Lewis base, consequently, a coordinate covalent bond is formed. This is essentially the neutralization reaction.

Electron dot formulas clarify the mechanism of forming a coordinate bond in the neutralization of a Lewis acid. Neutralization of a proton by an ammonia molecule is

$$H^+ + \overset{H}{\underset{H}{\ddot{H}}}:H \longrightarrow \left[H : \overset{H}{\underset{H}{\ddot{N}}} : H \right]^+$$

The acid-base pair Al^{3+} and H_2O undergoes Lewis neutralization by forming six coordinate bonds.

Many displacement reactions illustrate the relative strength of the Lewis acid or base. For example, the cyanide ion is a stronger base than the fluoride ion by virtue of its ability to displace the fluoride ion from the hexafluoroferrate(III) ion:

$$FeF_6^{3-} + 6CN^- \longrightarrow Fe(CN)_6^{3-} + 6F^-$$

There are many reactions that fit the Lewis acid-base concept. In contrast to proton loss or gain in the Brønsted-Lowry concept, the Lewis concept emphasizes the electron pair — a Lewis acid lacks an electron pair in an empty orbital, or has an orbital that can be vacated and a Lewis base has a nonbonding electron pair and can supply this pair to another substance lacking an electron pair.

It is evident that the Lewis concept applies not only to the chemical behavior correlated by the Brønsted-Lowry concept, but also to many chemical reactions that do not involve proton transfer, and for this reason it is most useful. Its generality precludes the establishment of a scale of acid and base strengths for all Lewis acids and bases, but comparisons can be made between selected substances.

The terms nucleophilic, and electrophilic are sometimes applied to bases and acids. Lewis acids that accept an electron pair are electrophilic, and the strength of Lewis bases is measured by their tendencies to supply electrons. For example, H_2O is a stronger base than Cl^- because H_2O displaces Cl^- in supplying an electron pair for the proton:

$$HCl + H_2O \longrightarrow H_3O^+ + Cl^-$$

Consequently, H_2O is a stronger nucleophilic agent than Cl^-.

New words and Expressions

donor['dəunə] n. 给(予)体
acceptor[ək'septə] n. 接受体
aggregate['ægrigit] n.a. 聚集(体);
　　　　　['ægrigeit] v. 总计,聚集
compete[kəm'pi:t] vi. 竞争

competition[ˌkɔmpi'tiʃən] n. 竞争
conjugate[ˌkɔndʒugit] a. 共轭的
　　　　　[ˌkɔndʒugeit] v. 结合,共轭,配对
subscript['sʌbskript] n. 下标(符)
donate[dəu'neit] v. 捐赠,给出

amphoteric [ˌæmfəˈterik] a. 两性的
function [ˈfʌŋkʃən] vi. 起作用
equilibrium [ˌiːkwiˈlibriəm]
　　（复 equilibria [ˌiːkwiˈlibriə]）n. 平衡
weak [wiːk] a. 弱的, 差的
extent [iksˈtent] n. 程度
fashion [ˈfæʃən] n. 方式, 样子
measure [ˈmeʒə] n. 量度, 标准, 尺度; v. 度量
relative [ˈrelətiv] a. 有关的, 相对的
distinguish [disˈtiŋgwiʃ] v. 区别; 识别
leveling effect [ˈlevəliŋ iˈfekt] 均衡效应
solvent [ˈsɔlvənt] n. 溶剂, a. 溶解的
electron pair [iˈlektrɔn pɛə] 电子对
coordinate covalent bond　配位共价键
coordinate bond　配位键
dot [dɔt] n. 圆点; v. 打点
mechanism [ˈmekənizəm] n. 机理, 历程
cyanide ion　氰离子
fluoride ion　氟离子
virtue [ˈvəːtjuː] n. 效力, 功效, 力
hexafluoferrate [ˈheksəˈflu(ː)ərəˈfereit]
ion　六氟合铁(Ⅲ)离子
fit [fit] v.a. 适合, 符合
lack [læk] vt. 没有, 缺少; 需要
empty [ˈempti] a. 空的, 未占用的
vacate [vəˈkeit] vt. 空出, 腾出
product [ˈprɔdəkt] n. 产物, 生成物; 乘积
generality [ˌdʒenəˈræliti] n. 概念, 普遍性
preclude [priˈkluːd] vt. 排除, 消除
nucleophilic [ˌnjuːkliəˈfilik] a. 亲核的,
　　亲质子的
nucleophilic agent　亲核试剂

前缀和后缀

dis- [dis-] 否定, 相反, 分开, 离去
　　displace, disappear, discharge discolor,
　　discontinuous
-phile [-fail]（名词词尾）亲, 爱好, 嗜
　　nucleophile, dienophile（亲双烯体）
-philic [-ˈfilik]（形容词词尾）亲, 嗜,
　　electrophilic, hydrophilic
-ize(-ise) [-aiz]（动词词尾）, 使……化
　　neutralize, oxidize, ionize, crystallize,
　　hybridize（杂化）

词　组

depend (up)on…　取决于……
that is…　也就是（插入语）
in the presence of …　在……存在下,
　　　　　　　　　　在……参加下
serve as …　用作, 起……作用
compete (with…) for…　（与……）竞争
lose out　输掉, 失败
enough (+a.) to (+inf.)　足够……（可以）……
derive…from…由　　推出, 由……得出
what else　还有什么
for this reason　为此, 因此之故
so called…　通常所说的
distinguish between…　区别……, 识别……
contrast to (with)…　与……对比,
　　　　　　　　　　与……相反

Notes

1. The first equation is a neutralization reaction according to the Arrhenius concept as well as Brönsted-Lowry but, since Arrhenius recognized only the base OH−, the other reactions are excluded according to his concept. 该句是 but 连接的并列复合句, 在第二个分句中主句是"the

other … to his concept", "since…base OH-"是它的原因状语从句。本句译文:"按阿累尼乌斯和布朗斯特概念,第一个方程是中和反应。但是由于阿累尼乌斯认为碱只是OH-,所以按他的概念,其它反应都被排斥在中和反应之外。"

2. Relative acid strength of proton donors is measured by the extent to which reactions with a common base proceed at equilibrium. 该句是带有定语从句的主从复合句,"Relative…by the extent"是主句,"to which…at equilibrium"是定语从句,这里which代表the extent作介词to的宾语,注意在指物时,介词后只能用which,不能用that,如用that,就必须把介词放到句子后面,即"……by the extent that reactions with a common base proceed to at equlibrium"。**译文**:"质子给体的相对酸强度是依靠它与同一碱反应进行到平衡的程度来度量。"

Exercises

1. Fill in the blacks with proper words or phrases.

   In 1923 ___ chemists Johannes Nicolaus Brønsted in Denmark and Thomas Martin Lowry in England both ___ proposed the theory that ___ their names. In the Brønsted–Lowry theory acids and bases are ___ by the way they react with each other, which allows for greater ___. The definition is expressed ___ an equilibrium expression.

   A. carries B. defined C. physical D. generality

   E. in terms of F. independently

2. Match the following explanation with the proper word.

   ___ donor, ___ conjugate, ___ amphoteric, ___ mechanism, ___ vacate

   A. the way that something works

   B. a matter, person, or group that gives something

   C. joined together especially in a pair or pairs

   D. to leave a seat or room so that someone or something else can use it

   E. having characteristics of both an acid and a base and capable of reacting as either

Lesson 9
THE COORDINATION COMPLEX

The chemical basis for the formation of a coordination complex is the coordinate bond. There must be an electron-pair acceptor and an electron-pair donor. Thus, coordination reactions are examples of Lewis acid-base neutralization. The central ion is the Lewis acid, or electron-pair acceptor, and the surrounding groups, called ligands, are Lewis bases or electron-pair donors. In general, this reaction may be described as

$$M^{n+} + x:L \rightleftharpoons M(:L)_x^{n+}$$

The forward reaction is coordination and the reverse reaction is dissociation. Both coordination and dissociation usually occur in stepwise fashion when x is greater than 1, so that the general equation above may be the sum of several stepwise equations.

The nature of the coordinate bond varies from essentially covalent, including double bond character in some cases, to essentially ionic. In the case of complexes such as $Na(H_2O)_x^+$ in aqueous solution, the sodium ion interacts with the coordinated water molecules much as a sodium ion in crystalline sodium chloride interacts with neighboring chloride ions. The bonding forces can be treated as electrostatic interactions. At the other extreme, complexes like $Fe(CN)_6^{4-}$ involve primarily covalent bonding. This variation in behavior is further illustrated by compounds listed in Table 9. The first four compounds listed are saltlike substances in which either the cation or anion is complex. This is apparent from[1] their composition in aqueous solution. However, these same ions exist in the solids, although their formulas may be written as double salts as in the table. In each of these cases the complex ion survives through the solution process and there is little evidence of free ligand or hydrated monatomic cation in solution. Note that the ligands may be molecular, as NH_3, or anionic, as CN^-, NO_2^-, or F^-. In the case of $VF_4(H_2O)_2^-$, vanadium is coordinated with both anionic and molecular ligands. The fifth compound, $Pt_2Cl_4 \cdot 4NH_3$, contains both complex cation and complex anion in solution and in the solid. The last compound has a crystal structure indicating the presence of $CuCl_4^{2-}$ complex ions. However, when the substance is dissolved in water, a mixture of hydrated monatomic ions results. This simply indicates that Cu

Table 9. Examples of Coordination Complexes

Solid	Solution in Water
$CoCl_3 \cdot 6NH_3$	$Co(NH_3)_6^{3+}$, $Cl^-(aq)$
$4KCN \cdot Fe(CN)_2$	$K^+(aq)$, $Fe(CN)_6^{4-}$
$3KNO_2 \cdot Co(NO_2)_3$	$K^+(aq)$, $Co(NO_2)_6^{3-}$
$NH_4F \cdot VF_3 \cdot 2H_2O$	$NH_4^+(aq)$, $VF_4(H_2O)_2^-$
$Pt_2Cl_4 \cdot 4NH_3$	$Pt(NH_3)_4^{2+}$, $PtCl_4^{2-}$
$2KCl \cdot CuCl_2 \cdot 2H_2O$	$K^+(aq)$, $Cu^{2+}(aq)$, $Cl^-(aq)$

(Ⅱ) forms a more stable complex with water ligands than with chloride ligands, and that a high degree of ionic character allows rapid conversion to hydrated Cu(Ⅱ) upon solution. The absence of $CuCl_4^{2-}$ in solution does not prove that it is not present in the solid.

Ligand field theory

A recent theory which has proven useful in connection with coordination compounds is the crystal field or ligand field theory, which states that ligands are held to the central ion of a coordination compound by electrostatic attractive forces due primarily to the charge of the central ion and the polar nature of the ligands[2]. The extent of the attractive forces determines the stability of the complex, and is dependent on the charge and size of the central ion and ligands.

Using this theory, it is possible to calculate the energy effects for many of the charge and size factors and finally to obtain values for bond energies. The latter are useful in predicting properties and types of configurations for a large number of compounds.

Ligand Groups

Ligands must be polar or polarizable (nonpolar molecules are poor coordinating groups), and they usually have unshared electron pairs that may form coordinate bonds with the central ion. Ligands may be classified (Table 10) on the basis of their ability to form one bond with the metal ion (unidentate) or their ability to form two bonds (bidentate). Ligands that are divalent or contain two donor atoms, like ethylenediamine ($H_2N-CH_2-CH_2-NH_2$), are bidentate because they contain two points or sites that enter into bonds with the central ion.

Ligands may be still more complex and many tridentate, tetradentate, and even hexadentate ligands are known. Polydentate ligands may have both neutral and anionic sites that coordinate with the central ion, such as the glycinate ion:

$$H-\underset{\underset{\uparrow}{H}}{\overset{H}{\underset{|}{N}}}-CH_2-\underset{\uparrow}{\overset{\overset{:O:}{\|}}{C}}-\ddot{\ddot{O}}:^-$$

Table 10. Some Ligand Groups

	Unidentate		
	Ions		Molecules
CN^-	Cyano	H_2O	Aquo
OH^-	Hydroxo	NH_3	Ammine
CH_3COO^-	Acetato	CO	Carbonyl
NO_3^-	Nitrato	CH_3NH_2	Methylamine
NO_2^-	Nitro		
F^-	Fluoro		
Cl^-	Chloro		
Br^-	Bromo		

	Bidentate		
$-OOC-COO-$	Oxalato	$H_2N-CH_2-CH_2-NH_2$	Ethylenediamine
CO_3^{2-}	Carbonato		

Here both the anionic oxygen and the nonionic nitrogen have unshared electron pairs and may coordinate with a given cation. An example is the complex $[Co(H_2NCH_2CO_2)_3]$, in which the cobalt is coordinated with the nitrogen and one of the oxygens of each glycinate. A polydentate ligand, when coordinated at two or more points to a central ion, forms a ring structure such as illustrated by the ethylenediamine complex of cobalt above[3]. This type of complex is called a chelate, and the polydentate ligand is a chelating group.

Naming and Examples of Coordination Compounds

1. Cations are named first, anions last.

2. The names of negative ion ligands end in o. Examples from Table 10 are cyano-, hydroxo-, chloro-, carbonato-. The order of ligands is alphabetic.

3. Neutral units have historical endings, such as ammine, aqua-(formerly aquo-), carbonyl, nitroso-.

4. The oxidation number of the central atom in the complex is specified by Roman numerals in parentheses following the name of the element. When a complex ion is negative, the name of the central element has -ate appended.

5. Neutral complexes are named as if they were cations[4].

6. If a complex is a positive ion, the names of the acid radicals not in the complex[5] complete the name for the compound.

These rules will be clarified by the examples below.

$K^+[BF_4]^-$ $\qquad$ $[Ag(NH_3)_2]^+Cl^-$

Potassium tetrafluoroborate(Ⅲ) $\qquad$ Diamminesilver(Ⅰ) chloride

$K_4^+[Fe(CN)_6]^{4-}$ $\qquad$ $[Cu(NH_3)_4]^{2+}SO_4^{2-}$

potassium hexacyanoferrate(Ⅱ) $\qquad$ Tetraamminecopper(Ⅱ) sulfate

(potassium ferrocyanide)

$[Co(H_2O)_2(NH_3)_2(CO_3)]^+NO_3^-$ $\qquad$ $[Co(NH_3)_3(NO_2)_3]^0$

Diamminediaquacarbonatocobalt(Ⅲ) $\qquad$ Triamminetrinitrocobalt(Ⅲ)

nitrate

New words and Expressions

ligand['laigənd] n. 配位体,配位基
forward['fɔːwəd] a. 向前的,正向的
~ reaction 正反应
reverse reaction 逆反应

double bond['dʌbl'bɔnd] 双键
interact[ˌintə'rækt] vi. 相互作用
neighboring['neibəriŋ] a. 邻近的
electrostatic[iˌlektrəu'stætik] a. 静电的

bonding force['bɔndiŋ 'fɔ:s]价键力
involve[in'vɔlv] vt.包含,包括
variation[ˌvɛəri'eiʃən] n.变化,变异
saltlike['sɔ:lt'laik] a.类似盐的,盐状的
double salt['dʌbl'sɔ:lt]复盐
anionic['ænai'ɔnik] a.阴离子的
ligand field['laigənd 'fi:ld]配位场
crystal field['kristl 'fi:ld]晶体场
polar['pəulə] a.极性的
factor['fæktə] n.因素
bond energy['enədʒi]键能
complete[kəm'pli:t] vt.填满,完成; a.完全的
polarizable['pəulə raizəbl] a.可极化的
unidentate[ˌju:ni'denteit] a.一齿(状)的
unidentate ligand 一齿配体
bidentate[ˌbai'denteit] a.二齿(状)的
ethylenediamine['eθili:ndai'æmi:n] n.乙二胺
attach[ə'tætʃ] v.连结,附加
glycinate['glaisineit] n.甘氨酸盐(酯)
chelate['ki:leit] n.螯合物;v.a.螯合(的)
potassium tetrafluoroborate(Ⅲ) [pə'tæsjəm 'tetrə'flu(:)ərə'bɔ:reit] 四氟合硼(Ⅲ)酸钾
potassium hexacyanoferrate(Ⅱ) [pə'tæsjəm 'heksə 'saiənəu 'fereit] 六氰合铁(Ⅱ)酸钾
potassium ferrocyanide [pə'tæsjəmˌferəu'saiənaid]亚铁氰化钾
triamminetrinitrocobalt(Ⅲ)[trai'æmi:nˌtrai'naitrəukə'bɔ:lt]三硝基三氨合钴(Ⅲ)

前 缀

cyano-['saiənəu(ə)-]氰合,氰基
chloro-['klɔ:rə-]氯合,氯代
carbonato-['kɑ:bəneitə-]碳酸合(根)
ammine-['æmi:n-]氨合
aquo(a)-['ækwə-]水(合)
carbonyl-['kɑ:bənil-]羰基合;羰基
nitroso-[nai'trəusəu(ə)-]亚硝基(合)
acetato-['æsiteitə-]醋酸合
nitrato-['naitreitə-]硝酸(合)
hydroxo-[hai'drɔksəu-]羟基合
oxalato-['ɔksəleitə-]草酸合
methylamine-['meθilə'mi:n-]甲胺合;甲胺

词 组

much as… 差不多就像……那样
a series of… 一系列……,许多……
lie in… 在于……
(be)due to + 名词 由于,归因于
enter into… 参加,成为……的一部分

Notes

1.from:according to

2.A recent theory which has proven useful in connection with coordination compounds is the crystal field or ligand field theory, which states that ligands are held to the central ion of a coordination compound by electrostatic attractive forces due primarily to the charge of the central ion and the polar nature of the ligands. 该句系复合句,主句是"A recent theory is the crystal field or ligand field theoty","which has…compounds"是定语从句,修饰主语theory,"which states…of the ligands"也是定语从句,修饰整个主句,此定语从句中还包括了一个宾语从句"that ligands are …, of the ligands",宾语从句中by引起介词短语作方式状语,"due primarily to…ligands"

作原因状语。译文:"已证实在配位化合物中有用的新理论是晶体场或配位场理论,该理论说明配位体被配位化合物的中心离子拉住,主要是由于中心离子的原有电荷和配位体的极性间的静电引力。"

3."A polydentate ligand, when coordinated at two or more points to a central ion, forms a ring structure such as illustrated by the ethylenediamine complex of cobalt above."是复合句,when引导时间状语,when后是过去分词短语;主句"A polydentate ligand forms a … of cobalt above"中such是代词,指"多齿配体与中心离子形成环状结构",as+过去分词短语作定语,修饰such。译文:"当一个多齿配体的二个或多个点配位于中心离子时,形成一环状结构,这一点已被上述钴的乙二胺络合物的例子说明"。

4."Neutral complexes are named as if they were cations."为复合句,as if引导带有虚拟语气的方式状语从句,意思为"好像…似的"。译文:"中性络合物像阳离子似的命名"。

5. not in the complex:which do not include in the complex.

6. 配合物中配体体位基(齿)数的表示是在形容词dentate(齿状的)前加数字词头构成,如unidentate,bidentate,tridentate,tetradentate,multidentate等。

Exercises

1. Fill in the blanks with proper words or phrases.

   Coordination ____ the "coordinate covalent bonds"(dipolar bonds) between the ____ and the ____ atom. Originally, a complex implied a reversible ____ of molecules, atoms, or ions through such weak ____ bonds. As applied to ____ chemistry, this meaning has evolved. Some metal complexes are formed virtually ____ and many are bound together by bonds that are quite ____.

   A. chemical   B. refers to   C. irreversibly   D. central,
   E. ligands   F. coordination   G. strong   H. association

2. Choose a proper prefix to make the follow words match the meanings.

   ____ polymer 共聚物, ____ molecular 分子间, ____ cover 回收, ____ oxide 过氧化物,
   ____ angle 三角, ____ wave 微波, ____ thermal 等温的, ____ phase 均相

   A. iso-   B. re-   C. co-   D. tri-
   E. inter-   F. micro-   G. homo-   H. per-

Lesson 10
ALKANES

Number of Isomers

The compounds now assigned the generic name alkane are also referred to as saturated hydrocarbons and as paraffin hydrocarbons. The word paraffin, from the Latin parum affinis (slight affinity) refers to the inert chemical nature of the substances and is applied also to the wax obtanable from petroleum and consisting of a mixture of higher alkanes.

Derivation of the formulas of the pentanes (3 isomers), hexanes (5), and heptanes (9) has already demonstrated the sharp rise in diversity with increasing carbon content.

Normal Alkanes

Successive members of the series differ in composition by the increment CH_2 and form a homologous series. Thus heptane and octane are homologous hydrocarbons; icosane is a higher homolog of methane.

Saturated Unbranched-Chain Compounds and Uuivalent Radicals

The first four saturated unbranched acyclic hydrocarbons are called methane, ethane, propane and butane. Names of the higher members of this series consist of a numerical term, followed by " - ane" with elision of terminal "a" from the numerical term. Examples of these names are shown in the table below. The generic name of saturated acyclic hydrocarbons (branched or unbranched) is "alkane."

Examples of names: (n=total number of carbon atoms)

n		n		n	
1	Methane	15	Pentadecane	29	Nonacosane
2	Ethane	16	Hexadecane	30	Triacontane
3	Propane	17	Heptadecane	31	Hentriacontane
4	Butane	18	Octadecane	32	Dotriacontane
5	Pentane	19	Nonadecane	33	Tritriacontane
6	Hexane	20	Icosane	40	Tetracontane
7	Heptane	21	Henicosane	50	Pentacontane
8	Octane	22	Docosane	60	Hexacontane
9	Nonane	23	Tricosane	70	Heptacontane
10	Decane	24	Tetracosane	80	Octacontane
11	Undecane	25	Pentacosane	90	Nonacontane
12	Dodecane	26	Hexacosane	100	Hectane

13	Tridecane	27	Heptacosane	132	Dotriacontahectane
14	Tetradecane	28	Octacosane		

Saturated branched acyclic hydrocarbon is named by prefixing the designations of the side chains to the name of the longest chain which is numbered from one end to the other by Arabic numerals, the direction being so chosen as to give the lowest numbers possible to the side chains. When series of locants containing the same number of terms are compared term by term, that series is "lowest" which contains the lowest number on the occasion of the first difference[1]. This principle is applied irrespective of the nature of the substituents.

Example:
$$\overset{6}{C}H_3-\overset{5}{C}H-\overset{4}{C}H_2-\overset{3}{C}H-\overset{2}{C}H-\overset{1}{C}H_3$$
$$\quad\quad\ \ |\quad\quad\quad\ \ |\quad\ \ |$$
$$\quad\quad CH_3\quad\quad\ CH_3\ CH_3$$

2,3,5-Trimethylhexane

The presence of identical unsubstituted radicals is indicated by the appropriate multiplying prefix di-, tri-, tetra-, penta-, hexa-, hepta-, octa-, nona-, deca, etc.

Example:
$$\quad\quad\quad\quad\quad CH_3$$
$$\quad\quad\quad\quad\quad\ |$$
$$\overset{5}{C}H_3-\overset{4}{C}H_2-\overset{3}{C}-\overset{2}{C}H_2-\overset{1}{C}H_3$$
$$\quad\quad\quad\quad\quad\ |$$
$$\quad\quad\quad\quad\quad CH_3$$

3,3-Dimethylpentane

Univalent radicals derived from saturated acyclic hydrocarbons by removal of hydrogen from a terminal carbon atom are named by replacing the ending "-ane" of the name of the hydrocarbon by "-yl". The carbon atom with the free valence is numbered as 1. As a class, these radicals are called normal, or unbranched chain, alkyls.

Examples:

Pentyl	$\overset{5}{C}H_3-\overset{4}{C}H_2-\overset{3}{C}H_2-\overset{2}{C}H_2-\overset{1}{C}H_2-$
1-Methylpentyl	$CH_3CH_2CH_2CH_2CH(CH_3)-$
2-Methylpentyl	$CH_3CH_2CH_2CH(CH_3)CH_2-$
5-Methylhexyl	$(CH_3)_2CHCH_2CH_2CH_2CH_2-$
Isobutyl	$(CH_3)_2CH-CH_2-$
sec-Butyl	CH_3-CH_2-CH- $\quad\quad\quad\quad\ \ |$ $\quad\quad\quad\quad CH_3$
tert-Butyl	$(CH_3)_3C-$
Neopentyl	$(CH_3)_3C-CH_2-$

Stabilty. Alkanes are relatively inert, chemically, since they are indifferent to reagents which react readily with alkenes or with alkynes. n-Hexane, for example, is not attacked by concentrated sulfuric acid, boiling nitric acid, molten sodium hydroxide, potassium permanganate, or chromic acid; with the exception of sodium hydroxide, these reagents all attack alkenes at room temperature. The few reactions of which alkanes are capable require a high temperature or special catalysis.

Halogenation. If a test tube containing n-hexane is put in a dark place and treated with a drop of bromine, the original color will remain undiminished in intensity for days. If the solution is exposed to sunlight, the color fades in a few minutes, and breathing across the mouth of the tube produces a cloud of condensate revealing hydrogen bromide as one reaction product. The reaction is a photochemical substitution:

$$C_6H_{14} + Br_2 \xrightarrow{Light} C_6H_{13}Br + HBr$$

Chlorination of alkanes is more general and more useful than bromination and can be effected not only photochemically but also by other methods.

Light initiates chlorination of an alkane by converting chlorine molecules into chlorine atoms by a process of homolysis, in which a covalent bond is severed and one electron is retained by each of the atoms forming the bond: $Cl:Cl \rightarrow Cl \cdot + Cl \cdot$. A chlorine atom has an odd, or unpaired electron and is a free radical. Because of the tendency of atoms to attain their normal valence shells, any free radical is a highly reactive species. Photochemical chlorination proceeds through a succession of free radicals; it is a free radical chain reaction. The chain initiating step (1), homolytic fission of chlorine molecules, produces chlorine free radicals; in chain propagating steps, a chlorine radical attacks a molecule of alkane to produce hydrogen chloride and an alkyl radical (2), which in turn attacks a chlorine molecule to produce a chloroalkane and a chlorine radical (3). Since chlorine radicals required in step (2) are regenerated in step (3), the two reactions together constitute a chain which, if both reactions proceeded with perfect efficiency, would be self-propagating without further requirement of light energy, The efficiency, however, is not perfect, for chlorine radicals can recombine (4), combine with alkyl radicals (5), or dissipate energy by collision with the flask walls. Hence continued radiation is required to maintain an adequate supply of initiating radicals. The chain initiating step requires input of light energy amounting to +242.8kJ/mole. Step (2), however, is exothermic, since the energy required to break the C-H bond is less than the bond energy of H-Cl. The second chain propagating step (3) is likewise exothermic, and indeed chlorination of an alkane can proceed explosively.

Cracking. Heated to temperatures in the range 500~700°, higher alkanes undergo pyrolytic rupture or cracking to mixtures of smaller molecules, some saturated and some unsaturated. Unsaturated hydrocarbons produced by selective cracking of specific petroleum fractions are useful in chemical synthesis. Cracking ruptures carbon-carbon rather than carbon-hydrogen bonds because the energy required to break the C-C bond is 247kJ/mol or kJ·mol^{-1}, whereas the C-H bond energy is 364kJ/mol or kJ·mol^{-1}.

Oxidation. The reaction of hydrocarbons with oxygen with the output of energy is the basis for use of gasoline as fuel in internal combustion engines. The energy release on burning a given hydrocarbon is expressed as the heat of combustion in terms of kJ/mol or kJ·mol^{-1}.

Incomplete combustion of gaseous hydrocarbons is important in the manufacture of carbon blacks, particularly lampblack, a pigment for ink, and channel black, used as a filler in rubber compounding. Natural gas is used because of its cheapness and availability; the yield of black varies with the type of gas and the manufacturing process but usually is in the range of 2~6% of the theoretical amount.

Partial air oxidation of a more limited extent is a means for production of specific oxygenated substances. Controlled air oxidation of high-boiling mineral oils and waxes from petroleum affords mixtures of higher carboxylic acids similar to those derived from fats and suitable for use in making soaps.

New words and Expressions

alkane['ælkein] n. 烷烃
isomer['aisəumə] n. 异构体,同分异构体
saturate['sætʃəreit] vt.a. 饱和,浸透
hydrocarbon['haidrou'kɑːbən] n. 烃,碳氢化合物
paraffin['pærəfin] n. 石蜡,链烷烃
derivation[ˌderi'veiʃən] n. 起源,由来
derive[di'raiv] vi. 衍生,派生
pentane['pentein] n. 戊烷
wax[wæks] n. 蜡
petroleum[pi'trəuljəm] n. 石油
hexane['heksein] n. 己烷
heptane['heksein] n. 庚烷
content['kɔntent] n. 含量
octane['ɔktein] n. 辛烷
nonane['nɔnein] n. 壬烷
decane['dekein] n. 癸烷
increment['inkrimənt] n. 增量,增加
homologous[hɔ'mɔləgəs] a. 同系列的
homolog['hɔməlɔg] n. 同系物
methane['meθein] n. 甲烷
branch['brɑːntʃ] v. 分支
univalent[juːni'veilənt] a. 一价的
acyclic[ei'saiklik] a. 无环的,非环状的
ethane['eθein] n. 乙烷
propane['prəupein] n. 丙烷
butane['bjuːtein] n. 丁烷
elision[iː'liʒən] n. 省略
undecane[ʌn'dekein] n. 十一(碳)烷
dodecane[dəu'dekein] n. 十二(碳)烷
tridecane[trai'dekein] n. 十三(碳)烷
icosane['aikəsein] n. 二十(碳)烷
henicosane[he'naikəsein] n. 二十一(碳)烷
docosane['dɔːkəsein] n. 二十二(碳)烷

tricosane['traikəsein] n. 二十三(碳)烷
triacontane[ˌtraiə'kɔntein] n. 三十(碳)烷
hentriacontane[ˌhentraiə'kɔntein] n. 三十一(碳)烷
dotriacontane[ˌdɔtraiə'kɔntein] n. 三十二(碳)烷
tritriacontane[ˌtraitraiə'kɔntein] n. 三十三(碳)烷
tetracontane[ˌtetrə'kɔntein] n. 四十(碳)烷
pentacontane[ˌpentə'kɔntein] n. 五十(碳)烷
hexacontane[ˌheksə'kɔntein] n. 六十(碳)烷
octacontane[ˌɔktə'kɔntein] n. 八十(碳)烷
nonacontane[ˌnɔnə'kɔntein] n. 九十(碳)烷
hectane['hektein] n. 一百(碳)烷
dotriacontahectane[ˌdɔtraiə'kɔntə'hektein] n. 一百三十二(碳)烷
number['nʌmbə] vt. 给…编号
locant['ləukənt] n. 位次
substituent[sʌbs'titjuənt] n. 取代基
isobutyl[ˌaisou'bjuːtil] n. 异丁基
tert.=tertiary['təːʃəri] a. 叔的,第三的,特的
sec.=secondary['sekəndəri] a. 仲的,第二的
indifferent[in'difrənt] a. 惰性的,无关紧要的
alkene['ælkiːn] n. 烯烃
alkyne['ælkain] n. 炔烃
attack[ə'tæk] v. 起(化学)反应,侵蚀,进攻
concentrate['kɔnsentreit] v. 提浓,浓缩
sodium hydroxide 氢氧化钠
catalysis[kə'tælisis] n. 催化(作用)
halogenation[ˌhælədʒə'neiʃən] n. 卤化(作用)
test tube[test tjuːb] 试管
intensity[in'tensiti] n. 强度
fade[feid] v. 褪色

condensate[kən'denseit] n. 冷凝物（液）
reveal[ri'vi:l] vt. 呈现,展现
hydrogen bromide['haidrədʒən 'brəumaid]
溴化氢
photochemical[ˌfəutəu'kemikəl] a. 光化学的
initiate[i'niʃieit] vt. 引发,发动
homolysis[hɔ'mɔlisis] n. 匀裂
sever['sevə] vi. 断裂,裂开
free radical　自由基,游离基
succession[sək'seʃən] n. 继续,继承,传递
chain reaction　链（锁）反应.
chloroalkane[ˌklɔ(:)rə'ælkein] n. 氯代烷
regenerate[ri'dʒenəreit] vt. 再生
efficiency[i'fiʃənsi] n. 效率
collision[kə'liʒən] n. 碰撞
radiation[ˌreidi'eiʃən] n. 辐射,照射,放射线
exothermic[ˌeksəu'θə:mik] a. 放热的
explosively[iks'plousivli] ad. 爆炸式地
pyrolytic[ˌpaiərə'litik] a. 热解的,高温分解的
rupture['rʌptʃə] n.v. 裂开,断裂
crack[kræk] v. 裂解,断裂
fraction['frækʃən] n. 馏分,部分,分数
synthesis['sinθisis]（复 syntheses
['sinθisi:z])n. 合成,综合
gasoline['gæsəli:n] n. 汽油
fuel[fjuəl] n. 燃料
internal combustion engine 内燃机
combustion[kəm'bʌstʃən] n. 燃烧
gaseous['geizjəs] a. 气体的,气态的
carbon black 炭黑
lampblack['læmp'blæk] n. 灯黑
pigment['pigmənt] n. 颜料,色料

channel['tʃænl] black 槽法炭黑
compounding[kəm'paundiŋ] n. 配料,配方
oxygenate[ɔk'sidʒineit] vt. 氧化
fat[fæt] n. 脂纺

前缀

a-(an-)[ei-,ə-、æ-(æn-)]不,无,非
　acyclic, asymmetric, amorphous
self-[self-] 自,自动,自发
　self-propagating, self-combustion
　self-feeding（自动加料）
neo-[ˌni:əu-] 新 neopentane

后缀

-ane[-ein]……烷（名词词尾)methane,
　decane
-yl[-il, -əl]……基（名词词尾)methyl,
　decyl

词组

apply to…　适合,适用于
differ in …(by…)　在……方面不同,在……
方面相（差）……
prefix A to B 把A放在B前面,给B加A前缀
term by term…　逐项,逐个
so…as to(+inf.)　如此……以便
on the occasion of…　在……时候,
　　　　　　　　　　　　值此……之际
express…in(+单位)　以(……)为单位表示
vary with…　随……而变(化)
suitable for(to)…　适合于,适于

Notes

1. "When series of locants containing the same number of terms are compared term by term, that series is "lowest" which contains the lowest number on the occasion of the first difference." 系主从复合句,"When series…term by term"是时间状语从句,其中分词短语"containing the same number of terms"作主语 series of locants 的定语, term by term 作状语,其中的 by 在此表示连续性,该复合句主句是"that series…first difference",其中"which contains…difference"是定语

从句,修饰 that series.**译文**:"当逐一比较包含着相同(碳原子)数目的词语的许多位次时,根据第一个位次差值的理由,含最低数字的是最低者"。

2. 温度和压力的表达方法:温度和压力一般使用介词短语结构来表达,如:

to

 Heated to temperatures in the range 500～700℃,…

 Cool the mixture to room temperature.

at

 The beaker of water is maintained at a temperature of 90～95℃.

 Zinc may be rolled into sheets at between 120° to 150℃.

 CuI undergo similar phase changes at high pressures.

 Diamond changes at a pressure of 600000 atm and 1000℃ from the four-neighbor diamond crystal to…

above

 Above 160℃ the S8 rings break open.

about

 It becomes brittle again about 200℃.

under

 under high pressures it is possible to convert many nonconductors to conductors.

Exercises

1. Fill in the blacks with proper words or phrases.

 Alkanes with more than three can be arranged in various different ways, forming ____ isomers. The ____ isomer of an alkane is the one in which the carbon atoms are ____ in a single chain with no ____. This isomer is sometimes called the n-isomer (n for "normal", although it is not necessarily the most common). However the ____ of carbon atoms may also be ____ at one or more points.

 A. chain, B.simplest, C.structural, D.arranged, E.carbon atoms,

 F.branched, G.branches

2. Match the following explanation with the proper word.

 ____ isomer, ____ homologous, ____ condensate, ____ exothermic, ____ oxygenate

 A. having the same evolutionary origin but serving different functions the wing of a bat and the arm of a man are homologous

 B. impregnate, combine, or supply with oxygen

 C. atmospheric moisture that has condensed because of cold

 D. (of a chemical reaction or compound) occurring or formed with the liberation of heat

 E. a compound that exists in forms having different arrangements of atoms but the same molecular weight

Lesson 11
UNSATURATED COMPOUNDS

Unsaturated Compounds and Univalent Radicals

Unsaturated unbranched acyclic hydrocarbons having one double or triple bond are named by replacing the ending "-ane" of the name of the corresponding saturated hydrocarbon with the ending "-ene" or "-yne". If there are two or more double or triple bonds, the ending will be "-adiene", "-atriene" or "-adiyne", "-atriyne", etc. The generic names of these hydrocarbons (branched or unbranched) are "alkene", "alkadiene", "alkatriene" or "alkyne", "alkadiyne", "alkatriyne", etc. The chain is so numbered as to give the lowest possible numbers to the double or triple bonds.

Examples:

2-Hexene $\overset{6}{C}H_3-\overset{5}{C}H_2-\overset{4}{C}H_2-\overset{3}{C}H=\overset{2}{C}H-\overset{1}{C}H_3$

1,4-Hexadiene $\overset{6}{C}H_3-\overset{5}{C}H=\overset{4}{C}H-\overset{3}{C}H_2-\overset{2}{C}H=\overset{1}{C}H_2$

The following non-systematic names are retained:

Ethylene $CH_2=CH_2$ Allene $CH_2=C=CH_2$ Acetylene $HC\equiv CH$

Isoprene $CH_2=CH-\underset{|}{\overset{CH_3}{C}}=CH_2$

The names of univalent radicals derived from unsaturated acyclic hydrocarbons have the endings "-enyl", "-ynyl", "-dienyl", etc., the positions of the double and triple bonds being indicated where necessary. The carbon atom with the free valence is numbered as 1.

Examples:	Ethynyl	$CH\equiv C-$
	2-propynyl	$CH\equiv C-CH_2-$
	1-Propenyl	$CH_3-CH=CH-$
	2-Butenyl	$CH_3-CH=CH-CH_2-$
	1,3-Butadienyl	$CH_2=CH-CH=CH-$

Exceptions:

The following names are retained:

Vinyl(for ethenyl) $CH_2=CH-$

Allyl(for 2-propenyl) $CH_2=CH-CH_2-$

Isopropenyl(for 1-methylvinyl) $CH_2=\underset{|}{\overset{}{C}}-$ CH_3 (for unsubstitued radical only)

Physical Properties. Alkenes are known also as ethylenic hydrocarbons and as olefins. The term olefin, meaning oil-forming, was applied by early chemists because the gaseous members of the series combine with chlorine and bromine to form oily addition products.

Alkenes are hardly distinguishable from the corresponding saturated hydrocarbons. The boiling points are no more than[1] a few degrees below those of alkanes of slightly higher molecular weight, and the densities are a few percent higher; in the first few members of the two series there is even a marked correspondence in the melting points. Cycloalkanes differ more from alkanes than alkenes do[2], and hence ring formation influences physical properties more than introduction of an ethylene linkage. The heat of combustion of 1-hexene is practically the same as[3] that of n-hexane on either a weight or volume basis.

Pyrolytic Dehydration. Elimination of water from alcohols is a useful method for the preparation of alkenes; thus on elimination of OH from one carbon atom and of H from another, ethanol yields ethylene.

One of several experimental procedures is catalytic dehydration. The alcohol is distilled through a tube packed with granules of alumina and maintained at a temperature of 350~400° in an electrically heated furnace. The reaction resembles pyrolysis of an alkane, since it involves production of an unsaturated product from a saturated one at an elevated temperature, but the pyrolysis temperature for an alcohol is distinctly lower, and the process is simpler and more uniform.

Sulfuric Acid Method. Sulfuric acid is a dihydroxy acid represented for simplicity by formula I. The substance may have one semipolar bond as in II, in which case sulfur has expanded its shell to accommodate twelve electrons, or it may have two double bonds (III).

$$\text{HO-SO}_2\text{-OH} \quad\quad \text{HO}-\overset{\overset{O^-}{|}}{\underset{\underset{O}{\|}}{S^+}}-\text{OH} \quad\quad \text{HO}-\overset{\overset{O}{\|}}{\underset{\underset{O}{\|}}{S}}-\text{OH}$$

$$\text{I} \quad\quad\quad \text{II} \quad\quad\quad \text{III}$$

When concentrated sulfuric acid is added gradually to ethanol with ice cooling, water is eliminated from the two components and ethylsulfuric acid is formed. The reaction proceeds to completion because the water formed is absorbed by the concentrated acid; the process, however, is an equilibrium, and can be reversed by treatment of the product with a large excess of water. Ethylsulfuric acid is the mono ester of the inorganic diacid; it is a strong acid soluble in sulfuric acid as well as in water. It is stable at a low temperature but decomposes when heated. This chief organic product of the decomposition is ethylene, formed by loss of OSO_2OH from one carbon atom and of hydrogen from the adjacent position to produce sulfuric acid. A side reaction consists in formation of diethyl ether (ordinary ether) by the action of alcohol on ethylsulfuric acid. This reaction can be operated for preparation of ether by adjusting the proportions of reagents and maintaining a temperature of 140°. The difference in the optimum temperatures for ethylene and ether formation is so slight that each product is a by-product of the production of the other. A third product, diethyl sulfate, can be prepared by heating ethylsulfuric acid at a temperature below 140° at a pressure sufficiently reduced to cause diethyl sulfate to distill from the nonvolatile acids. Diethyl sulfate is the normal ester, or di-ester, of sulfuric acid. It is useful as an ethylating reagent, sometimes as an alternative to[4] an ethyl halide. Dimethyl sulfate (b.p.188.5°) is used similarly.

Acid - Catalyzed Dehydration. Dehydration can be effected also with hydrochloric acid, phosphoric acid, potassium bisulfate, or oxalic acid, $(COOH)_2$. These are strong acids capable of effecting elimination in catalytic amounts, and some are incapable of forming intermediate esters.

Ease of Dehydration. – Among alcohols having no activating group, the ease of dehydration depends upon the alcohol type and is in the following order: tertiary>secondary>primary.[5] The differences are illustrated in the accompanying examples. The case of ethanol, a primary alcohol, has been cited; the acid strength is 96% and the temperature 170°. The secondary alcohol is dehydrated by 62% acid at the temperature of the steam bath, whereas the tertiary alcohol affords an alkene at the same temperature on reaction with acid of only 46% strength.

Direction of Dehydration. – The structure of an alcohol may be such that two routes of dehydration are open. 2-Pentanol offers the possibility for elimination of hydrogen from either the 1- or the 3- postition, along with the adjacent hydroxyl group, but the 3-hydrogen is utilized almost exclusively and the product is 2-pentene.

New words and Expressions

alkadiene[ˌælkəˈdaiiːn] n. 二烯
alkatriene[ˌælkətraiiːn] n. 三烯
cyclic[ˈsaiklik] a. 环的,环状的
substiution[ˌsʌbstiˈtjuːʃən] n. 取代
2-hexene[tuːˈheksiːn] n. 2—已烯
1,4-hexadiene[ˈwʌn fɔːˌheksəˈdaiiːn] n. 1,4—已二烯
ethylene[ˈeθiliːn] n. 乙烯;1,2—亚乙基,乙撑—CH_2CH_2-
allene[ˈæliːn] n. 丙二烯
triple[ˈtripl] a. 三重的
  ~ bond 三键
alkatriyne[ˌælkəˈtraiain] n. 三炔
alkadiyne[ˌælkəˈdaiain] n. 二炔
acetylene[əˈsetiliːn] n. 乙炔
isoprene[ˈaisəupriːn] n. 异戊二烯
ethynyl[eˈθainil] n. 乙炔基
2-propynyl[ˈprəupainil] n. 2—丙炔基
1-propenyl[ˈprəupənil] n. 1—丙烯基
2-butenyl[ˈbjuːtənil] n. 2—丁烯基

1,3-butadienyl[ˌbjuːtəˈdaiiː(ə)nil] n. 1,3—丁二烯基
vinyl[ˈvainil] n. 乙烯基
ethenyl[ˈeθənil] n. 乙烯基
allyl[ˈælil] n. 烯丙基
isopropenyl[ˌaisəˈprəupənil] n. 异丙烯基,1—甲基乙烯基
ethylenic[ˌeθiˈlenik] a. 烯的,乙烯的
olefin[ˈəuləfin] n. 烯烃
oil-forming[ɔilˈfɔːmiŋ] 生成石油,成油
oily[ˈɔili] a. 油的,油状的,含油的
addition[əˈdiʃən] n. 加成,加合
distinguishable[disˈtiŋgwiʃəbl] a. 可区别的,可辨别的
density[ˈdensiti] n. 密度,比重
cycloalkane[ˌsaikləuˈælkein] n. 环烷(烃)
influence[ˈinfluəns] n. 影响
linkage[ˈliŋkidʒ] n. 键(合)
heat of combustion 燃烧热
volum[ˈvɔljum] n. 卷,册,体积

synthetic[sin'θetik] a. 合成的,综合的
dehydration[ˌdiːhai'dreiʃən] n. 脱水
eliminate[i'limineit] vt. 消去,消除
elimination[iˌlimi'neiʃən] n. 消去,消除
ethanol['eθənɔl] n. 乙醇
distil[dis'til] v. 蒸馏
pack['pæk] vt. 填充,塞满
granule['grænjuːl] n. 颗粒
alumina[ə'ljuːminə] n. 氧化铝,钒土
pyrolysis[pai'rɔlisis] n. 热解,高温分解
elevate['eliveit] vt. 提高,抬高
ethylsulfuric acid 乙基硫酸,硫酸单乙酯
diacid[dai'æsid] n. 二元酸
decompose[diːkəm'pəuz] v. 分解
adjacent[ə'dʒeisənt] a. 邻近的,相邻的
side reaction 副反应
ether['iːθə] n. 醚,乙醚
diethyl ether[dai'əθil'iːθə] 二乙基醚,乙醚
adjust[ə'dʒʌst] vt. 调整,调节
diethyl sulfate[dai'eθil'sʌlfeit] 硫酸二乙酯
ester['estə] n. 酯
normal ester 等当量酯,中性酯
ethylating[ˌeθi'leitiŋ] n.a. 乙基化
ethyl halide['eθil'hælaid] 卤乙烷
potassium bisulfate 酸式硫酸钾
oxalic acid[ɔk'sælik'æsid] 草酸,乙二酸
intermediate[ˌintə(ː)'midjət] n. 中间体;
　　　　　　　　　　　　　a. 中间的

前　缀

alk-['ælk-] 烃类
　　alkane, alkene, alkyl
　　alkoxyl, alkylation
de-['diː-]
　①脱, dehydration, dewax
　　decolorize, decarboxylation;
　②去,除 detergent, deshield
　　deodorant, defroster(除霜剂)
　③解,减,消 depolymerize, detoxication,
　　dehumidifier(减湿剂),
　　defrother(消泡剂)

后　缀

-ene[iːn]……烯
-adiene[ə'daiiːn]……二烯
-atriene[ə'traiiːn] n.……三烯
-yne[ain]……炔
-adiyne[ə'daiain]……二炔
-atriyne[ətraiain]……三炔
-enyl[ənil]……烯基
-ynyl[ainil]……炔基
-dienyl[daiənil]……二烯基

词　组

no more than　不过是,仅仅
differ from　不同于
consist in　是,在于

Notes

1. no more than: merely
2. do: differ from alkanes
3. same as: 作表语, as that of 即"as is the heat of combustion of", 它是 same 的定语从句。
4. as an alternative to: which may be used in place of…
5. tertiary>secondary>primary: 该句中">"读作 greater than, 请注意以下读法:
　　A>B——A is greater than B
　　A<B——A is less than B

A=B——A equals B; A is equal to B

A≈B——A approximately equals to B

A_2——A sub two

6.程度的表达

程度从句(clauses of degree)给出了与某些标准相比较的作用强度或状态的信息,其强度可以等于此标准,与它成比例或与其不同(大于或小于它)。

(a)相等强度:相等强度通常以"the same as""so … as"或"as … as"表示,如:

The heat of combustion of 1-hexene is practically the same as that of n-hexane on either a weight or volume basis.

Cellulose is not hydrolysed so easily as starch.

These metals are almost as reactive as the alkali metals.

注意the same as 用于名词,如第一句中指与the heat of combustion of 1-hexene相比较的强度,而so … as用于动词,如在第二句中是hydrolysed,而so … as之间的词是副词或形容词。as … as是so … as的另一种选择,它一般用于肯定句,而so … as通常用于否定句中。

(b)比例强度:比例可通过许多方法表达,但是引入比例程度从句中的连接词是短语"in proportion as (to)",如:

Newton had demonstrated clearly that an elastic fluid is constituted of small particles or atoms of matter, which repel each other by a force increasing in proportion as their distance diminishes.

(c)不同强度:程度的不同可通过"形容词比较形式+than"来表示,如:

Cycloalkanes differ more from alkanes than alkenes do, and hence ring formation influences physical properties more than introduction of an ethylene linkage.

也可利用"less(more)+a.than"的形式,如:

The group IIB elements are less reactive than are those of group IIA, but are more reactive than the neighboring elements of group IB.

Exercises

1. Fill in the blacks with proper words or phrases.

Unsaturated compounds are ____ compounds whose ____ structure contains one or more double bonds or triple bonds. These can also be ____. Examples are the unsaturated fatty acids or unsaturated ____ (alkenes and alkynes). Many natural products are unsaturated compounds. Unsaturated compounds are ____ more reactive than ____ compounds. Triglycerides (rapeseed oil, linseed oil, olive oil, etc.) with a high content of unsaturated fatty acid residues turn more rapidly rancid than those with a high ____ of saturated fatty acid residues, as for example coconut fat.

  A. carbon-carbon B. saturated C. generally   D. conjugated,

  E. molecular   F. proportion G. organic-chemical H. hydrocarbons

2. Choose a proper prefix to make the following words to its antonyms.

  ____ hydration, ____ place, ____ cyclic reaction, ____ appear, ____ tergent, ____ odic

  A. de-   B. peri-   C. dis-

Lesson 12
THE NOMENCLATURE OF CYCLIC HYDROCARBONS

The names of saturated monocyclic hydrocarbons (with no side chains) are formed by attaching the prefix "cyclo" to the name of the acyclic saturated unbranched hydrocarbon with the same number of carbon atoms. The generic name of saturated monocyclic hydrocarbons (with or without side chains) is "cycloalkane". Exa. cyclopropane, cyclohexane.

Univalent radicals derived from cycloalkanes (with no side chains) are named by replacing the ending "-ane" of the hydrocarbon name by "-yl", the carbon atom with the free valence being numbered as 1. The generic name of these radicals is "cycloalkyl". Exa. cyclopropyl, cyclohexyl.

The names of unsaturated monocyclic hydrocarbons (with no side chains) are formed by substituting "-ene", "-adiene", "-atriene", "-yne", "-adiyne", etc., for "-ane" in the name of the corresponding cycloalkane. The double and triple bonds are given numbers as low as possible. Exa. cyclohexene, 1,3-cyclohexadiene.

The following names for monocyclic aromatic hydrocarbons are retained: benzene, toluene, xylene, mesitylene, cumene, styrene, etc.

The names of univalent radicals derived from unsaturated monocyclic hydrocarbons have the endings "-enyl", "-ynyl", "-dienyl", etc. The carbon atom with the free valence is numbered as 1. Example: 2-cyclopenten-1-yl, 2,4-cyclopentadien-1-yl.

The following trivial names for radicals having a single free valence are retained: phenyl, tolyl, xylyl, mesityl, cumenyl, benzyl, styryl, etc.

Since the name phenylene (o-, m-, or p-) is retained for the radical -C_6H_4-, bivalent radicals formed from substituted benzene derivatives and having the free valences at ring atoms are named as substituted phenylene radicals. The carbon atoms having the free valences are numbered 1, 2-, 1, 3-, or 1, 4-as appropriate.

The names of polycyclic hydrocarbons with maximum number of non-cumulative double bonds end in "-ene". The following list contains the names of polycyclic hydrocarbons which are retained, such as naphthalene, anthracene, phenanthrene, indene etc.

The names of "ortho-fused" or "ortho- and peri-fused" polycyclic hydrogenated hydrocarbons with less than maximum number of non-cumulative double bonds are formed from a prefix "dihydro-", "tetrahydro-", etc., followed by the name of the corresponding unreduced hydrocarbon. The prefix "perhydro-" signifies full hydrogenation. Exa. 1,4-dihydronaphthalene, tetradecahydroanthracene or perhydroanthracene.

Univalent radicals derived from "ortho-fused" or "ortho- and peri-fused" polycyclic hydrocarbons with names ending in "-ene" by removal of a hydrogen atom from an aromatic or alicyclic ring are named in principle by changing the ending "-ene" of the names of the hydrocarbons to "-enyl".

Examples: 2-indenyl.

Exceptions: naphthyl, anthryl, phenanthryl.

Bicyclic Bridged Hydrocarbons

Saturated alicyclic hydrocarbon systems consisting of two rings only, having two or more atoms in common, take the name of an open chain hydrocarbon containing the same total number of carbon atoms preceded by the prefix "bicyclo-"[1]. The number of carbon atoms in each of the three bridges① connecting the two tertiary carbon atoms is indicated in brackets in descending order. The system is numbered commencing with one of the bridgeheads, numbering proceeding by the longest possible path to the second bridgehead; numbering is then continued from this atom by the longer unnumbered path back to the first bridgehead and is completed by the shortest path from the atom next to the first bridgehead.[2]

Examples:

$$\overset{7}{C}H_2 - \overset{1}{C}H - \overset{2}{C}H_2 \qquad \overset{2}{C}H_2 - \overset{3}{C}H_2 - \overset{4}{C}H_2 - \overset{5}{C}H_2$$

Bicyclo[3.2.1]octane   Bicyclo[4.3.2]undecane

Note: Longest path 1,2,3,4,5

Next longest path 5,6,7,1

Shortest path 1,8,5

Radicals derived from bridged hydrocarbons are named by replacing the ending "ane" of respective bridged hydrocarbon name by "yl". The numbering of the hydrocarbon is retained and the point or points of attachment are given numbers as low as is consistent with the fixed numbering of the saturated hydrocarbon[3].

Polycyclic Systems

Cyclic hydrocarbon systems consisting of three or more rings may be named in accordance with the naming principles stated in bicyclic system. The appropriate prefix "tricyclo-" "tetracyclo-", etc., is substituted for "bicyclo-" before the name of the open-chain hydrocarbon containing the same total number of carbon atoms. Radicals derived from these hydrocarbons are named

---

* A bridge is a valence bond or an atom or an unbranched chain of atoms connecting two different parts of a molecule. The two tertiary carbon atoms connected through the bridge are termed "bridgeheads".

according to the naming principles set forth in radicals derived from bridged hydrocarbons.

When there is a choice, the following criteria are considered in turn, until a decision is made:

(a) The main ring shall contain as many carbon atoms as possible, two of which must serve as bridgeheads for the main bridge.

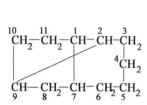
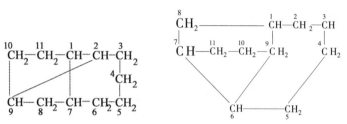

Tricyclo[5.4.0.0^{2,9}]undecane   Tricyclo[4.2.1.2^{7,9}]undecane

Correct numbering   Incorrect numbering

(b) The main bridge shall be as large as possible.

(c) The main ring shall be divided as symmetrically as possible by the main bridge.

(d) The superscripts locating the other bridges shall be as small as possible.

Hydrocarbon Bridges

Polycyclic hydrocarbon systems which can be regarded as "ortho-fused" or "ortho- and peri-fused" systems and which, at the same time, have other bridges, are first named as "ortho-fused" or "ortho- and peri-fused" systems. The other bridges are then indicated by prefixes derived from the name of the corresponding hydrocarbon by replacing the final "-ane", "-ene", etc., by "-ano", "-eno", etc., and their positions are indicated by the points of attachment in the parent compound. If bridges of different types are present, they are cited in alphabetical order. Examples of bridge names:

Butano	—CH$_2$—CH$_2$—CH$_2$—CH$_2$—
Benzeno(o-, m-, p-)	—C$_6$H$_4$—
Ethano	—CH$_2$—CH$_2$—
Etheno	—CH=CH—
Methano	—CH$_2$—
Propano	—CH$_2$—CH$_2$—CH$_2$—

New words and Expressions

monocyclic [ˌmɔnəˈsaiklik] a. 单环的
cyclopropane [ˌsaikləˈprəupein] n. 环丙烷
cyclohexane [ˌsaikləˈheksein] n. 环己烷
cycloalkyl [ˌsaikləˈælkil] n. 环烷基
cyclopropyl [ˌsaikləˈprəupil] n. 环丙(烷)基
cyclohexyl [ˌsaikləˈheksil] n. 环己(烷)基
1,3-cyclohexadiene [ˌsaikləˈheksədaii:n] n. 1,3—环己二烯
2-cyclopenten-1-yl [ˌsaikləˈpentin(wʌn)il] n. 环戊-2-烯-1-基

phenyl[ˈfenil] n. 苯基
cumene[ˈkjuːmiːn] n. 异丙(基)苯
mesitylene[miˈsitəliːn] n.
1,3,5—三甲苯,对称三甲苯
styrene[ˈstaiəriːn] n. 苯乙烯
toluene[ˈtɔljuiːn] n. 甲苯
xylene[ˈzailiːn] n. 二甲苯
cumenyl[ˈkjuːmiːnil] n. 异丙苯基
mesityl[ˈmesitil] n. 对称三甲苯基
tolyl[ˈtɔlil] n. 甲苯基
xylyl[ˈzailil] n. 二甲苯基
phenylene[ˈfeniliːn] n. 苯撑,亚苯基
($-C_6H_4-$)
benzyl[ˈbenzil] n. 苄基,苯甲基
styryl[ˈstaiəril] n. 苯乙烯基
polycyclic[ˌpɔliˈsaiklik] a. 多环的,聚集的
cumulative[ˌkjuːmjulətiv] a. 累积的,稠环的
indene[ˈindiːn] n. 茚
naphthalene[ˈnæfθəliːn] n. 萘
phenanthrene[fəˈnænθriːn]菲
anthracene[ˈænθrəsiːn] n. 蒽
hydrogenate[haiˈdrɔdʒineit] vt. 氢化,加氢
ortho-fused 邻位稠单边稠(并)
peri-fused 迫位稠,萘环之[1,8]或[4,5]
位稠(并)
perhydro-[pəˈhaidrɔ](前缀)全氢化
1,4-dihydronaphthalene[daiˈhaidrɔˈ
næʃθəliːn] n.1,4—二氢化萘
tetradecahydroanthracene=perhydro
anthracene n. 十四氢化蒽(=全氢化蒽)
alicyclic[æliˈsaiklik] a. 脂环的
2-indenyl[ˈindiːnil] n.2—茚基
naphthyl[ˈnæfθil] n. 萘基
anthryl[ˈænθril] a. 蒽基
phenanthryl[fəˈnænθril] n. 菲基
bicyclic[baiˈsaiklik] a. 双环的
precede[ˈpriˈsiːd] vt. 在前,在先,先于
bicyclo[3,2,1]octane n. 双环[3.2.1]辛烷
bridgehead[ˈbridʒˈhed] n. 桥头
bicyclo[4,3,2]undecane n. 双环[4.3.2]
十一(碳)烷
superscript[ˈsjuːpəskript] n. 上标符
criterion[kraiˈtiəriən](复 criteria[kraiˈ
tiəriə]) n. 准则,标准
butano[ˈbjuːtənə]桥丁撑,桥亚丁基
ethano[ˈeθənə]桥乙撑,桥亚乙基
benzeno[ˈbenzinə]桥苯撑,桥亚苯基
etheno[ˈeθinə]桥乙烯撑,桥亚乙烯基
methano[ˈmeθənə]桥甲撑,桥亚甲基
propano[ˈproupənə]桥丙撑,桥亚丙基

前　缀

cyclo-[ˌsaikləu(ə)]环 cyclopentane;
para-[ˌpærə]①对(p-)
p-xylene,parathion(对硫磷)
②顺-,paramagnetic resonance;
③聚-,paraformaldehyde
ortho-[ˌɔːθəu](o-)邻-
o-xylene,
o-dichlorobenzene;
正-,ortho-phosphoric acid;
原-,orthoformate(原甲酸酯)
meta-[ˌmetə](m-)
间-,m-xylene,m-dinitrobenzene;
偏-,meta-aluminic acid;
(位)变,metathesis 复分解(位变作用)
peri-[ˌpiəri-,peri-]周(围),包围,迫位
periodic,pericyclic reaction;
(周环反应)peri-position

词　组

substitute(…) for…
以(……) 代替……
in common 共同的,共同地
set forth… 陈述,宣布
consistent with 和……一致
in turn 依次
(…)preceded by… 在(……)前加上……

Notes

1. Saturated alicyclic……by the prefix bicyclo-. 该句是简单句,主语是 Saturated alicyclic hydrocarbon systems,consisting 引导的分词短语作主语的后置定语,having 引导的短语作时间状语,containing 引导的分词短语为 open chain hydrocarbon 的定语,过去分词 preceded 引导的短语作宾语 name 的定语。译文:"只有两个环的饱和脂环烃体系,有两个或多个共用原子时,采用含相同碳原子总数的开链烃的名称,加'双环'前缀"。

2. The system is numbered……next to the first bridgehead. 该句为并列句,在第一分句"The system …to the second bridgehead"中,"commencing with one of the bridgeheads"是分词短语,作方式状语,其中 with 意即 from,"numbering proceeding …bridgehead"是独立主格结构,作状语。"numbering is then… 到句末"为第二个分句。译文:"此体系的编号是从其中一个桥头开始,沿着最长路线到第二桥头,然后再继续从第二桥头原子,沿着较长的未编号路线返回第一桥头,并沿着紧靠第一桥头原子的最短路线结束"。

3. "The numbering of the …… of the saturated hydrocarbon." 系并列复合句,在第二分句"the point … hydrocarbon"中,attachment 在这里指的是 radical,"as is …hydrocarbon"是比较状语从句,第二个 as 在从句中作主语。译文:"保留烃的编号,给基的一个或多个点以最低编号,且与饱和烃原编号相一致。"

Exercises

1. Fill in the blacks with proper words or phrases.

   Cyclic hydrocarbon, also known as closed-chain ____, is the general term of hydrocarbons with cyclic structure. It ____ to that a hydrocarbon has its

   ____ carbons attached to each other to form a cyclic hydrocarbon.

   ____ to the structure and ____, it can be ____ into alicyclic hydrocarbons and aromatic hydrocarbons. But general cyclic refers to alicyclic hydrocarbons. According to the number of cylinders, it can be further divided into ____, bicyclic,____and other cyclic hydrocarbons.

    A. divided   B. According   C. properties   D. refers,

    E. hydrocarbon F. inter-molecular G.tricyclic   H. monocyclic

2. Match the following explanation with the proper word.

    ____ bicyclic, ____ criterion, ____ hydrogenate, ____ pyrolysis, ____ decompose

    A. add hydrogen to the molecule of (an unsaturated organic compound)

    B. separate (substances) into constituent elements or parts

    C. transformation of a substance produced by the action of heat

    D. having molecules consisting of two fused rings

    E. the ideal in terms of which something can be judged

Lesson 13
SUBSTITUTIVE NOMENCLATURE

Compulsory prefixes

10.1-The characteristic groups listed in Table 11 are always cited by prefixes, as given in the Table, to the name of the parent compound[1]. Multiplying affixes and locants are added as necessary.

Example:1,2-dichlorocyclohexane

Principal Group for Citation as Suffix

10.2 - Characteristic groups other than those listed in Table 11 may be cited as suffixes or prefixes to the name of the parent compound.

Table11. Characteristic groups cited only as prefixes in substitutive nomenclature

Characteristic group	Prefix	Characteristc group	Prefix
$-Br$	Bromo	$-IX_2$	X may be halogen or a radical, and the prefix names are dihalogenoiodo, etc., for radicals, patterned on diacetoxyiodo
$-Cl$	Chloro		
$-ClO$	Chlorosyl		
$-ClO_2$	Chloryl		
$-ClO_3$	Perchloryl		
$-F$	Fluoro		
$-I$	Iodo	$-N_2$	Diazo
$-IO$	Iodosyl	$-N_3$	Azido
$-IO_2$	Iodyl(replacing iodoxy)	$-NO$	Nitroso
$-I(OH)_2$	Dihydroxyiodo	$-NO_2$	Nitro
		$=N(O)OH$	aci-Nitro
		$-OR$	R-oxy
		$-SR$	R-thio (similarly R-seleno and R-telluro)

10.3-If any characteristic groups other than those in Table 11 are present, one kind must be cited as suffix, but only one kind[2]. That kind of group is termed the principal group. When a compound contains more than one kind of group not listed in Table 11, the principal group is that which characterizes the class occurring as high as possible in Table 12; all other characteristic groups are then cited as prefixes. Some suffixes and prefixes to be used with the general classes given in Table 12 are listed in Table 13. Multiplying affixes and locants are added as necessary. If, but only if, the complete suffix(that is, the suffix itself plus its multiplying affix, if any[3]) begins with a vowel, a terminal "e" (if any) of the preceding parent name is elided. Elision or retention of the terminal "e" is independent of the presence of numerals between it and the following letter.

Table 12. Some general classes of compound in the order in which the characterisitc groups have decreasing priority for citation as principal group[4].

1. Onium and similar cations
2. Acids: in the order COOH, C(=O)OOH, then successively their S and Se derivatives, followed by sulfonic, sulfinic acids, etc.
3. Derivatives of acids: in the order anhydrides, esters, acyl halides, amides, hydrazides, imides, amidines, etc.
4. Nitriles (cyanides), then isocyanides
5. Aldehydes, then successively their S and Se analogues; then their derivatives
6. Ketones, then their analogues and derivatives, in the same order as for aldehydes
7. Alcohols and phenols; then their S, Se and Te analogues; then neutral esters of alcohols and phenols with inorganic acids, except hydrogen halides, in the same order.
8. Hydroperoxides
9. Amines; then imines, hydrazines, etc.
10. Ethers; then successively their S and Se analogues
11. Peroxides

Examlpes:

$\overset{2}{C}H_3-\overset{1}{C}H_2OH$ and $HO\overset{2}{C}H_2-\overset{1}{C}H_2OH$

Suffix for principal group, OH: -ol
Parent name for CH_3-CH_3:   Ethane
Full names: Ethanol
                  1,2-Ethanediol

$HO\overset{7}{C}H_2-\overset{6}{C}H_2-\overset{5}{C}H_2-\overset{4}{C}H_2-\overset{3}{C}H_2-\overset{2}{C}O-\overset{1}{C}H_3$

Principal group of class higher in Table 12: $\diagdown$(C)=O. denoted by suffix -one

Parent name	Heptane
Name based on:	Heptanone
Full name:	7-Hydroxy-2-heptanone

$\overset{5}{C}H_2=\overset{4}{C}H-\overset{3}{C}H_2-\overset{2}{C}H_2-\overset{1}{C}H_2OH$ 4-Penten-1-ol

Suffix: -ol
Parent name: Pentane, altered to pentene for unsaturation
Full name: 4-Penten-1-ol

10.4 1-Derivative groups have priority for citation as principal group after the respective parents of their general class, as indicated in Table 12.

Example: 2-Hydroxy-1-cyclohexanecarboxamide

The compound is both an amide and an alcohol. Amides, being derivatives of acids, are higher than alcohols in Table 12. The amide group is therefore cited as suffix and the hydroxyl group as prefix.

There are two methods of using suffixes in substitutive nomenclature for aliphatic carboxylic acids and their derivatives, for aliphatic nitriles, and for aliphatic aldehydes; of these methods,

Table 13. Suffixes and prefixes used for some important groups in substitutive nomenclature (see RuleC-10.3)

Class	Formula*	Prefix	Suffix
Carboxylic acid	-COOH	Carboxy	-carboxylic acid
	-(C)OOH	-	-oic acid
Sulfonic acid	-SO$_3$H	Sulfo	-sulfonic acid
Salts	-COOM	-	Metal···carboxylate
	-(C)OOM	-	Metal···oate
Esters	-COOR	R···oxycarbonyl	R···carboxylate
	-(C)OOR	-	R···oate
Acid halides	-CO-Halogen	Haloformyl	-carbonyl halide
	-(C)O-Halogen	-	-oyl halide
Amides	-CO-NH$_2$	Carbamoyl	-carboxamide
	-(C)O-NH$_2$	-	-amide
Amidines	-C(=NH)-NH$_2$	Amidino	-carboxamidine
	-(C)(=NH)-NH$_2$	-	-amidine
Nitriles	-C≡N	Cyano	-carbonitrile
	-(C)≡N	-	-nitrile
Aldehydes	-CHO	Formyl	-carbaldehyde
	-(C)HO	Oxo	-al
Ketones	\(C)=O /	Oxo	-one
Alcohols	-OH	Hydroxy	-ol
Phenols	-OH	Hydroxy	-ol
Thiols	-SH	Mercapto	-thiol
Hydroperoxides	-O-OH	Hydroperoxy	-
Amines	-NH$_2$	Amino	-amine
Imines	=NH	Imino	-imine
Ethers	-OR	R-oxy	-
Sulfides	-SR	R-thio	-
Peroxides	-O-OR	R-dioxy	-

*carbon atoms enclosed in parentheses are included in the name of the parent compound and not in the suffix or prefix.

one involves modification of the preceding rules, the other does not, as illustrated for carboxylic acids by Rules C-11.11(b) and (a), respectively, below. However, when the characteristic group concerned is attached to a ring, only the method of Rule C-11.11(b) is used on systematic nomenclature.

Carboxylic Acids

C-11.1-(a) The atom =O and the group -OH on one and the same carbon atom are together denoted by a suffix "-oic" attached to the name of the parent aliphatic chain and the word "acid" is added thereafter. Thus the change "-ane" to "-anoic acid" denotes the change of -CH$_3$ to -COOH. For numbering see Rule C-401 1.

Example: $\overset{6}{C}H_3-\overset{5}{C}H_2-\overset{4}{C}H_2-\overset{3}{C}H_2-\overset{2}{C}H_2-\overset{1}{C}OOH$ Hexanoic acid

(b) The group -COOH is treated as a complete subsituent, denoted by the ending, "carboxylic acid", which includes the carbon atom of the carboxyl group.

Examples: 1,3,5-Pentanetricarboxylic acid, cyclohexanecarboxylic acid

Derivatives and Radicals from Carboxylic Acids

C-11.2-The two methods described in Rule C-11.1 apply to all derivatives of aliphatic acids and to acyl radicals; choice is made between endings such as $-CO-NH_2$ -amide or -carboxamide, $-CO-NH-NH_2$ -ohydrazide or -carbohydrazide, $-COCl$ -oyl chloride or -carbonyl chloride, $-COOR$ (for esters) -oate or -carboxylate, etc., and for radicals R-CO- -oyl or -carbonyl.

Aldehydes and Nitriles

11.31-The two methods described in Rule C-11.1 apply in principle also to aliphatic aldehydes and their derivatives and to nitriles; choice is made between the endings:

$$-(C)\overset{O}{\underset{H}{\diagdown}} \quad \text{-al or} \quad -C\overset{O}{\underset{H}{\diagdown}} \quad \text{-carbaldehyde}$$

$-(C)\equiv N-$ nitrile or $-C\equiv N-$ carbonitrile

Examples: Hexanal, 1,3,5-Pentanetricarbaldehyde, Hexanenitrile,
1,3,5-Pentanetricarbonitrile

Amines

Amines are named by traditional methods, details of which are given in Subsection C-8.1; the following are the two main principles.

11.41-For simple amines RNH_2, R^1R^2NH, and $R^1R^2R^3N$, the names of the radicals are attached in front of the word "amine".

Examples: Butylamine, N-Methylbutylamine, Triethylamine

In this method, amine can be considered to replace ammonia as parent compound.

11.42-Primary amines may also be named by adding the suffix "-amine" to the name of the parent compound (with elision of terminal "e", if present).

Examples: 2-Hexanamine, 2-Furanamine

New words and Expressions

substitutive ['sʌbstitju:tiv] a. 取代的，取代物的
parent ['pɛərənt] n.a. 母体，起源
characterize ['kæriktəraiz] vt. 表示的…特征，成为…的特征
terminal ['tə:minl] n. 词尾, a. 末端的
elide [i'laid] vt. 取消，省略，删去

onium ['əuniəm] n. 鎓离子；有机阳离子的词尾
sulfonic [sʌl'fɔnik] acid 磺酸
sulfinic [sʌl'finik] acid 亚磺酸
acyl halide [æsil'hælaid] 酰卤
amide ['æmaid] n. 酰胺
hydrazide ['haidrəzaid] n. 酰肼

imide['imaid] n. 酰亚胺
amidine['æmidi:n] n. 脒
nitrile = nitril['naitril] n. 腈
isocyanide[,aisə'saiənaid] n. 异氰
amine['æmi:n] n. 胺
imine[i'mi:n] n. 亚胺
hydrazine['haidrəzi:n] n. 肼
1,2-ethanediol['eθein'daiəul] n.1,2-乙二醇
7-hydroxy-2-heptanone[hai'drɔksi'heptənoun] n.7-羟基庚-2-酮
4-penten-1-ol['penti:nɔl] n.4-戊烯-1-醇
alter['ɔ:ltə] vt.改变,更改
thiol['θaiəul] n.硫醇
1,3,5-pentanetricarboxylic acid 1,3,5-三羧基戊烷,戊烷-1,3,5-三甲酸
cyclohexanecarboxylic acid 环己烷基甲酸
hexanal['heksənəl] n.己醛
1,3,5-pentanetricarbaldehyde n.1,3,5-戊烷三甲醛
hexanenitrile[,heksein'naitril] n.己腈
1,3,5-pentanetricarbonitrile n.戊烷-1,3,5-三甲腈,1,3,5-三氰基戊烷
butylamine[,bju:tə'læmi:n] n.丁胺
N-methylbutylamine[,meθil'bju:tə'læmi:n] n.N-甲基丁胺
triethylamine[trai,eθil'æmi:n] n.三乙胺
2-hexanamine[,heksə'næmi:n] n.2-氨基己烷
2-furanamine[,fjuərə'næmi:n] n.2-氨基呋喃

前　缀

bromo-[,brəumə]　溴代
chloro-['klɔ:rəu(ə)]　氯代
chlorosyl-['klɔ:rəsil]　亚氯酰
chloryl-['klɔ:ril]　氯氧基
perchloryl-[pə,klɔ:ril]　三氧氯基,过氧氯基,氯过氧基
fluoro-[flu:ərə]　氟代
iodo-[ai'ɔdə]　碘代

iodosyl['aiədəsil]=iodoso-[ai'ɔdəsou]　亚碘酰
iodyl-['aiədil](= iodoxy-[aiə'dɔksi])　碘酰基
dihydroxyiodo-[,dai'haidrɔksiai'ɔdə]　二羟基碘代
diazo-[dai,æzəu]　重氮,偶氮
azido-['eizidə]　叠氮基
nitroso-[nai,trəusəu]　亚硝基
nitro-['naitrə(əu)]　硝基
aci-nitro-[,æsi'naitrə]　异硝基,酸式硝基
R-oxy-['ɔksi]　R-氧基
R-thio-[θaiəu]　R-硫代
diacetoxyiodo-[dai,æsitɔksiai'ɔdə]　二乙酰氧基碘代
seleno-[si'li:nə]　硒基Se-
telluro-[te'ljuərə]　碲基Te-
carboxy-[kɑ:'bɔksi]　羧基
sulfo-[,sʌlfəu(ə)]　磺(酸)基
R-oxycarbonyl[,ɔksi'kɑ:bənil]　R-氧羰基
haloformyl-[,hælə'fɔmil]　卤甲酰基
carbamoyl-[kɑ:'bæmɔwil]　酰胺基
amidino-[ə'midinə]　脒基
formyl-['fɔmil]　甲酰基,甲醛基
oxo-['ɔksəu]　氧代
mercapto-[mə'kæptə]　巯基,氢硫基
amino-['æminəu,ə'mi:nəu]　氨基
imino-['imənəou(ə)]　亚氨基

后　缀

-carboxylic acid　甲酸
-oic acid[-əuik'æsid]酸
(metal)…oate……酸金属盐
R…carboxylate[kɑ:'bɔksileit]……羧酸R酯
R…oate……酸R酯
-carbonyl halide 甲酰卤
-oyl halide[-ɔil'hælaid]酰卤
-carboxamide[,kɑ:bɔk'sæmaid]甲酰胺
-carboxamidine['kɑ:bɔk'sæmidi:n]甲脒
-carbonitrile['kɑ:bə'naitril]甲腈

-carbaldehyde[kɑːˈbældi(ə)haid]甲醛
-al[æl,əl]醛
-one[əun]酮
-ol[ɔl,ɔːl,əul]醇
-ohydrazide[ɔˈhaidrəzaid]甲酰肼
-carbohydrazide[ˈkɑːbɔˈhaidrəzaid]甲酰肼
-oyl[əuil,əwil]酰(基)

-carbonyl[ˈkɑːbənil]酰,羰(基)

词 组

in principle　原则上
one and the same　同一个
attach to…　与……相连,接到……上

Notes

1. "The characteristic groups listed in Table 11 are always cited by prefixes, as given in the Table, to the name of the parent compound." 系主从复合句,"as given in the Table"为非限制性定语从句,as后省去了is,as在句中作主语,主句中"to the name …compound"是prefixes的定语。**译文**："表11中列出的特征性基团,对母体化合物名称来说总是按表中所给出的前缀加以引用。"

2. "If any characteristic groups other than those in Table 11 are present, one kind must be cited as suffix, but only one kind." 为复合句,其中"but only one kind"是插入语。**译文**："除了表11的基团外,如果存在着其它特征官能团,一种,而且只有一种,必定作为后缀引用。"

3. if any: if there is any

4. Some general classes of… as principal group. **译文**："化合物的一般分类,按特征官能团优先引用作主要基团的递减顺序排列"。

Exercises

1. Fill in the blanks with proper words or phrases.

In ____ chemistry and biochemistry, a substituent is an atom or group of ____ which replaces one or more ____ atoms on the parent chain of a hydrocarbon, ____ a moiety of the resultant new molecule. The terms substituent and ____ group, as well as other ones (e.g. side chain, pendant group) are used almost interchangeably to describe ____ from a parent structure, though certain distinctions are made in the context of ____ chemistry. In polymers, side chains extend from a backbone structure. In proteins, side chains are ____ to the alpha carbon atoms of the amino acid backbone.

　　A. becoming　　B. attached　　C. organic　　D. polymer,
　　E. functional　　F. branches　　G. atoms　　H. hydrogen

2. Choose a proper prefix to make the follow words match the meanings.

　　____ -benzene 苯甲酸, ____ -methionine 甲酰蛋氨酸, ____ -acetonitrile 氯代乙腈,
　　____ -benzene 氨基苯, ____ -cellulose 羧基纤维素, ____ -acetate 氟化乙酸盐,
　　____ -methane 溴化甲烷, ____ -glycerin 硝酸甘油

　　A. bromo-　　B. amino-　　C. carboxy-　　D. sulfo-
　　E. nitro-　　F. chloro-　　G. fluoro-　　H. formyl-

Lesson 14
THE COMPOUNDS CONTAINING OXYGEN

Alcohols and phenols. In substitutive and conjunctive nomenclature of alcohols the hydroxyl group(OH)as principal group is indicated by a suffix"-ol", with elision of terminal"e" (if present) from the name of the parent compound, for example, methanol, 2-propanol, triphenylmethanol, etc.

The following are examples of trivial names which are retained: allyl alcohol, tert-butyl alcohol, benzyl alcohol, ethylene glycol, glycerol, etc.

Hydroxy derivatives of benzene and other aromatic carbocyclic systems are named by adding the suffix "-ol", "-diol", etc., to the name of the hydrocarbon, with elision of terminal "e" (if present) before "ol", e.g., 1,2,4-benzenetriol. The following are examples of trivial names of aromatic hydroxy compounds which are retained, for example, phenol, cresol, naphthol, pyrocatechol, resorcinol, hydroquinone, etc.

Radicals RO- are named by adding "oxy" as a suffix to the name of the radical R, e.g., pentyloxy, allyloxy, benzyloxy, etc. Only the following contractions for oxygen-containing radical names are recommended as exceptions to this rule: methoxy, ethoxy, propoxy, butoxy, phenoxy, isopropoxy.

Except when forming part of a ring system, bivalent radicals of the form -O-X-O- are named by adding "dioxy" to the name of the bivalent radical -X-, e.g., ethylenedioxy, trimethylenedioxy.

Salts. Anions derived from alcohols or phenols are named by changing the final "-ol" of the name of the alcohol or phenol to "-olate". This applies to substitutive, radicofunctional, and trivial names, e.g., potassium methanolate, sodium phenolate.

Ethers. Compounds R^1—O—R^2 have the generic name "ethers" and may be named by either the substitutive or the radicofunctional method.

Substitutive names of unsymmetrical ethers are formed by using names of radicals R^1O- as prefixes to the names of the hydrocarbons corresponding to the second radical R^2. The senior component is selected as the parent compound. Radicofunctional names of ethers are formed by citing the names of the radicals R^1 and R^2 followed by the word "ether", for example, 1-isopropoxypropane, ethyl methyl ether, diethyl ether, ethyl vinyl ether.

Aldehydes. The term "aldehyde" is applied to compounds which contain the group -C(=O)H attached to carbon. Aldehydes are named by means of the suffixes "-al", "-aldehyde", or "-carbaldehyde", or by the prefix "formyl-"〔representing the -C(=O)H group when present as the terminal group of a carbon chain], or, in connexion with trivial names, by the prefix "oxo-" (representing=O). The name of an unbranched acyclic mono-or di-aldehyde is formed by adding the suffix "-al" (for a monoaldehyde), with elision of a terminal "e" (if present), or "-dial" (for a dialdehyde) to the name of the hydrocarbon containing the same number of carbon atoms, e.g., ethanal, hexanal.

Trivial Names. When the corresponding monobasic acid has a trivial name, the name of the aldehyde may be formed from the trivial name of the acid by changing the ending "-ic acid" or "-oic acid" to "-aldehyde", for example, formaldehyde, acetaldehyde, propionaldehyde, acrylaldehyde(or acrolein), benzaldehyde, cinnamaldehyde, furfural.

Ketones. The generic name "ketone" is given to compounds containing an oxygen atom doubly bound to a single carbon atom with the carbonyl group $\diagdown$C=O joined to two carbon atoms[1]. Ketones are named by means of the suffix "-one", the prefix "oxo-", the functional class name "ketone", or, in special cases, the suffix "-quinone". In substitutive nomenclature the name of an acyclic ketone is formed by adding the suffix "-one" or "-dione", etc. to the name of the hydrocarbon corresponding to the principal chain, with elision of terminal "e" (if present) before "one", e.g., 2-butanone, 2,4-pentanedione.

Radicofunctional names of ketones, R^1COR^2, are formed, by citing the names of the radicals R^1 and R^2 followed by the word "ketone", for example, ethyl methyl ketone, diethyl ketone.

Carboxylic acids. Carboxyl groups COOH replacing CH_3 at the end of the main chain of an acyclic hydrocarbon are denoted by adding "-oic acid" or "-dioic acid" to the name of this hydrocarbon, with elision of terminal "e" (if present) before "oic". Alternatively, the name may be formed by substitutive use of the suffix "-carboxylic acid". In the aliphatic series the numbering of the chain dose not then include the carbon atom of the carboxyl group, e.g., heptanoic acid, heptanedioic acid, cyclohexanecarboxylic acid.

The name of a univalent or bivalent acyl radical formed by removal of hydroxyl from all the carboxyl groups is derived from the name of the corresponding acid by changing the ending "-oic" to "-oyl".

Systematic and trivial names for some acids and their acyl radicals are given in Table 14.

Neutral salts of carboxylic, amic, imidic, carboximidic, hydroxamic, carbohydroxamic, etc., acids are named by citing the cation(s) and then the anion or, when the carboxyl groups of the acid are not all named as affixes, by use of a periphrase, such as "(metal) salt of (the acid)"[2], e.g., sodium heptanoate, potassium acetate.

Neutral esters of carboxylic acids, etc., are named in the same way as their neutral salts except that (a) the name of the alkyl or aryl, etc., radical replaces the name of the cation and (b) a periphrase such as "(alkyl or aryl) ester" replaces "(metal)salt", e.g., ethyl acetate, diethyl malonate, ethyl chloroformate.

Symmetrical anhydrides of monocarboxylic acids, when unsubstituted, are named by replacing the word "acid" by "anhydride", e.g., acetic anhydride, phthalic anhydride.

Acyl halides, that is, compounds in which the hydroxyl group of a carboxyl group is replaced by halogen, are named by placing the name of the corresponding halide after that of the acyl radical, e.g., acetyl chloride, benzoyl iodide.

Table 14. The names of some acids and their radicals*

Names of Acids		Acyl radicals	
Systematic	Trivial name	Trivial name	Formula
methanoic**	formic	formyl	HCO—
ethanoic**	acetic	acetyl	CH_3—CO—
propanoic**	propionic	propionyl	CH_3—CH_2—CO—
butanoic**	butyric	butyryl	CH_3—$(CH_2)_2$—CO—
2-methylpropanoic**	isobutyric*	isobutyryl*	$(CH_3)_2$CH—CO—
pentanoic**	valeric	valeryl	CH_3—$(CH_2)_3$—CO—
dodecanoic	lauric*	lauroyl*	CH_3—$(CH_2)_{10}$—CO—
hexadecanoic	palmitic*	palmitoyl*	CH_3—$(CH_2)_{14}$—CO—
octadecanoic	stearic*	stearoyl*	CH_3—$(CH_2)_{16}$—CO—
ethanedioic**	oxalic	oxalyl	—CO—CO—
propanedioic**	malonic	malonyl	—CO—CH_2—CO—
butanedioic**	succinic	succinyl	—CO—$(CH_2)_2$—CO—
hexanedioic**	adipic	adipoyl	—CO—$(CH_2)_4$—CO—
propenoic**	acrylic	acryloyl	CH_2=CH—CO—
2-methylpropenoic**	methacrylic	methacryloyl	CH_2=C(CH_3)—CO—
cis-butenedioic**	maleic	maleoyl	CH—CO— ‖ CH—CO—
benzenecarboxylic	benzoic	benzoyl	C_6H_5—CO—
1,2-benzenedicarboxylic	phthalic	phthaloyl	⌬(—CO—)(—CO—)
1,3-benzenedicarboxylic	isophthalic	isophthaloyl	⌬(—CO—)(—CO—)
1,4-benzenedicarboxylic	terephthalic	terephthaloyl	—OC—⌬—CO—

*Systematic names are recommended for derivatives formed by substitution on a carbon atom.

**The trivial name is normally preferred.

New words and Expressions

methanol[meθənɔl]n.甲醇
2-propanol['prəupənɔl]n.2-丙醇
triphenylmethanol['traifenil'meθənɔl]n.
　三苯甲醇
allyl['ælil] alcohol烯丙(基)醇
tert-butyl['tə:ʃəri'bju:til] alcohol
　第三丁醇,叔丁醇
benzyl['benzil] alcohol苯甲醇,苄醇
ethylene glycol['eθili:n 'glaikɔl]乙二醇
glycerol['glisərɔl]n.甘油,丙三醇
carbocyclic['kɑ:bəu'saiklik]a.碳环的
cresol['kri:sɔl]n.甲酚
naphthol['næfθɔl]n.萘酚
pyrocatechol['pairə'kætəkəul]n.
　邻苯二酚,儿茶酚
resorcinol[rez'ɔ:sinɔl]n.间苯二酚
hydroquinone['haidrəukwi'nəun]n.
　对苯二酚,氢醌
potassium methanolate[meθə'nɔleit]
　甲醇钾
component[kəm'pəunənt]n.成分,组分;
　　　　　　　　　　　　　a.组成的
1-isopropoxypropane['aisəprə'pɔksi
　'prəupein]n.1-异丙氧基丙烷
aldehyde['ældihaid]n.醛,乙醛
ethyl methyl ether['eθil'meθil'i:θə]
　甲乙醚,甲氧基乙烷
ethyl vinyl ether['eθil'vainil'i:θə]
　乙基乙烯基醚
ethanal['eθənəl]n.乙醛
undecanedial[ʌn,dekein'daiəl]n.
　十一碳二醛
monobasic[,mɔnəu'beisik] acid一元酸
formaldehyde[fɔ:'mældihaid]n.甲醛
acetaldehyde['æsi'tældihaid]n.乙醛
propionaldehyde[,prəupiə'nældihaid]n.
　丙醛
acrylaldehyde[,ækrə'lældihaid]n.丙烯醛
acrolein[ə'krəuliin]n.丙烯醛
cinnamaldehyde[,sinə'mældihaid]n.肉桂醛
2-furaldehyde[fju'rældihaid]n.
　呋喃醛,糠醛
furfural['fə:fərəl]n.呋喃醛
2-butanone['bju:tənəun]n.2-丁酮
2,4-pentanedione n.2,4-戊二酮
ethyl methyl ketone['eθil'meθil'ki:təun]
　甲乙酮
diethyl ketone[dai'eθil'ki:təun]二乙酮
heptanoic['heptə'nəuik] acid 庚酸
heptanedioic['hepteindai'əuik] acid
　庚二酸
methanoic[meθə'nəuik] acid甲酸
formic['fɔ:mik] acid甲酸,蚁酸
ethanoic[eθə'nəuik] acid乙酸
acetyl['æsitil]乙酰(基)
propanoic[,prəupə'nəuik] acid丙酸
propionic['prəupi'ɔnik] acid丙酸
propionyl['prəupiəni(ə)l]n.丙酰(基)
butanoic[bju:tə'nəuik] acid丁酸
butyryl['bju:təril]n.丁酰(基)
butyric[bju:'tirik] acid丁酸
valeric[və'lɛərik] acid戊酸
valeryl[və'lɛəril]n.戊酰(基)
lauric['lɔ:rik] acid月桂酸,十二(碳烷)酸
lauroyl['lɔ:rəuil]n.月桂酰(基)
palmitic[pæl'mitik] acid
　软脂酸,十六(碳烷)酸
palmitoyl[pæl'mitəuil]n.十六(碳)酰(基)
stearic['stirik] acid
　硬脂酸,十八(碳烷)酸
stearoyl['stirəuil]n.十八(碳)酰(基)
oxalyl[ɔk'sælil,'ɑ:ksəlil]n.草酰(基),

乙二酰(基)
malonic[məˈlɔnik] acid 丙二酸
malonyl[ˈmælənil]n. 丙二酰(基)
succinic[səkˈsinik] acid 丁二酸,琥珀酸
succinyl[ˈsʌksinəl]n. 丁二酰(基)
adipic[əˈdipik] acid 己二酸
adipoyl[əˈdipəuil]n. 己二酰(基)
acrylic[əˈkrilik] acid 丙烯酸
acryloyl[əˈkriləuil,əˈkriləwil]n. 丙烯酰(基)
methacrylic[ˈmeθəˈkrilik] acid 甲基丙烯酸
methacryloyl[ˈmeθəˈkriləuil]n. 甲基丙烯酰(基)
maleic[məˈliːik] acid 顺丁烯二酸,马来酸
maleoyl[məˈliːəuil]n. 顺丁烯二酰(基)
benzoic[benˈzəuik] acid 苯甲酸,安息香酸
benzoyl[benˈzəuil]n. 苯甲酰(基),安息香酰(基)
phthalic[ˈfθælik] acid (邻)苯二甲酸
phthaloyl[fˈθæləuil,fˈθæləwil]n. (邻)苯二甲酰(基)
isophthalic[ˈaisəfˈθælik] acid 间苯二甲酸
isophthaloyl[ˈaisəfˈθæləuil]n. 间苯二甲酰(基)
terephthalic[ˈterefˈθælik]acid 对苯二甲酸
terephthaloyl[ˈterefˈθæləuil]n. 对苯二甲酰(基)
amic[ˈæmik] acid 酰胺酸
imidic[iˈmidik] acid 亚胺酸
carboximidic[ˈkɑːbɔksiˈmidik] acid 甲亚胺(基)酸
hydroxamic[ˌhaidrəukˈsæmik] acid 异羟肟酸
carbohydroxamic acid 甲酰异羟肟酸
sodium heptanoate[heptəˈnəueit]庚酸钠
potassium acetate 醋酸钾

ethyl acetate 醋酸乙酯
diethyl malonate[məˈlɔneit] 丙二酸二乙酯
ethyl chloroformate 氯甲酸乙酯
acetic anhydride[əˈsitik ænˈhaidraid] 醋酸酐,乙酸酐
phthalic anhydride (邻)苯二甲酸酐
acetyl chloride[ˈæsitilˈklɔːraid]乙酰氯
benzoyl iodide[ˈbenzəuilˈaiədaid] 苯甲酰碘

前　缀

pentyloxy-[ˌpentiˈlɔksi]戊氧基
allyloxy-[ˌæliˈlɔksi]烯丙氧基
benzyloxy-[ˌbenziˈlɔksi]苄氧基
methoxy-[meˈθɔksi]甲氧基
ethoxy-[eˈθɔksi]乙氧基
propoxy-[prəˈpɔksi]丙氧基
butoxy-[bjuˈtɔksi]丁氧基
phenoxy-[fiˈnɔksi]苯氧基
isopropoxy-[ˌaisəprɔˈpɔksi]异丙氧基
ethylenedioxy-[ˌeθiliːnˈdaiɔksi] 亚乙基二氧基,乙撑二氧基
trimethylenedioxy- 亚丙二氧基,丙撑二氧基
trans-[træns]反式
cis-[sis]顺式

后　缀

-olate[ɔ(ə)leit]……醇(酚)……(盐)
-ketone[ˈkiːtəun]……酮
-quinone[kwiˈnəun]……醌;n. 苯醌
-dione[daiəun]二酮

词　组

by means of…借助于,通过

Notes

1. The generic name "ketone" is given to compounds containing an oxygen atom doubly bound to a single carbon atom with the carbonyl group >C=O joined to two carbon atoms.",该句为简单句,其中"containing…carbon atoms"分词短语是compounds的定语。"with the …atoms"介词短语是single carbon atom的定语。**译文**:"含有与一个碳原子双重键合的氧原子,此羰基的碳原子又与两个碳原子相接的化合物通称为酮"。

2. Neutral salts of carboxylic, amic, imidic, carboximidic, hydroxamic, carbohydroxamic, etc., acids are named by citing the cation(s) and then the anion or, when the carboxyl groups of the acid are not all named as affixes, by use of a periphrase, such as "(metal)salt of(the acid).",该句是主从复合句,主句主语是"Neutral salt of…acids",谓语是are named…,有两个方式状语,即"by citing…anion"和"by use of…salt of (the acid)",第二个方式状语带有一个时间状语从句,即"when the…affixes"。**译文**:"羧酸、酰胺酸、亚胺酸、甲亚胺酸、异羟肟酸、甲酰异羟肟酸等的中性盐的命名是先引用阳离子,后引用阴离子,当酸的羧基不是全部作为词缀而命名时,通过繁琐的(某)酸(某金属)盐的说法而加以命名"。

Exercises

1. Fill in the blanks with proper words or phrases

   Methanol, also known as ____, is a chemical with the formula CH3OH. Methanol is the simplest alcohol,____ of a methyl group linked to a hydroxyl group. It is a light, ____, colorless, flammable liquid with a distinctive ____ similar to that of ethanol (drinking alcohol). Methanol is however far more toxic than ethanol. At room ____, it is a polar liquid. With more than 20 million tons produced ____, it is used as a ____ to other commodity chemicals, including ____, acetic acid, methyl tert-butyl ether, as well as a host of more specialized chemicals.

   A. temperature   B. consisting   C. methyl alcohol   D. precursor
   E. annually   F. formaldehyde G. odor   H. volatile

2. Match the following explanation with the proper word

   ____ carbocyclic, ____ glycerol, ____ component, ____ addition,

   A. something determined in relation to something that includes it

   B. a component that is added to something to improve it

   C. having or relating to or characterized by a ring composed of carbon atoms

   D. a sweet syrupy trihydroxy alcohol obtained by saponification of fats and oils

Lesson 15
PREPARATION OF A CARBOXYLIC ACID BY THE GRIGNARD METHOD

IIntroduction

Some of the common methods for the preparation of carboxylic acids include the oxidation of alcohols, the carbonation of Grignard reagents, and the hydrolysis of nitriles.

Because a solid carboxylic acid is much easier to isolate and purify than a liquid acid, so benzoic acid, m.p.121℃, rather than a liquid aliphatic acid, has been selected for this assignment. It will first be prepared by treatment of phenylmagnesium bromide with solid carbon dioxide (Dry Ice) and subsequent acidification of the reaction mixture by the addition of hydrochloric acid.

The Grignard reagent, phenylmagnesium bromide, is to be prepared at the start of the present period[1] by reaction of bromobenzene with metallic magnesium. The entire sequence of steps employed in this synthesis of benzoic acid is shown in the equations at the beginning of this experiment.

Since most of the common aliphatic carboxylic acids are liquids at room temperature, such acids are purified by distillation. If one were to prepare such an acid by the Grignard method, not only would the purification by distillation have to be carried out, but also it would be necessary to prepare at least one solid derivative of the acid in order to identify the product in a reasonably certain manner. Hence the entire experiment could not conveniently be completed in a regular 3-hour laboratory period.

However, in the actual practice of organic chemistry in industrial or institutional laboratories, where there are no arbitrary time limits for completing an experiment, the syntheses of certain aliphatic carboxylic acids can better be achieved by the Grignard process than by other methods. For example, trimethylacetic acid, $(CH_3)_3CCOOH$, is readily prepared from t-butyl chloride, $(CH_3)_3CCl$, by the Grignard method, whereas the nitrile route would be of no use whatsoever; treatment of t-butyl chloride with sodium cyanide gives sodium chloride, hydrogen cyanide and isobutylene rather than trimethylacetonitrile, $(CH_3)_3CCN$.

Anhydrous ether is generally used as the solvent in the preparation of a Grignard reagent. Actually, the ether plays a more important role in the reaction than merely that of solvent. Ether molecules combine with the various components of a Grignard reagent to form complex etherates. For example, one of the components present in phenylmagnesium bromide-ether solution is a complex.

As implied in the previous paragraph, a Grignard reagent is actually an equilibrium mixture of different molecular species. One of the equilibria thought to exist in any Grignard reagent is shown in the following equation, which has been simplified in that the ether molecules coordinated with the magnesium are not shown:[2]

$$2RMgX \rightleftharpoons R_2Mg + MgX_2$$

Addition of anhydrous dioxane to an ether solution of a Grignard reagent causes $RMgX$ and MgX_2 to precipitate, leaving R_2Mg in solution. Occasionally, for certain specific reactions, it is preferable to use the filtrate from such a mixture rather than the Grignard mixture itself; i.e., there are certain Grignard reactions in which the use of R_2Mg is preferable to the use of the equilibrium mixture of $RMgX$, MgX_2, and R_2Mg. However, no such complication exists in today's experiment.

Grignard reagents can be prepared successfully only in a completely anhydrous medium and in an atmosphere that is free of oxygen. The presence of water causes hydrolysis of the reagent, and the presence of oxygen causes loss of the reagent by oxidation.

$$RMgX + H_2O \rightarrow RH + Mg(OH)X$$
$$2RMgX + O_2 \rightarrow 2ROMgX$$

Therefore, in carrying out a Grignard reaction, one must take care to dry the apparatus and all of the reagents (magnesium metal, ether, and organic halide) carefully and also to provide an inert atmosphere over the reaction mixture. The latter condition is best realized by passing highly purified nitrogen over the surface of the liquid.

However, in today's experiment, the oxygen-free atmosphere will be attained by keeping the ether solution warm during the preparation of the Grignard reagent. Ether is so highly volatile (b.p.35°) that a blanket of ether vapor over the warm solution keeps the reagent reasonably well insulated from contact with the air.

Experimental

Fit a dry, 200-ml. round bottomed flask with a dry, water-cooled condenser. A drying tube filled with calcium chloride is attached to the top of the condenser. In the flask place 2.4g. (0.10 mole) of dry magnesium turnings, a crystal of iodine, 30 ml. of anhydrous ether, and 10 ml (15.7 g.0,10mole) of anhydrous bromobenzene. If reaction does not start immediately, warm the flask on the steam bath so that the ether refluxes gently and then remove the bath. This will usually initiate the reaction. The disappearance of the iodine color, the production of a cloudiness in the solution, and gentle boiling of the ether are all indications that the reaction has started. Once the reaction has begun it will proceed of its own accord causing the ether to boil gently 35~40 minutes.

You are now ready for the carbonation of the Grignard reagent by reaction with Dry Ice. Place about 15g. of crushed Dry Ice in a 250-ml. beaker and gradually pour into this mixture the solution of phenylmagnesium bromide. The reaction is vigorous, and the mixture sets to a stiff paste. Continue stirring the mixture until all the excess carbon dioxide has evaporated. Take the beaker to the hood and add 50~60 ml. of hot water in order to evaporate the small remaining quantity of ether. Acidify the contents of the beaker with dilute hydrochloric acid in order to dissolve the magnesium salts and liberate the benzoic acid. Stir the warm mixture well and set it in a pan of ice so that the portion of benzoic acid that is dissolved in the warm water may crystallize. When the solution is cold, collect the benzoic acid on the Büchner funnel.

The benzoic acid that you have now obtained is usually contaminated with a little biphenyl that has been formed as a by-product according to the following equation:

$$C_6H_5MgBr + C_6H_5Br \longrightarrow C_6H_5-C_6H_5 + MgBr_2$$

In as much as biphenyl is but slightly soluble in hot water, you may purify your sample of benzoic acid by recrystallizing it from that solvent.

New words and Expressions

carbonation[ˌkɑːbəˈneiʃən] n. 碳酸;(盐)化;羧基化
Grignard reagent [ˌgriːˈnjɑːriˈeidʒənt] 格氏试剂
purify[ˈpjuərifai] vt. 纯化,净化
phenylmagnesium bromide[ˈfenil mæˈgniːziəmˈbrəumaid] 溴化苯基镁
dry ice [drai ais] 干冰,固体二氧化碳
acidification[əˌsidifiˈkeiʃən] n. 酸化
bromobenzene[ˌbrəuməˈbenziːn] n. 溴苯
laboratory[ləˈbɔrətəri] n. 实验室
institutional[ˌinstiˈtjuːʃənl] a. 规定的,学校的
arbitrary[ˈɑːbitrəri] a. 武断的,随意的
trimethylacetic[traiˈmeθiləˈsiːtik] acid 三甲基乙酸
t-butyl chloride[ˈtəːʃəriˈbjuːtilˈklɔːraid] 第三氯丁烷
hydrogen cyanide[ˈhaidrɔdʒənˈsaiənaid] 氰化氢
trimethylacetonitrile[ˈtraiˈmeθilˈæsitɔˈnaitril] 三甲基乙腈
etherate[ˈiːθəreit] n. 醚化物
dioxane[ˌdaiˈɔksein] n. 二氧杂环己烷
filtrate[ˈfiltreit] v. 过滤 n. 滤液
loss[lɔːs] n. 损失
blanket[ˈblæŋkit] n. 毛毯,掩盖,覆盖层
round bottomed flask[raundˈbɔtəmd flɑːsk] 圆底烧瓶
condenser[kənˈdensə] n. 冷凝器
mole[məul] n. 摩尔
turning[ˈtəːniŋ] n. 削屑
steam bath 蒸气浴
gentle[ˈdʒentl] a. 温和的,轻轻的
cloudiness[ˈklaudinis] n. 阴暗,朦胧
crush[krʌʃ] vt. 压碎,磨细
beaker[ˈbiːkə] n. 烧坏
vigorous[ˈvigərəs] a. 有力的,活泼的,激烈的
stiff[stif] a. 浓粘的,硬的
hood[hud] n. 通风橱
acidify[əˈsidifai] v. 酸化
pan[pæn] n. 盘
Büchner funnel[ˈbuːknəˈfʌnl] 瓷漏斗,布氏漏斗
biphenyl[ˈbaiˈfenil] n. 联苯
recrystallize[riːˈkristəlaiz] v. 再结晶,重结晶

词 组

preferable to… 比……更好,优于
free of (from)… 无……的,除……的
take care to (+ inf.) 一定……,务必…
insulate from… 与……隔离,与……绝缘
of one's (its) own accord 自动地,自然而然地
in as much as… 因为,由于……缘故

Notes

1. at the start of the present period: at the beginning of this experiment
2. "One of the equilibria thought to exist in any Grignard reagent is shown in the following equation, which has been simplified in that the ether molecules coordinated with the magnesium are not shown." 是主从复合句,主句是"one of the …equation", which引导的是非限定性定语从句,此从句中又包括了一个介词宾语从句,即"in that the ether…not shown"。**译文**:"认为存在于任何格氏试剂中的平衡之一示于下面的方程式中,此方程已简化为没有示出与镁配位的醚分子"。
3. 处所的表达(位置的状语修饰语):

    用来引导状语修饰语的介词有完整的语义,也就是说它具有很深的意思,而许多其它词只不过是结构符号,在很多情况下特定介词的用法都由它后面所跟的名词来决定,我们很容易区分"in the corner of a room"与"at the corner of a street",但不易理解为什么英国人用"in the rain",而法国人用"under the rain"。在另一些情况下动词决定选择的介词。

    鉴于上述情况,我们把一些词语按习惯用法排列如下:

(a)介词 + 名词

in

  in(…)nucleus   in (writting) equation
  in nature   in the periodic table
  in solvent   in the reactivity series
  in(…)solution   in a reaction
  in an atom   in the (…) process
  in (4f) orbitals   in an experiment
  in (an open) conta   inerin quantity
  in the tower   in the preparation of

at

  at the anode   at the cathode
  at the positive electrode   at(ordinary, or room) temperature
  at a pressure of (+数字)atm.
  at (high) pressure   at the mouth
  at 200°   at the bottom of …
  at equilibrium   at a range of …

on

  on the bottom of (marine craft)   on all plates
  on the fingers   on ions
  on the surface of   on the negative electrode

under

  under high pressure   under ethanol
  under water

above, below

  above (or below)boiling point, melting point, 160℃ etc.

outside

  outside the nucleus

(b)动词 + 介词

with

    react with   compete with…(for…)

    combine with   charge with

    mix with   compare with

    consistent with

to

    add to   convert A to B

    apply to   adjacent to

    expose to   correspond to

    attach to   pass to

    give…to…

on

    act on   expand on

    base on   depend on

through

    sweep through   pass through

from

    proceed from   insulate from

    protect A from B   result from

    follow from   separate from

    differ from   suffer from

    arise from   transfer from

between

    distinguish between

Exercises

1. Fill in the blanks with proper words or phrases.

    Grignard reagent, any of ___ organic derivatives of ___ (Mg) commonly represented by the general ___ RMgX (in which R is a hydrocarbon radical: CH_3, C_2H_5, C_6H_5, etc.; and X is a__ ___, ___ chlorine, bromine, or iodine). They are called ___ after their discoverer, French chemist Victor Grignard, who was a ___ of the 1912 ___ for Chemistry for this work.

    A. magnesium   B. numerous   C. Nobel Prize   D. Grignard reagents,

    E. corecipient   F. halogen atom   G. usually   H. formula

2. Match the following explanation with the proper word.

    ___ carbonation, ___ purify, ___ acidification, ___ filtrate(n.), ___ recrystallize

    A. remove impurities from, increase the concentration of, and separate through the process of distillation

    B. to dissolve and subsequently crystallize (a substance) from the solution, as in purifying chemical compounds, or (of a substance) to crystallize in this way

    C. the process of becoming acid or being converted into an acid

    D. saturation with carbon dioxide (as soda water)

    E. the product of filtration; a gas or liquid that has been passed through a filter

Lesson 16
THE STRUCTURES OF COVALENT COMPOUNDS

Most of us realize from our earlier chemical training that each covalent chemical compound has a structure—that is, a definite arrangement of its constituent atoms in space. The concept of covalent compounds as three-dimensional objects was developed in the latter part of the last century. Chemists who lived before that time regarded covalent compounds as shapeless groups of atoms held together in a rather undefined way by poorly understood electrical forces. Although the currently accepted structural characteristics of organic compounds were first suggested in 1874, these postulates were based on indirect chemical and physical evidence. Until the early twentieth century no one knew whether they had any physical reality, since scientists had no techniques for viewing molecules at the atomic level of resolution. Thus, as recently as the second decade of the twentieth century investigators could ask two questions: (1) Do organic molecules have specific geometries and, if so, what are they? (2) Can we develop simple principles to predict molecular geomerty?

A. Methods for Determining Molecular Structure

Among the greatest developments of chemical physics in the early twentieth century were the discoveries of ways to peer into molecules and deduce the arrangement in space of their constituent atoms. Most information of this type today comes from three sources; X-ray crystallography, electron diffraction, and microwave spectroscopy. The arrangement of atoms in the crystalline solid state can be determined by ***X-ray crystallography***. This technique, discovered in 1915 and revolutionized in recent years by the availability of high-speed computers, uses the fact that X rays are diffracted from the atoms of a crystal in precise patterns that can be translated into a molecular structure. In 1930 another technique, **electron diffraction**, was developed from the observation that electrons are scattered by the atoms in molecules of gaseous substances. The diffraction patterns resulting from this scattering can also be used to deduce the arrangements of atoms in molecules. Following the development of radar in World War II came **microwave spectroscopy**, in which the absorption of microwave radiation by gaseous molecules provides structural information.

Most of the details of atomic structure in this book are derived from gas-phase methods—electron diffraction and microwave spectroscopy. For molecules that are not readily studied in the gas phase, X-ray crystallography is the most important source of structural information. There is no comparable method that allows the study of structures in solution, a fact that is unfortunate because most chemical reactions take place in solution. The consistency of gas-phase and crystalline structures suggests, however, that molecular structures in solution probably differ little from those of molecules in the solid or gaseous state.

B. Prediction of Molecular Geometry

The geometry of a simple covalent molecule is defined by two quantities, bond length and bond angle. The **bond length** is defined as the distance between the centers of bonded nuclei. Bond length is usually measured in angstroms; $1 Å = 10^{-10}$ meter $= 10^{-8}$ centimeter. Bond angle is the angle between two bonds to the same atom. Consider, for example, the compound methane, CH_4. When the C—H bond length and H—C—H bond angles are known, we know the structure of methane.

Molecular structrure is important because the way a molecule reacts chemically is closely linked to its structure. From the many structures that have been determined, we can now make several generalizations about the structures of covalent compounds.

Bond Length   The following generalizations can be made about bond length:
1. Bond lengths between atoms of a given type decrease with the amount of multiple bonding. Thus, bond lengths for carbon-carbon bonds are in the order C—C>>C=C>C≡C.
2. Bond lengths tend to increase with the size of the bonded atoms. This effect is most dramatic as we proceed down the periodic table. Thus, a C—H bond is shorter than a C—F bond, which is shorter than a C—Cl bond. Since bond length is the distance between the centers of bonded atoms, it is reasonable that larger atoms should form longer bonds.
3. When we make comparisons within a given row of the periodic table, bonds of a certain type (single, double, or triple) between a given atom and a series of other atoms become shorter with increasing electronegativity. Thus, the C—F bond in H_3C—F is shorter than the C—C bond in H_3C—CH_3. This effect occurs because a more electronegative atom has a greater attraction for the electrons of the bonding partner, and therefore "pulls it closer," than a less electronegative atom.

Bond Angle   The bond angles within a molecule determine its shape—whether it is bent or linear. Two generalizations allow us to predict the approximate bond angles, and therefore the general shapes, of many simple molecules. The first is that the groups bound to a central atom are arranged so that they are as far apart as possible. For methane, CH_4, the central atom is carbon and the groups are the four hydrogens. The hydrogens of methane are farthest apart when they occupy the vertices of a tetrahedron centered on the carbon atom. Because the four C—H bonds of methane are identical, the hydrogen atoms of methane lie at the vertices of a regular tetrahedron (a tetrahedron with equal sides). The tetrahedral shape of methane requires an H—C—H bond angle of 109.5°.

In applying this rule for the purpose of predicting bond angles, we regard all groups as identical. Thus the groups that surround carbon in H_3CCl (methyl chloride) are treated as identical, even though in reality the C—Cl bond is considerably longer than the C—H bonds.

Although the bond angles show minor deviations from the exact tetrahedral bond angle of 109.5°, methyl chloride in fact has the general tetrahedral shape.

The tetrahedral structure, then, is assumed by molecules when four groups are arranged about a central atom. Since carbon is tetravalent, this is an extremely important geometry for many organic compounds. Since we shall see this geometry repeatedly, it is worth the effort to become familiar with it. Tetrahedral carbon is often represented, as shown below in the structure of methylene chloride, CH_2Cl_2, with two of its bound groups in the plane of the page. One of the remaining groups, indicated with a dashed line, is behind the page, and the other, indicated with a wedge-shaped heavy line, is in front of the page.

You should make a molecular model of a simple tetrahedral molecule such as CH_2Cl_2 and relate the three-dimensional model to its two-dimensional representation above.

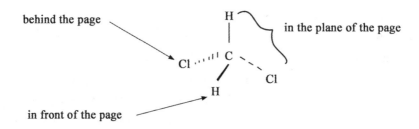

When three groups surround a central atom, the groups are as far apart as possible when all bonds lie in the same plane with bond angles of 120° in boron trichloride, the central atom (in this case boron) is said to have trigonal geometry.

When a central atom is surrounded by two groups, maximum separation of the groups demands a bond angle of 180°. This is the situation with each carbon in acetylene, H—C≡C—H. Each carbon is surrounded by two groups, a hydrogen and another carbon. (It makes no difference that the carbon has a triple bond.) Because of the 180° bond angle at each carbon, acetylene is a linear molecule.

The second generalization about molecular structure applies to molecules with unshared electron pairs. In predicting the geometry of a molecule, an unshared electron pair can be considered as a bond without a nucleus at one end. This rule allows us to handle, for example, the geometry of ammonia. In view of this rule, ammonia, $:NH_3$, has four groups about the central nitrogen, three hydrogens and an electron pair. To a first approximation, these groups adopt the tetrahedral geometry with the electron pair occupying one corner of the tetrahedron. (This geometry is sometimes called **pyramidal**, or pyramid-like, geometry.) We can refine our prediction of geometry even more if we recognize that an electron pair without a nucleus at one end has an especially repulsive interaction with electrons in adjacent bonds. As a result, the bond angle between the electron pair and the N-H bond is a little larger than tetrahedral, leaving the H—N—H angle a little smaller than tetrahedral; in fact, the H—N—H angle is 107.3°.

If we examine the structure of a molecule that is somewhat more complex than those we have just discussed, we discover one further aspect of structure. Consider, for example, the structure of ethylene, $H_2C=CH_2$. Each carbon of ethylene is bound to three groups: two hydrogens and the other carbon. Our rules for predicting bond angle require, then, that the geometry at each carbon should be trigonal. The structure of ethylene shows that the H—C—H bond angle, 117°, indeed approaches the idealized trigonal value of 120° quite closely, and that the three groups bonded to either of the carbons lie in the same plane. However, ethylene poses a new problem of geometry. Imagine sighting along the carbon-carbon double bond from one end of the molecule. The resulting angle between the C—H bonds on adjacent carbons is called the dihedral angle. The value of the dihedral angle determines the **conformation** of ethylene: the spatial relationship of the groups on one carbon to those on the other.

In summary, then, the structure of a molecule is completely determined by three elements: its bond lengths, its bond angles, and its conformation. There are simple molecules whose structures are completely determined by bond lengths and bond angles. We have learned what trends in bond lengths to expect and how to predict the shapes of molecules from bond angles. Conformation enters the picture for more complex molecules. We shall learn at various stages of our study some of the principles that determine molecular conformation.

New words and Expressions

arrangement[əˈreindʒmənt]n. 排列
three-dimensional[-diˈmenʃənəl]a. 立体的，体型的，三维的
shapeless[ˈʃeiplis]a. 不定形的，无形状的
postulate[ˈpɔstjuleit]vt. 假定，n. 假说
resolution[ˌrezəˈljuːʃən]n. 溶解，分辨率
investigator[inˈvestigeit]n. 研究者
peer[piə]vi. 凝视
deduce[diˈdjuːs]vt. 推论
crystallography[ˌkristəˈlɔgrəfi]n. 结晶学
diffraction[diːˈfrækʃən]n. 衍射
spectroscopy[spekˈtrɔskəpi]n. 波谱学
crystalline[ˈkristəlain]n. 结晶质，结晶体
scatter[ˈskætə]vt. 散射
radar[ˈreidə]n. 雷达
absorption[əbˈsɔːpʃən]n. 吸收
angstrom[ˈæŋstrəm]n. 埃(Å，长度单位)
multiple bond 重键
vertex[ˈvəːteks]n.(pl. -tices[-tisiːz])顶点
tetrahedron[ˌtetrəˈhedrən]n.(pl. -dra[drə])四面体
deviation[ˌdiːviˈeiʃən]n. 偏离，偏差
wedge[wedʒ]n. 楔形
trigonal[ˈtrigənl]a. 三角形的
methylene[ˈmeθiliːn]n. 甲撑，亚甲
pyramidal[piˈræmidl]a. 锥体的
adjacent[əˈdʒeisənt]a. 相邻的
pose[pəuz]vt. 提出(问题)
imagine[iˈmædʒin]vt.vi 想象
dihedral[daiˈhedrəl]a. 二面的
conformation[ˌkɔnfɔːˈmeiʃən]n. 构型
spatial[ˈspeiʃəl]a. 空间的

后缀

- graphy[-grəfi] 一种方法(名词词尾)
  crystallography, chromatography, photography

- hedron[-ˈhedrən] …面体，…面晶体(名词词尾) tetrahedron, hexahedron, ployhedron

词 组

peer(vi.)into… 凝视
be derived from… 由…派生出(衍生出)
result from… 由…而造成
come from… 得自…
be linked to… 被连结到,与…有关
in the order of… 依…序
become familiar with… 熟悉…
in view of… 考虑到,由…看来

Notes

1. 化学英语中复合词的组成形式主要有以下几种方式:
   (1)形容词 + 名词,如:high-temperature;high-speed;
   (2)名词 + 动名词或现在分词,如:oxygen-containing;
   (3)名词 + 过去分词,如:home-made;
   (4)副词 + 过去分词,如:so-called,well-known;
   (5)副词 + 现在分词,如:fast-spreading;
   (6)名词 + 名词,如:rate constant;
   (7)名词 + 形容词,如:ice-cold;water-free;
   (8)形容词 + 现在分词,如:fine-looking;
   (9)动名词或现在分词 + 名词,如:melting point;
   (10)形容词 + 形容词,如:dark–blue;
   (11)介词 + 名词,如:online;
   (12)复合动词,如:overload.

2. Radar是一个以缩写的形式构成的词汇,它取自:radio detecting and ranging,类似的词汇还有,如:sonar(sound operation navigation and range)声纳;laser(light amplification by simulated emission of radiation)激光;e.g.(for example);i.e.(that is to say);etc[it'setrə]等等。

3. 一些由名词簇集组成的化学术语:

electron diffraction	电子衍射
microwave spectroscopy	微波波谱学
gas phase	气相
bond length	键长
bond angle	键角
shared electron pair	共享电子对
electron pair	电子对
repulsive interaction	推斥相互作用
carbon–carbon double bond	碳—碳双键
alkali metal	碱金属
energy level	能级
valence shell	价电子层
activity coefficient	活度系数
oxidation level	氧化水平

oxidation number	氧化数
oxidation state	氧化态
rate constant	速率常数
reaction mechanism	反应机理
transition state	过渡态
hydrogen bond	氢键
energy barrier	能垒
quantum mechanics	量子力学
state function	状态函数
heat capacity	热容
potential energy	势能
vapor pressure	蒸气压
partition ratio	分配比
retention volume	保留体积
diffusion coefficient	扩散系数
buffer solution	缓冲溶液
ion atmosphere	离子氛
ground state	基态
data base	数据库

Exercises

1. Fill in the blanks with proper words or phrases.

   Chemical compounds are generally grouped into one of two ___: covalent compounds and compounds. Ionic compounds are made up of electrically ___ atoms or molecules as a result of ___ or losing electrons. Ions of ___ charges form ionic compounds, usually as a result of a metal reacting with a nonmetal. Covalent, or molecular, compounds generally result from two ___ reacting with each other. The ___ form a compound by sharing electrons, resulting in an electrically ___ molecule.

    A. nonmetals B. elements C. opposite   D. ionic
    E. gaining   F. neutral G. categories H. charged

2. Match the following explanation with the proper word.

   ___ shapeless, ___ crystallography, ___ postulate(n.), ___ spectroscopy, ___ trigonal

    A. a proposition that is accepted as true in order to provide a basis for logical reasoning

    B. the branch of science that studies the formation and structure of crystals

    C. having threefold symmetry

    D. the use of spectroscopes to analyze spectra

    E. having no definite form or distinct shape

Lesson 17
OXIDATION AND REDUCTION IN ORGANIC CHEMISTRY

The conversion of alcohols into carbonyl compounds is an important reaction of primary and secondary alcohols, and is one of many examples in organic chemistry of oxidation. How do we know when an organic compound has been oxidized? In the last section, we recognized that conversion of an alcohol to a ketone is an oxidation because it is brought about by the reduction of Cr(Ⅵ). But there are other oxidations in which the oxidizing agent is less obvious. Our goal in this section is to be able to recognize an oxidation or reduction merely by examining the transformation of the organic compound itself. The procedure for doing this involves three steps:

Step 1. Assign an **oxidation level** to each carbon atom in reactant and product. (It is only necessary to assign an oxidation level to carbons that undergo some chemical change during the transformation; other carbons may be ignored.) The oxidation level of a particular carbon is assigned by considering the relative electronegativities of the groups bound to the carbon, as follows.

(a) For every bond to an element less electronegative than carbon (including hydrogen), and for every negative charge on the carbon, assign a −1.

(b) For every bond to another carbon atom, and for every unpaired electron on the carbon, assign a zero.

(c) For every bond to an element more electronegative than carbon, and for every positive charge on the carbon, assign a +1.

(d) Add the numbers assigned in (a), (b), and (c) to obtain the oxidation level for the carbon atom under consideration.

Let us apply this first step to the transformation of isopropyl alcohol to acetone.

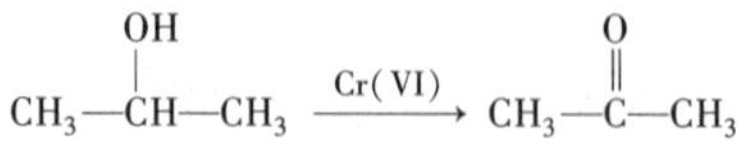

$$CH_3-\underset{\underset{\text{isopropyl alcohol}}{}}{\overset{OH}{\underset{|}{CH}}}-CH_3 \xrightarrow{Cr(VI)} CH_3-\underset{\underset{\text{acetone}}{}}{\overset{O}{\underset{\|}{C}}}-CH_3 \qquad (17\text{—}1)$$

Since the carbon atoms of the two methyl groups do not change, we do not need to assign oxidation levels to these carbons. Notice in the treatment of acetone that the C=O double bond is counted as two bonds; +1 for each bond gives a total of +2 for the double bond.

reactant: $CH_3 \overset{0}{-} \overset{\overset{+1}{OH}}{\underset{\underset{CH_3}{0}}{C}} \overset{-1}{-} H$ product: $CH_3 \overset{0}{-} \overset{\overset{+2}{O}}{\underset{}{C}} \overset{0}{-} CH_3$

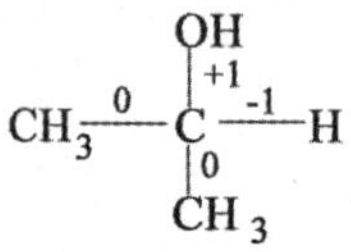

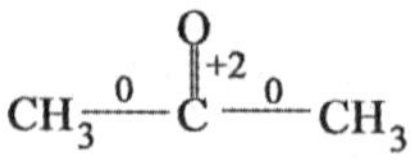

Sum: 0+0+(+1)+(−1)=0   Sum: 0+0+(+2)=+2

Step 2. The oxidation number N_{ox} for each compound is computed by adding the oxidation levels of all carbons. In the structures above, only one carbon has changed its oxidation level, so the N_{ox} values of the reactant and product are simply equal to the respective oxidation levels of this carbon. Therefore, the oxidation level of the reactant is 0 and that of the product is +2. In other reactions involving more than one carbon atom, N_{ox} is computed by summing the oxidation levels of all carbon atoms that undergo a chemical change.

Step 3. Compute the difference.

$$N_{ox}(\text{product}) - N_{ox}(\text{reactant})$$

If this difference is positive, the transformation is an oxidation. If this difference is negative, the transformation is a reduction. If the difference is zero, neither an oxidation nor a reduction has taken place. For the reaction of Eq. 17-1 this difference is +2-0=+2. This transformation is thus an oxidation.

Although the oxidation-number formalism is very useful, we should not lose sight of the following two general characteristics of organic oxidations and reductions. These two points can enable us to spot an oxidation or reduction at a glance.

1. In most oxidations of organic compounds, either hydrogen in a C—H bond or carbon in a C—C bond is replaced by a more electronegative element, such as halogen or oxygen. The converse is true for reductions.

2. The oxidation state of a molecule is determined from the oxidation states of its individual carbon atoms.

The oxidation number concept can be simply related to a definition of oxidation that is often used in inorganic chemistry. According to this definition, oxidation is the loss of electrons and reduction is the gain of electrons. To see how this definition applies to organic compounds, let us consider as an example the oxidation of ethanol to acetic acid:

$$CH_3-CH_2-OH \longrightarrow CH_3-\overset{\overset{O}{\|}}{C}-OH \qquad (17-2)$$
$$\text{ethanol} \qquad \text{acetic acid}$$

We can write this oxidation as a balanced half-reaction using H_2O to balance missing oxygens, protons to balance missing hydrogens, and "dummy electrons" to balance charges.[1]

$$CH_3-CH_2-OH + H_2O \longrightarrow CH_3-\overset{\overset{O}{\|}}{C}-OH + 4H^+ + 4e^- \qquad (17-3)$$

According to this half-reaction, four electrons are lost from the ethanol molecule when acetic acid is formed. (Since this is only a half-reaction, a corresponding number of electrons must be gained by the species that brings about the oxidation.) It can be said that the oxidation of ethanol to acetic acid is a four-electron oxidation. This type of terminology, which is frequently used in biochemistry, comes from the half-reaction formalism.

If we compute the oxidation numbers of ethanol and acetic acid, we can see that the change in oxidation number for Eq. 17-2 is +4 (verify this statement). This example illustrates the following point: the change in oxidation number is equal to the number of electrons lost. If the change in oxidation number is negative, the reaction is a reduction, and the number corresponds to electrons gained.

Oxidations and reductions, like acid-base reactions, always occur in pairs. Therefore, whenever something is oxidized, something else is reduced. When an organic compound is oxidized, the reagent that brings about the transformation is called an **oxidizing agent**. Likewise, when an organic compound is reduced, the reagent that effects the transformation is called a **reducing** agent. For example, suppose that chromate ion (CrO_4^{2-}) is used to bring about the oxidation of ethanol to acetic acid in Eq.17-2 in this reaction, chromate ion is reduced to Cr^{3+}.

$$8H^+ + 3e^- + CrO_4^{2-} \longrightarrow Cr^{3+} + 4H_2O \qquad (17\text{—}4)$$

Three electrons are gained in the reduction of chromate to Cr^{3+}. Since four electrons are lost in the oxidation (Eq.17-3), stoichiometry requires that for every three ethanol molecules oxidized to acetic acid (twelve electrons lost), four CrO_4^{2-} are reduced (twelve electrons gained).

By considering the change in oxidation number for a transformation, we can tell whether an oxidizing or reducing agent is required to bring about the reaction. For example, the following transformation is neither an oxidation nor a reduction (verify this statement):

$$CH_3-\underset{\underset{OH}{|}}{\overset{\overset{CH_3}{|}}{C}}-\underset{\underset{OH}{|}}{\overset{\overset{CH_3}{|}}{C}}-CH_3 \longrightarrow CH_3-\underset{\underset{CH_3}{|}}{\overset{\overset{CH_3}{|}}{C}}-\overset{\overset{O}{\|}}{C}-CH_3 \qquad (17\text{—}5)$$

Although one carbon is oxidized, another is reduced. Even though we might know nothing else about the reaction, it is clear that an oxidizing or reducing agent alone would not effect this transformation. (In fact, the reaction is the pinacol rearrangement, which is brought about by mineral acid.)

The oxidation-number concept can be used to organize organic compounds into functional groups with the same oxidation level. Compounds within a given box are generally interconverted by reagents that are neither oxidizing nor reducing agents. For example, we know that alcohols can be converted into alkyl halides with HBr, which is neither an oxidizing nor a reducing agent. On the other hand, conversion of an alcohol into a carboxylic acid involves a change in oxidation level, and indeed this transformation requires an oxidizing agent. There are a greater number of possible oxidation states for carbons with larger numbers of hydrogens. Thus, a tertiary alcohol cannot be oxidized at the α-carbon (without breaking carbon-carbon bonds) because this carbon bears no hydrogens. Methane, on the other hand, can be oxidized to CO_2. (Of course, any hydrocarbon can be oxidized to CO_2 if carbon-carbon bonds are broken.)

New words and Expressions

ignore [igˈnɔː] vt. 忽视
oxidation level n. 氧化水平,氧化级(数)
isopropyl [ˌaisəuˈprəupil] n. 异丙基
count [kaunt] vt. 算为

oxidation number n. 氧化值
formalism [ˈfɔːməlizəm] n. 拘泥形式,形式主义
spot [spɔt] vt. 找到,认准

dummy[dʌmi]a. 假的,虚的,设定的
terminology[tə:mi'nɔlədʒi]n. 术语
verify['verifai]vt. 证明
stoichiometry[stɔiki'ɔmitri]n. 化学计量学
pinacol['pinəkɔl]n. 口片呐醇,2,3-二基-2,3-丁二醇
rearrangement[ri:ə'reindʒmənt]n. 重排
bear[bɛə]vt. 携带
transformation[ˌtrænsfə'meiʃən]n. 转化

后　缀

-metry[-mitri]测定,度量的方法(名词词尾)stoichiometry, geometry spectrophotometry
-ism[-izəm]表示"特性、状态、主义、体系"之义（名词词尾）formalism, mechanism, alcoholism(酒中毒)

短　语

assign…to…　　把…指认为……
under consideration　　所考虑的,所讨论的
neither…nor　　既不…又不
at a glance　　一看就
either…or…　　要么…或者…
relate to…　　关联于
oxidation of…to…　　…氧化…为…
be equal to…　　等于…
in pairs　　成对地
be converted into…　　被转变为…
(be)characteristic of…　　有…特色的

Notes

1. "we can write……to balance charges". 该句为简单句,其中"as a balanced…句末",为as引出的修饰宾语this oxidation的宾语补语。**译文**:"我们能利用H₂O平衡失去的氧,用质子平衡失去的氢和用设定的电子平衡电荷,从而把该氧化反应写成一个平衡的半反应"。

Exercises

1. Fill in the blacks with proper words or phrases.
   Oxidation−reduction ____, also called ____, any chemical reaction in which the ____ number of a participating chemical species changes. The term covers a large and diverse body of processes. Many oxidation−reduction reactions are as common and familiar as fire, the rusting and dissolution of ____, the ____ of fruit, and ____ and photosynthesis—basic life functions.
   　A. oxidation　　B. reaction　　　C. browning　　D. redox reaction,
   　E. metals,　　F. respiration
2. Match the following explanation with the proper word.
   ____ oxidation, ____ terminology, ____ stoichiomtery, ____ rearrangement
   　A. the relation between the quantities of substances that take part in a reaction or form a compound (typically a ratio of whole integers)
   　B. a system of words used to name things in a particular discipline
   　C. changing an arrangement
   　D. the process of oxidizing; the addition of oxygen to a compound with a loss of electrons; always occurs accompanied by reduction
   　E. having characteristics of both an acid and a base and capable of reacting as either

Lesson 18
SYNTHESIS OF ALCOHOLS AND DESIGN OF ORGANIC SYNTHESIS

A. Synthesis of Alcohols and Glycols

Let us review the methods we have learned for the synthesis of alcohols and glycols. The methods we have studied so far involve alkenes as starting materials:

1. Hydroboration-oxidation of alkenes.

$$6RCH=CH_2 + B_2H_6 \longrightarrow 2(RCH_2CH_2)_3B \xrightarrow[OH^-]{H_2O_2} 6RCH_2CH_2OH$$

2. Oxymercuration-reduction of alkenes.

$$R-CH=CH_2 + Hg(OAc)_2 \xrightarrow{H_2O} R-\underset{OH}{CH}-CH_2-HgOAc \xrightarrow{NaBH_4} R-\underset{OH}{CH}-CH_3$$

Acid-catalyzed hydration of alkenes is used industrially to prepare certain alcohols, but is not an improtant laboratory method. In principle, the S_N2 reaction of ^-OH with an alkyl halide can also be used to prepare alcohols from alkyl halides. However, since alkyl halides are generally prepared from alcohols themselves, this method is of little practical improtance[1].

Some of the most improtant methods for the synthesis of alcohols involve the reduction of carbonyl compounds (aldehydes, ketones, or carboxylic acids and their derivatives), as well as the reactions of carbonyl compounds with Grignard or organolithium reagents.

We have learned only one method for preparing glycols, namely, oxidation of alkenes with either OsO_4 or $KMnO_4$.

$$2KMnO_4 + 4H_2O + 3RCH=CHR \longrightarrow 3RCH-CHR + 2MnO_2 + 2KOH$$
$$||$$
$$OHOH$$

$$RCH=CHR \xrightarrow{OsO_4} \xrightarrow{NaHSO_3} RCH-CHR$$
$$\phantom{RCH=CHR \xrightarrow{OsO_4} \xrightarrow{NaHSO_3} R}||$$
$$\phantom{RCH=CHR \xrightarrow{OsO_4} \xrightarrow{NaHSO_3} R}OHOH$$

The glycols can also be prepared from epoxides.

B. Design of Organic Synthesis

We have now studied a number of reactions that can transform one functional group into another. These reactions can be used to prepare organic compounds from readily available starting materials. The preparation of an organic compound by the use of one or more reactions is called an **organic synthesis**.

Although many useful syntheses consist of only one reaction, more typically it is necessary to use several reactions to complete the synthesis of an organic compound. Let us examine the logic used in planning such a multistep synthesis.

We shall call the molecule we desire to synthesize the **target molecule**. In order to assess the best route to the target molecule from the starting material, we take the same approach that a military officer might take in planning his assault on an objective: namely, we work backward from the target towards the starting material. Just as the officer considers secondary objectives—a hill here, a tree there—from which the final assault on the target can be launched, in planning a synthesis we first assess what compound can be used as the immediate precursor of the target. We then continue to work backward from this precursor step-by-step until the route from the starting material becomes clear. Sometimes more than one synthetic route will be possible. In such a case, we evaluate each synthesis in terms of yield, limitations, expense, and so on.

Let us illustrate the strategy we have just discussed with a sample problem. Suppose we are asked to prepare hexanal from 1-hexene:

$$CH_3CH_2CH_2CH_2CH=CH_2 \longrightarrow CH_3CH_2CH_2CH_2CH_2CH=O$$
$$\text{1-hexene} \qquad\qquad\qquad \text{hexanal}$$

The first question we ask is whether we know any ways to prepare aldehydes directly from alkenes. The answer is that we do. Ozonolysis can be used to transform alkenes into aldehydes and ketones. However, ozonolysis breaks a carbon-carbon double bond, and certainly would not work for preparing an aldehyde from an alkene with the same number of carbon atoms, because at least one carbon is lost when the double bond is broken. We have studied no other ways to prepare aldehydes directly from alkenes. The next step is to ask what ways we know to prepare aldehydes from other starting materials. The answer is that we know two: cleavage of glycols and oxidation of primary alcohols. The cleavage of glycols, like ozonolysis, breaks a carbon-carbon bond, and requires that we lose at least one carbon atom. However, the oxidation of a primary alcohol would be a satisfactory last step in our synthesis.

$$CH_3CH_2CH_2CH_2CH_2CH_2-OH \xrightarrow[\text{pyridine}]{CrO_3} CH_3CH_2CH_2CH_2CH_2CH=O$$
$$\text{1-hexanol}$$

We now ask whether it is possible to prepare 1-hexanol from 1-hexene; the answer is yes. Hydroboration-oxidation will convert 1-hexene into 1-hexanol. Our synthesis is now complete.

$$CH_3CH_2CH_2CH_2CH=CH_2 \xrightarrow{B_2H_6} \xrightarrow{H_2O_2/OH^-} CH_3CH_2CH_2CH_2CH_2CH_2-OH \quad \text{1-hexanol}$$

$$\downarrow CrO_3/\text{pyridine}$$

$$CH_3CH_2CH_2CH_2CH_2CH=O \qquad\qquad \text{hexanal}$$

Notice how we worked backward from the target molecule one step at a time.

Working problems in organic synthesis is one of the best ways to master organic chemistry[2].

It is akin to mastering a language: it is rather easy to learn to read a language (be it a foreign language, English, or even a computer language), but we have to understand it more thoroughly in order to write it. Similarly, it is relatively easy to follow individual organic reactions, but to integrate them and use them out of context requires more understanding. One way to bring together organic reactions and study them systematically is to go back through the text and write a representative reaction for each of the methods we have studied for preparing each functional group. For example, which reactions can be used to prepare alkanes? Carboxylic acids? Then jot down some notes describing the stereochemistry of each reaction (if known) as well as its limitations—that is, the situations in which the reaction would not be expected to work. For example, dehydration of tertiary and secondary alcohols is a good laboratory method for preparing alkenes, but dehydration of primary alcohols is not. (Do you understand the reason for this limitation?) This type of study can be continued throughout future chapters.

We have begun this process for you by summarizing in Appendix IV the reactions discussed in this text that can be used to prepare compounds containing each functional group. It is now time to begin to integrate what you have learned about organic chemistry, and problems in organic synthesis will help you achieve that goal.

New words and Expressions

synthesize['sinθisaiz] vt. 合成
glycol[glaikɔl] n. 1,2-乙二醇, 二元醇
hydroboration['haidrəbɔː'reiʃən] n. 硼氢化(作用)
oxymercuration['ɔksiməːkju'reiʃən] n. 羟汞化(作用)
catalyze['kætəlaiz] vt. 催化
ketone['kiːtəun] n. 酮类
carboxylic[kɑːbɔk'silik] a. 羧基的, 羧酸的
organolithium[ˈɔːgənəuˈliθiəm] n. 有机锂化合物
appendix[ə'pendiks] n. 附录
transform[trænsˈfɔːm] vt. 使转变
assess[ə'ses] vt. 评价
assault[ə'sɔːlt] n. 袭击
lauch[lɔːtʃ] vt. 发起
immediate[i'miːdjət] a. 最接近的
target['tɑːgit] n. 靶, 目标
illustrate['iləstreit] vt. (用图解)说明

strategy['strætidʒi] n. 战略
ozonolysis[əuzə'nɔlisis] n. 臭氧分解(作用)
systematical[sisti'mætikəl] a. 系统的
dehydration[dihai'dreiʃən] n. 脱水(作用)

前缀和后缀

organo-[ˈɔːgənəu-] 有机, organolithium, organometallic
-lysis[-lisis, -'ləsis] 分解, 消散(名词词尾) ozonolysis, analysis, pyrolysis
-tion[-ʃən] (或作-ation, -cion, -xion, -ion, -sion) 动作、状态、行为之结果(名词词尾) hydroboration, oxidation, isomerization

词　　组

prepare A from B　　由B制A
(be)akin to…　　与…类似(近似)

Notes

1. be + 介词短语与 be + 名词或形容词类似，均作表语，如 be of use = be useful，即 is of… importance = is improtant.
2. "Working problems in……organic chemistry." 系简单句，其中 Working 引导的动名词短语作主语，to master 引导的不定式短语作表语 "is one…ways" 的定语。**译文**："解决有机合成的难点是精选有机化学的诸多最好方法中的一个"。

Exercises

1. Fill in the blacks with proper words or phrases.

   Organic synthesis is a special branch of chemical synthesis and is concerned with the ____ construction of ____ compounds. Organic____ are often more ____ than ____ compounds, and their ____ has developed into one of the most important branches of organic chemistry. There are several ____ areas of research within the general area of organic synthesis: total synthesis, ____, and methodology.

   A. intentional   B. molecules   C. main   D. inorganic
   E. semisynthesis   F. synthesis   G. organic   H. complex

2. Choose a proper prefix to make the follow words match the meanings.

   ____ organic 无机的，____ arrangement 重新排列，____ synthesis 半合成，
   ____ oxide 过氧化物，____ cyclic 双环的，____ hydration 脱水，____ act 相互作用

   A. re-   B. in-   C. inter-   D. bi-
   E. per-   F. de-   G. semi-

Lesson 19
ORGANOMETALLICS—METAL π COMPLEXES

Metal π complexes are characterized by a type of direct carbon-to-metal bonding that is not a classical ionic, σ, or π bond. Numerous molecules and ions, eg, mono- and diolefins, polyenes, arenes, cyclopentadienyl ions, tropylium ions[1], and π-allylic ions, can form metal π complexes with transition-metal atoms or ions. These are classified as organometallic complexes, because of their direct carbon-metal bond, and as coordination complexes, because the nature and characteristics of the π ligands are similar to those in coordination complexes. In 1827, Zeise reported that ethylene reacts with platinum (II) chloride to form a salt $K(C_2H_4)PtCl_3(1)$, but it was not until after the elucidation of the structure of ferrocene (2) in 1953 that attention was redirected to Ziese's salt, which was the first reported metal π complex.

Generally, metal π complexes can be classified into three main groups: olefin-, cyclopentadienyl-, and arene-metal π complexes; mixed complexes are categorized according to structural or chemical analogies within these groups. Allyl π complexes are designated as olefin π complexes in this review. Study of metal π complexes has contributed to the elucidation of the mechanisms of Ziegler-Natta polymerization, the oxo reaction, and catalytic hydrogenation, and to the development of the Wacker process which is used for the oxidation of olefins[2].

The following nomenclature for metal π complexes is used: (1) Organic π ligands precede the metal atom. (2) Organic π ligands precede inorganic π ligands. (3) Inorganic π ligands, eg, carbonyl or nitrosyls, generally follow the metal atom; halides also follow the metal but precede carbonyls or nitrosyls. (4) A prefix, eg, di, is preferred rather than bis in describing sandwich-type π complexes, eg, dibenzenechromium. (5) The symbol π can be used preceding a ligand in order to distinguish π-complex bonding from σ, ionic, or other bonding. The symbol η (eta or hapto) precedes a ligand and indicates the number of C—M bonds in the ligand.

Monoolefins, dienes, polyolefins, and acetylenes serve as ligands to transition metals and form olefin π complexes. Typical examples of olefin π complexes are monoolefin ligands, eg, potassium $η^2$-ethyleneplatinum trichloride(1); and cyclopentadienylium.-$η^3$-cycloheptatrienylium molybdenum dicarbonyl (3); diene ligands, eg, η^4-butadieneiron tricarbonyl(4).

Certain of the delocalized π-electron ring systems of aromatic molecules overlap with d_{xy} and d_{y3} metal orbitals as do the π electrons of alkenes with metal d orbitals[3]. The following aromatic rings can form π complexes:

The $C_5H_5^-$, C_6H_6, and C_8H_8 arenes are the most common in arene π complexes that are characterized by π-bonded rings alone or π-bonded rings that are associated with one ring and other ligands, eg, halogens, CO, RNC, and R_3P. Typical examples are the di-η^5-cyclopentadienyl complexes, ie, metallocenes, eg, di-η^5-cyclopentadienyliron(2). In di-η^5-cyclopentadienyliron, ie, ferrocene, the 6-π-electron system of the $C_5H_5^-$ ion is bonded to the metal. Other aromatic ring systems are mono-η^5-cyclopentadienylmetal nitrosyl and carbonyl complexes.

Properties

The π-Complex Bond. Metal π complexes are among those that are least satisfactorily described by crystal-field theory (CFT) or valence-bond theory (VBT). The nature of the bonding can be treated more completely and quantitatively by molecular-orbital theory (MOT) or ligand-field theory (LFT). The ligand-field theory originally was advanced as a corrected CFT. The LFT relies on the use of molecular orbitals and often is used interchangeably with the MOT. The usual approach is to use the linear combination of atomic orbitals (LCAO) method. It is assumed that when an electron in a molecule is near a particular nucleus, the molecular wave function is approximately an atomic orbital that is centered at the nucleus. The molecular orbitals are formed by adding or subtracting the appropriate atomic orbitals. For transition metals, the $3d$, $4s$, and $4p$ orbitals are the atomic orbitals of interest. The ligands may have σ- and π-valence orbitals. Once the appropriate atomic orbitals have been selected for the metal and ligands, the proper linear combination of valence atomic orbitals is determined for the molecular orbitals. The determination of orbital overlaps that are possible, ie, meet inherent symmetry requirements, is done by application of the principles of group theory. At this point, the procedure becomes arbitrary in that approximate wave functions must be selected for use in the calculations of the overlap integrals and coulomb integrals[4]. Finally, an arbitrary charge distribution is chosen and the orbital energies and interaction energies are calculated, and a solution of the secular equation for the energies and coefficients of the atomic wave functions can be determined. A new initial charge distribution is repeated until consistent values are obtained.

Reactions

Metal π complexes react with a wide range of chemical reagents. However, the reactions of the π-olefin-, π-cyclopentadienyl-, and π-arene-metal complexes are distinctly characteristic of each group. π Cyclopentadienyl complexes, ie, metallocenes, exhibit a high degree of aromaticity and undergo many typical aromatic substitution reactions. However, the π arene complexes do not exhibit a discernible degree of aromaticity.

Although most physical properties, particularly the structure of metal π complexes, are interpreted by use of the basic principles of coordination chemistry, these established principles do not explain suitably some reaction anomalies of the different groups of metal π complexes.

Olefin π Complexes. Reactions involving olefin π complexes similarly are characteristic of uncomplexed and complexed olefinic functions. Generally, reactions involving the former are not very different from those observed for free olefins. However, reactions of the latter are altered significantly by π-complex formation. Among the reactions of interest are addition, elimination, and substitution.

Cyclopentadienyl π Complexes. The most significant feature of the reactions of π-cyclopentadienyl complexes in general and ferrocene in particular involves their aromatic nature. The resonance stabilization energy for ferrocene is 210 kJ/mol (50 kcal/mol). Ferrocene undergoes a large number of typical ionic aromatic substitution reactions, eg, Friedel–Crafts acylation, alkylation, metalation, sulfonation, and aminomethylation.

Friedel - Crafts Acylation. The acylation of metallocenes proceeds easily. The equimolar reaction of ferrocene and acetyl chloride in the presence of aluminum chloride yields monoacetylferrocene almost exclusively. When an excess of acetyl chloride and aluminum chloride is used, a mixture of two isomeric diacetylferrocenes is produced. The heteroannular disubstituted derivative 1,1′-diacetylferrocene and the homoannular isomer 1,2-diacetylferrocene are obtained in a ratio of 60:1. The first acetyl group deactivates the π-cyclopentadienyl ligand toward further electrophilic substitution. Thus, the second acetyl group enters the other ring.

Sulfonation. Ferrocene can be sulfonated readily by sulfuric acid or cholrosulfonic acid in acetic anhydride to form ferrocenesulfonic acid and heteroannular disulfonic acid. π-Cyclopentadienylrhenium tricarbonyl can be sulfonated with concentrated sulfuric acid in acetic anhydride; the product is isolated as the p-toluidine salt.

Formylation. Ferrocene is formylated with N-methylformanilide in the presence of phosphorus oxychloride. This reaction also is characteristic of highly reactive aromatic rings.

Arylation. The most significant radical substitution reaction of ferrocene is its reaction with aryl diazonium salts giving an arylation product.

Arene–Metal π Complexes. Generally, arene π complexes do not undergo the reactions that are characteristic of benzene and its derivatives. However, arene π complexes do undergo a limited number of substitution, addition, expansion, and condensation reactions.

Uses

Catalysis Involving Metal π-Complex Intermediates. Many metal-catalyzed reactions proceed by way of a substrate metal π-complex intermediate. Commercially, the most significant of these include the polymerization of ethylene, the hydroformylation of olefins yielding aldehydes, ie, the oxo process (qv), and the air oxidation of ethylene-producing acetaldehyde (qv), ie, the Wacker process.

Polymerization of Olefins. Ziegler - Natta Process. During the 1950s, ethylene was polymerized using a Ziegler-Natta catalyst, ie, a mixture of transition metal halides, eg, titanium halides, and trialkylaluminum (triethylaluminum commonly is used). The use of trialkylaluminum stimulated research into the use of organometallic compounds in general. It has been determined that the Ziegler-Natta process involves a metal π-complex intermediate. A plausible mechanism for the polymerization can be formulated by applying typical organometallic and coordination reactions.

Oxidation of Olefins. Wacker Process. The oxidation of ethylene exclusively to acetaldehyde and of other straight-chain olefins to ketones is achieved by the catalytic reaction of ethylene in an aqueous solution by palladium(II) or by oxygen in the presence of palladium(II) chloride, copper(II) chloride, or iron(III) chloride. Generally, the oxidation of olefins by other metal ions, eg, Hg(II), Th(III), and Pb(IV), yields glycol derivatives as well as carbonyl products. The mechanism for the oxidation is postulated to include π-σ rearrangements.

Addition of Carbon Monoxide. Oxo Reaction. The oxo process has been developed extensively to produce primary alcohols by the reduction of the aldehydes which are formed in the process.

Health and Safety Factors

Some metal π complexes are air-sensitive and, therefore, their preparation requires an air-free reaction system. Their toxicity usually is based on the metal; however, organometallic compounds generally exhibit greater toxicities than their corresponding inorganic salts. The alkyl derivatives tend to be more toxic than the aryl complexes, which exhibit toxicities similar to those of the corresponding inorganic compounds.

New words and Expressions

organometallics[ˌɔːgənəumiˈtæliks] n. 金属有机化学(化合物)
polyene[ˈpɔliːn] n. 多烯，聚烯
arene[ˈæriːn] n. 芳烃
tropylium[trɔˈpiliəm] n. 䓬䓬，环庚三烯芳香型阳离子
elucidation[iˌljuːsiˈdeiʃən] n. 阐明
analogy[əˈnælədʒi] n. 类似
oxo[ˈɔksəu] reaction 羰基合成
polymerization[pɔliməraiˈzeiʃən] n. 聚合作用
catalytic[ˌkætəˈlitik] a. 催化的
nitrosyl[ˈnaitrəusil] n. 亚硝酰(基)
sandwich-type[ˈsænwidʒ taip] 夹心式的，三明治型的
dibenzenechromium[daibenˈziːnˈkrəumiəm] n. 二苯铬
ethyleneplatinum trichloride 乙烯合三氯化铂
cyclopentadienylium-η^3-cycloheptatrienylium molybdenum dicarbonyl 环戊二烯基合-η^3-环庚三烯基合二羰基合钼
butadieneiron tricarbonyl 丁二烯合三羰基合铁
di-η^5-cyclopentadienyliron = ferrocene 二茂铁
delocalize[diːˈləukəlaiz] vt. 离域，非定域
molecular-orbital theory = MOT 分子轨道理论
linear combination of atomic orbitals = LCAO 原子轨道的线性组合

metallocene[ˈmetələsiːn]n. 二茂金属化合物
inherent[inˈhiərent]a. 固有的，内在的
group theory　群论
wave function　波函数
integral[ˈintigrəl]n. 积分
overlap[əuvəˈlæp] integral 重叠积分
coulomb[ˈkuːlɔm] integral 库仑积分
secular[ˈsekjulə] equation 久期方程
distribution[ˌdistriˈbjuʃən]n. 分布，分配
symmetry[ˈsimitri]n. 对称
aromaticity[ˌærəməˈtisiti]n. 芳香性
discernible[diˈsəːnibl]a. 可辨别的
acylation[æsiˈleiʃən]n. 酰基化(作用)
alkylation[ˌælkiˈleiʃən]n. 烷基化(作用)
metalation[metəˈleiʃən]n. 金属化(作用)
sulfonate[ˈsʌlfəˌneit]n. 磺酸盐(酯)
aminomethylation[ˈæminəumeθiˈleiʃən] n. 氨基甲基化(作用)
equimolar[ˌiːkwiˈməulə]a. 等摩尔的
annular[ˈænjulə]a. 环形的
deactivate[diːˈæktiveit]vt. 使失活
chlorosulfonic[ˌklɔːrəsʌlˈfɔnik]a. 氯磺酸的
p-toluidine[təˈluːidin]n. 对甲苯胺
formylate[ˈfɔːmiˌleit]vt. 甲酰化
formanilide[fɔmˈænilaid]n. N-甲酰苯胺
oxychloride[ˌɔksiˈklɔːraid]n. 氯氧化物
aryl[ˈæril]n. 芳基
diazonium[ˌdaiəˈzəuniəm]n. 重氮化
stimulate[ˈstimjuleit]vt. 刺激
plausible[ˈplɔːzəbl]a. 似乎有道理的，似合理的
toxicity[ˈtɔkˈsisiti]n. 毒性
expansion[iksˈpænʃən]n. 膨胀

词　组

be characterized by…　突出的表现为…
be classified as…　归类为
be classified into…　分类成
be designated as…　被叫做…
in a ratio of…　以…的比例
by way of…　经由……
stimulate…into…　激励…到…

Notes

1. Tropylium ion: In organis chemistry, the tropylium ion is an aromatic species with a formula of $[C_7H_7]^+$. Its structure shows as following:

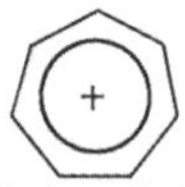

2. "Study of metal……of olefins"，该句为复合句，主句是"study of…the Wacker process"，主句中谓语 contributed to 有两个宾语，即 the elucidation…和 the development…。"which…句末"为定语从句，修饰 the Wacker process.
3. "Certain of the ……metal d orbitals"，为复合句，其中"as…句末"为 as 引导的非限制性定语从句，as 代替整个主句，从句中谓语 do 意即主句中的 overlap，从句中主谓倒装。
4. "At this point，……and coulomb integrals"，该句为复合句，"At this……arbitrary"是主句，"in that……句末"为定语从句，修饰主句主语 the procedure。**译文**："在(使用群论)这一点上，使用近视波函数来计算重叠积分和库仑积分的方法便成为专用的。"
5. 常用的前辍：

acet-	乙酰	hetero-	异，杂
acyl-	酰基	homo-	同，均
alk-	烷烃类	hydro-	氢，水

amino-	氨基	iso-	等,异
azo-	偶氮	nitro-	硝基
bromo-	溴	organo-	有机的
chloro-	氯	oxo	-氧化
electro-	电	poly-	多,聚
ferro-	亚铁	stereo-	立体
halo-	卤素		

6. 常用的后缀

-al	醛类	-ite	亚酸盐
-amide	酰胺	-ium,-um	金属元素词尾
-amine	胺	-ocene	二茂(金属)化合物
-ane	饱和烷烃	-ol	醇,酚
-ate	含氧酸盐,酯	-one	酮,砜
-ene	烯,苯系烃	-ous	亚
-fold	倍	-oxide	氧化物
-form	仿	-philic	亲近
-free	无	-yl	基
-ide	某化物,无氧酸盐	-ylene	亚基
-ide	酐		

Exercises

1. Fill in the blanks with proper words or phrases.

   Organometallic chemistry is the study of organometallic compounds, chemical compounds containing at least one chemical bond between a carbon ___ of an organic molecule and a metal, including alkaline, alkaline ___, and transition metals, and sometimes ___ to include metalloids like boron, ___, and tin, as well. Aside from bonds to organyl ___ or molecules, bonds to '___' carbon, like carbon monoxide (metal carbonyls), cyanide, or carbide, are generally considered to be organometallic as well. Some related such as ___ hydrides and metal phosphine complexes are often included in discussions of organometallic compounds, though strictly speaking, they are not necessarily organometallic.

   A. earth   B. broadened   C. atom   D. compounds
   E. transition metal   F. inorganic   G fragments   H. silicon

2. Match the following explanation with the proper word.

   ___ polymerization, ___ catalytic, ___ delicalize, ___ annular, ___ toxicity

   A. relating to or causing or involving catalysis

   B. shaped like a ring

   C. remove from the proper or usual locality

   D. a chemical process that combines several monomers to form a polymer or polymeric compound

   E. the degree of strength of a poison

Lesson 20
THE ROLE OF PROTECTIVE GROUPS IN ORGANIC SYNTHESIS

Properties of a Protective Group

When a chemical reaction is to be carried out selectively at one reactive site in a multifunctional compound, other reactive sites must be temporarily blocked. Many protective groups have been, and are being, developed for this purpose. A protective group must fulfill a number of requirements. It must react selectively in good yield to give a protected substrate that is stable to the projected reactions. The protective group must be selectively removed in good yield by readily available, preferably nontoxic reagents that do not attack the regenerated functional group. The protective group should form a crystalline derivative (without the generation of new chiral centers) that can be easily separated from side products associated with its formation or cleavage. The protective group should have a minimum of additional functionality to avoid further sites of reaction.

Historical Development

Since a few protective groups cannot satisfy all these criteria for elaborate substrates, a large number of mutually complementary protective groups are needed and, indeed, are becoming available. In early syntheses the chemist chose a standard derivative known to be stable to the subsequent reactions. Other classical methods of cleavage include acidic hydrolysis (eq. 1), reduction (eq. 2), and oxidation (eq. 3):

(1) ArO—R→ArOH

(2) RO—CH_2Ph→ROH

(3) RNH—CHO→[RNHCOOH]→RNH_2

Some of the original work in the carbohydrate area in particular reveals extensive protection of carbonyl and hydroxyl groups. For example, a cyclic diacetonide of glucose was selectively cleaved to the monoacetonide. A more recent summary describes the selective protection of primary and secondary hydroxyl groups in a synthesis of gentiobiose, carried out in the 1870s, as triphenylmethyl ethers.

Development of New Protective Groups

As chemists proceeded to synthesize more complicated structures, they developed more satisfactory protective groups and more effective methods for the formation and cleavage of protected compounds. At first a tetrahydropyranyl acetal was prepared, by an acid-catalyzed reaction with dihydropyran, to protect a hydroxyl group. The acetal is readily cleaved by mild acid hydrolysis, but formation of this acetal introduces a new chiral center. Formation of the 4-methoxytetrahydropyranyl acetal eliminates this problem.

Catalytic hydrogenolysis of an O-benzyl protective group is a mild, selective method introduced by Bergmann and Zervas to cleave a benzyl carbamate ($>$NCO—OCH$_2$C$_6$H$_5$→$>$NH) prepared to protect an amino group during peptide syntheses. The method also has been used to cleave alkyl benzyl ethers, stable compounds prepared to protect alkyl alcohols; benzyl esters are cleaved by catalytic hydrogenolysis under neutral conditions.

Three selective methods to remove protective groups are receiving much attention: "assisted," electrolytic, and photolytic removal. Four examples illustrate "assisted removal" of a protective group. A stable allyl group can be converted to a labile vinyl ether group (eq.4); a β-haloethoxy (eq.5) or a β-silylethoxy (eq.6) derivative is cleaved by attack at the β-substituent; and a stable o-nitrophenyl derivative can be reduced to the o-amino compound, which undergoes cleavage by nucleophilic displacement:

(4) ROCH$_2$CH=CH$_2$ $\xrightarrow{\text{O-t-Bu}}$ [ROCH=CHCH$_3$] $\xrightarrow{\text{H}_3\text{O}^+}$ ROH

(5) RO—CH$_2$—CCl$_3$ + Zn → RO$^-$ + CH$_2$=CCl$_2$

(6) RO—CH$_2$—CH$_2$—SiMe$_3$ $\xrightarrow{\text{F}^-}$ RO$^-$ + CH$_2$=CH$_2$ + FSiMe$_3$

R=alkyl, aryl, R'CO—, or R'NHCO—

The design of new protective groups that are cleaved by "assisted removal" is a challenging and rewarding undertaking.

Removal of a protective group by electrolytic oxidation or reduction can be very satisfactory. The equipment required ranges from a minimum of two electrodes, a potentiostat, and a source of DC current to quite sophisticated systems. A suitable electrolyte/solvent system is needed, and the deprotected product must not undergo further electrochemistry under the experimental conditions. The use and subsequent removal of chemical oxidants or reductants (e.g., Cr or Pb salts; Pt—or Pd—C) are eliminated. Reductive cleavages have been carried out in high yield at —1 to —3 V (vs. SCE) depending on the group; oxidative cleavages in good yield have been realized at 1.5—2 V (vs. SCE). For systems possessing two or more electrochemically labile protective groups, selective cleavage is possible when the half-wave potentials, $E_{1/2}$, are sufficiently different; excellent selectivity can be obtained with potential differences on the order of 0.25 V. Protective groups that have been removed by electrolytic oxidation or reduction are described at the appropriate places in this book; a review article by Mairanovsky discusses electrochemical removal of protective groups.

Photolytic cleavage reactions (e.g., of o-nitrobenzyl, phenacyl, and nitrophenyl-sulfenyl derivatives) take place in high yield on irradiation of the protected compound for a few hours at 254~350 nm. For example, the o-nitrobenzyl group, used to protect alcohols, amines, and carboxylic acids, has been removed by irradiation. Protective groups that have been removed by photolysis are described at the appropriate places in this book; in addition, the reader may wish to consult three review articles.

Selection of a Protective Group from This Book

To select a specific protective group, the chemist must consider in detail all the reactants, reaction conditions, and functionalities involved in the proposed synthetic scheme. First he or she must evaluate all functional groups in the reactant to determine those that will be unstable to the desired reaction conditions and require protection. The chemist should then examine reactivities of possible protective groups, listed in the Reactivity Charts, to determine compatibility of protective group and reaction conditions. The protective groups listed in the Reactivity Charts have been used most widely; consequently, considerable experimental information is available for them. He or she should consult the complete list of protective groups in the relevant chapter and consider their properties. It will frequently be advisable to examine the use of one protective group for several functional groups (i.e., a 2,2,2-trichloroethyl group to protect a hydroxyl group as an ether, a carboxylic acid as an ester, and an amino group as a carbamate). When several protective groups are to be removed simultaneously, it may be advantageous to use the same protective group to protect different functional groups (e.g., a benzyl group, removed by hydrogenolysis, to protect an alcohol and a carboxylic acid). When selective removal is required, different classes of protection must be used (e.g., a benzyl ether, cleaved by hydrogenolysis but stable to basic hydrolysis, to protect an alcohol, and an alkyl ester, cleaved by basic hydrolysis but stable to hydrogenolysis, to protect a carboxylic acid).

If a satisfactory protective group has not been located, the chemist has a number of alternatives: rearrange the order of some of the steps in the synthetic scheme so that a functional group no longer requires protection or a protective group that was reactive in the original scheme is now stable; redesign the synthesis, possibly making use of latent functionality (i.e., a functional group in a precursor form; e.g., anisole as a precursor of cyclohexanone). Or, it may be necessary to include the synthesis of a new protective group in the overall plan.

A number of standard synthetic reference books are available. A review article by Kossell and Seliger discusses protective groups used in oligonucleotide syntheses, including protection for the phosphate group, which is not included in this book; and a series of articles describe various aspects of protective group chemistry.

New words and Expressions

protective[prəˈtektiv] a. 保护的
  ~ group 保护基
temporarily [ˈtempərərili] ad. 临时地

yield[ji:ld] vt. 产出, n. 收率, 产率
substrate[ˈsʌbstreit] n. 底物, 基质
functional[ˈfʌŋkʃənl] group 官能团

chiral['tʃairəl]a. 手性的
cleavage['kli:vidʒ]n. 分解，裂解
elaborate[i'læbəreit]a. 精心的，复杂的
complementary[ˌkɔmpli'mentəri]a. 补充的
hydrolysis[hai'drɔlisis]n. 水解作用
carbohydrate[ˌkɑ:bəu'haidreit]n. 碳水化合物
acetonide['æsitəunaid]n. 丙酮化物
glucose['glu:kəus]n. 葡萄糖
gentiobiose[ˌdʒentiə'baiəz]n. 龙胆二糖
acetal['æsitæl]n. 乙缩醛，乙醛缩二乙醇
tetrahydropyranyl n. 四氢吡喃基
carbamate['kɑ:bəmeit]n. 氨基甲酸酯
amino['æminəu]a. 氨基的
peptide['peptaid]n. 肽
alkyl['ælkil]n. 烷基
labile['leibail]a. 不稳定态的
haloethoxy['hæləu'iθɔksi]n. 卤代乙氧基
silylethoxy['silil'iθɔksi]n. 甲硅烷基乙氧基
nitrophenyl['naitrəfenil]n. 硝基苯基
challenge['tʃælindʒ]n.v. 挑战
reward[ri'wɔ:d]vt. 值得做
potentiostat[pə'tenʃiəˌstæt]n. 恒电位器
half-wave potential 半波电位
phenacyl[fi'næsil]n. 苯甲酰甲基
nitrophenylsulfenyl['naitrəfenil'sʌlfenil] n. 硝基苯基亚磺酰基
photolysis[fəu'tɔlisis]n. 光解作用
photolytic[ˌfəutə'litik]a. 光解的
electrolytic[iˌlektrəu'litik]a. 电解的
symmetrical[si'metrikəl]a. 对称的

compatibility[kəmˌpæti'biliti]n. 适合，兼容
alternative[ɔ:l'tə:nətiv]a. 选择的
　　　　　　　　　　　n. 可供选择的对象
latent['leitənt]a. 潜在的
precursor[pri(:)'kə:sə]n. 前体
anisole['ænisoul]n. 茴香醚
review[ri'vju:]n. 综述，述评
oligonucleotide[ˌɔligəu'nju:kli:ətaid] n. 低聚核苷酸

前缀和后缀

-stat['stæt]恒，不动（名词词尾）
　　potentiostat, thermostat
hydro-['haidrə-]水，氢 hydrolyze,
　　hydrolysis, hydroxyl, hydrogenolysis
carb(o)-[kɑ:bə-]碳，羰 carbohydrate,
　　carbonation, carbocyclic, carbonyl
photo-[fəutə-]光，照像 photolytic,
　　photochemical, photometry, photograph
olig(o)-['ɔləg(əu)-,'ɔlig(ə)]少，低，齐
　　oligonucleotide, oligomer（低聚物）
　　oligomerization（低聚，齐聚）
　　oligodynamics（微动力学）

词组

1. be protected from…免于…
2. be separated from……从…被分离出
3. range from…to…范围从…到…；在…与…间变化
4. in detail 详细地

Exercises

1. Fill in the blanks with proper words or phrases.

   A protecting group or ____ is introduced into a molecule by chemical modification of a functional group to obtain __ in a subsequent chemical reaction. It plays an important __ in multistep organic ____. Protecting groups are more commonly used in small-scale laboratory work and initial development than in ___ production processes because their use adds ___ steps and material costs to the process. However, the ___ of a cheap chiral building block can ___ these additional costs (e.g. shikimic acid for oseltamivir).

A. role　　　　　B. availability　C. synthesis　　D. chemoselectivity

E. protective group　F. overcome　　G industrial　　H. additional

2. Match the following explanation with the proper word.

___ substrate, ___ functional group, ___ hydrolysis, ___ photolysis, ___ electrolysis

A. the substance upon which an enzyme acts

B. the process of passing an electric current through a substance in order to produce chemical changes in the substance.

C. chemical decomposition caused by light or other electromagnetic radiation

D. a chemical reaction in which a compound reacts with water to produce other compounds

E. the group of atoms in a compound, such as the hydroxyl group in an alcohol, that determines the chemical behaviour of the compound

Lesson 21
HETEROCYCLIC CHEMISTRY

The known organic compounds have an enormous diversity of structure. Many of these structures contain ring systems. If the ring system is made up of atoms of carbon and at least one other element, the compound can be classed as heterocyclic. The elements that occur most commonly, together with carbon, in ring systems are nitrogen, oxygen and sulfur. About half of the known organic compounds have structures that incorporate at least one heterocyclic component.

Uses of Heterocyclic Compounds

Heterocyclic compounds have a wide range of applications: they are predominant among the types of compounds used as pharmaceuticals, as agrochemicals and as veterinary products. They are used as optical brightening agents, as antioxidants, as corrosion inhibitors and as additives with a variety of other functions. Many dyestuffs and pigments have heterocyclic structures.

One of the reasons for the widespread use of heterocyclic compounds is that their structures can be subtly manipulated to achieve a required modification in function. Many heterocycles can be fitted into one of a few broad groups of structures that have overall similarities in their properties but significant variations within the group. Such variations can include differences in acidity or basicity, different susceptibility to attack by electrophiles or nucleophiles, and different polarity. The possible structural variations include the change of one heteroatom for another in a ring and the different positioning of the same heteroatoms within the ring. An example of the way in which heterocycles are used is provided by an account of the development of a new systemic fungicide, 1. The pyrimidine ring was incorporated into this structure because a related compound, 2, proved to be too lipophilic. The water solubility and the transport of the fungicide through the plant are improved by replacing a benzene ring by the more polar heterocycle.

Another important feature of the structure of many heterocyclic compounds is that it is possible to incorporate functional groups either as substituents or as part of the ring system itself. For example, basic nitrogen atoms can be incorporated both as amino substituents and as part of a ring. This means that the structures are particularly versatile as a means of providing, or of mimicking, a functional group. An example is the use of the 1H-tetrazole ring system 3 as a mimic of a carboxylic acid function 4, because of its similarity in acidity and in steric requirement. One of the main purposes of the later chapters of this book is to provide a framework for understanding and predicting the effects of structure on properties. Armed with this understanding, the organic chemist can 'tailor' a structure to meet a particular need by modifying the heterocyclic component.

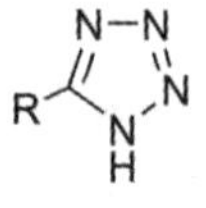

Heterocyclic compounds are also finding an increasing use as intermediates in organic synthesis. Very often this is because a relatively stable ring system can be carried through a number of synthetic steps and then cleaved at the required stage in a synthesis to reveal other functional groups. Many examples are given in later chapters, but the principle is illustrated by the frequent use of the furan ring system as a 'masked' 1,4-dicarbonyl compound (Fig. 1.): the furan ring can be cleaved by acid to reveal this functionality.

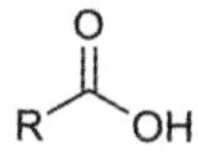

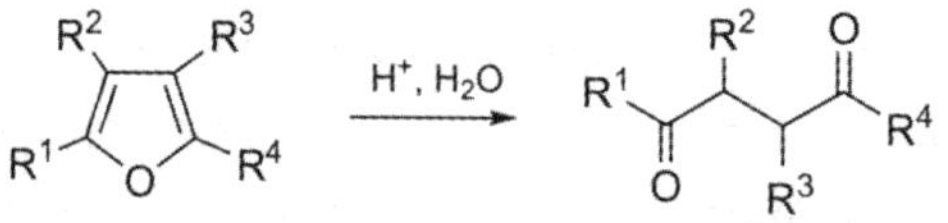

Figure 1. Furans as masked 1,4-dicarbonyl compounds.

Pyridines, imidazoles and other heterocyclic compounds have frequently been used as ligands for metals but their design is increasingly being tailored to specific applications. For example, heterocyclic compounds are now often used as chiral ligands for transition metals and the resulting complexes act as catalysts in a variety of asymmetric synthetic reactions. Thus, the disubstituted pyridine 5 has been incorporated as a tridentate ligand into the ruthenium complex 6. A useful reaction of diazo esters, N_2CHCO_2R, is their catalyzed decomposition, with the loss of nitrogen. When this decomposition is carried out in the presence of alkenes, cyclopropanes are produced. The chiral complex 6 not only catalyses the decomposition of the diazo esters but it enables the cyclopropanes to be isolated with a high degree of optical purity.

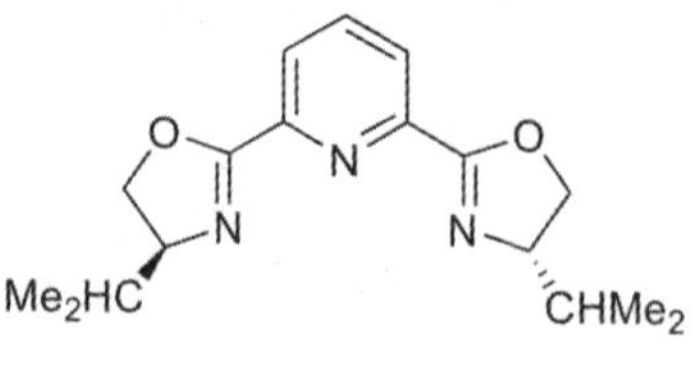

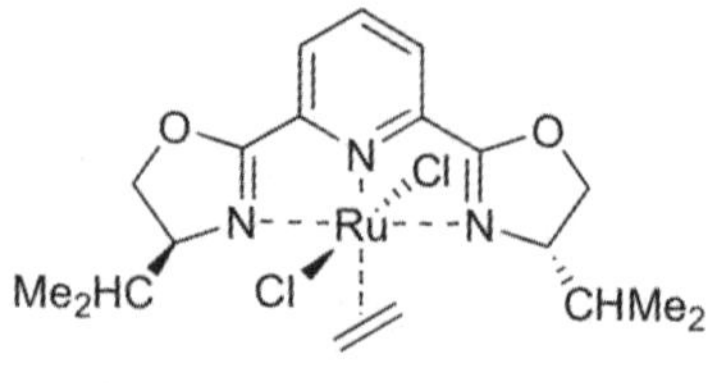

Heterocyclic compounds are widely distributed in nature. Many are of fundamental importance to living systems: it is striking how often a heterocyclic compound is found as a key component in biological processes. We can, for example, identify the nucleic acid bases, which are derivatives of the pyrimidine and purine ring systems, as being crucial to the mechanism of replication. Chlorophyll and heme, which are derivatives of the porphyrin ring system, are the components required for photosynthesis and for oxygen transport in higher plants and in animals, respectively. Essential diet ingredients such as thiamin (vitamin B1), riboflavin (vitamin B2), pyridoxol (vitamin B6), nicotinamide (vitamin B3) and ascorbic acid (vitamin C) are heterocyclic compounds. Of the twenty amino acids commonly found in proteins, three, namely histidine, proline and tryptophan, are heterocyclic. It is not surprising, therefore, that a great deal of current research work is concerned with methods of synthesis and properties of heterocyclic compounds.

Many of the pharmaceuticals and most of the other heterocyclic compounds with practical applications are not extracted from natural sources but are manufactured. The origins of organic chemistry do, however, lie in the study of natural products. These have formed the basis for the design of many of the useful compounds developed subsequently: examples are the early development of vat dyes based on the structure of indigo and the continuing invention of new antibacterial agents based on the ? - lactam structure of penicillin. Three different groups of pharmaceuticals with structures related to natural products are described briefly below as illustrations of the ways in which natural products chemistry interacts with synthetic heterocyclic chemistry.

Heterocyclic Bases

One particular group of hetero cyclic compounds are the heterocyclic bases. These examples are extremely important compounds. Look at the names of these compounds. If you have studied any biology, you will probably recognize these names as being very important parts of DNA and RNA molecules. The ones shown here are uracil, thymine, and cytosine.

In the biochemistry lesson on nucleic acids, you will learn about how these heterocyclic bases bond to sugar molecules and phosphate groups to form DNA and RNA. You will also learn about how they can bond to one another to pull the two strands of the DNA together.

For now, we will limit our consideration of these compounds to three things. The first is why they are called bases. Essentially, it is the presence of the unbonded pair of electrons on each nitrogen atom that gives these compounds their basic properties. Their ability to provide a pair of electrons in order to attract a proton makes these compounds basic.

Second, you should recall from when we dealt with amines, the presence of this unbonded electron pair also allowed for hydrogen bonding to occur to the nitrogen atom. This, of course, can happen only if there is another polar molecule nearby which has a hydrogen available for hydrogen bonding.

If you look carefully at these diagrams, you can see that each of these compounds has nitrogen or oxygen available to receive hydrogen for hydrogen bonding. Remember: unbonded electron pairs are also available on the oxygen atoms that are contained in these compounds. You should also notice that each of these compounds contains hydrogen atoms bonded to nitrogen. These hydrogen atoms are the ones that can form hydrogen bonds to other molecules. For future reference you should remember that there are groups on each of these heterocyclic bases which can form hydrogen bonds both by having the hydrogen available for the hydrogen bond and having a bonding site available to hydrogen from other molecules. We will go into the details of the hydrogen bonding patterns of these molecules in one of the biochemistry lessons.

Third, you should notice that each of these compounds has an amino group which can form amide bonds to other types of compounds. Look for such amide bonds in the structural formulas of such substances as DNA, RNA, and ATP in the biochemistry lesson on nucleic acids.

New words and Expressions

heterocyclic[ˌhetərəˈsaiklik] a. 杂环的
diversity[daiˈvəsiti] n. 差异,多样性
pharmaceuticals[ˌfɑːməˈsjuːtikəls][复] n. 医药品
agrochemicals[ˌægrouˈkemikəls][复] n. 农用化学品
veterinary[ˈvetərinəri] a. 兽医的
~ product 兽医用产品
optical[ˈɔptikəl] a. 光学的
~ purity 光学纯度
brightening agent 光亮剂
antioxidant[ˌæntiˈɔksidənt;ˌæntiˈɑksədənt] n. 抗氧化剂,防老剂
corrosion inhibitor 缓蚀剂,腐蚀抑制剂
function[ˈfʌŋkʃən] n. 官能(团)
dyestuff[ˈdaistʌf] n. 染料
pigment[ˈpigmənt] n. 颜料,色素
subtly[ˈsʌtli] adv. 精细的,巧妙的
manipulate[məˈnipjuleit] vt. 操作,处理,改造
modify[ˈmɔdifai] v. 修饰,修改,变更
heterocycle[ˌhetərəˈsaikl] n. 杂环
acidity[əˈsiditi] n. 酸度,酸性
basicity[beiˈsisiti] n. 碱度,碱性
susceptibility[səˌseptəˈbiliti] n. 敏感度,感受性
electrophile[iˈlectrəfail] n. 亲电试剂
nucleophile[ˈnjuːkliəfail] n. 亲核试剂
heteroatom[ˌhetərəˈætəm] n. 杂原子
fungicide[ˈfʌndʒisaid] n. 杀(霉)菌剂
pyrimidine[paiˈrimidiːn] n. 嘧啶
lipophilic[ˌlipəˈfilik] a. 亲油的,亲脂的

versatile[ˈvəːsətail] a. 多方面的,万能的,易变的
mimic[ˈmimik] vt.a. 模拟,模仿
tetrazole[ˈtetrəzɔ(əu)l] n. 四氮唑
tailor[ˈteilə] vt. 裁制,裁剪
 n. 裁缝
mask[mɑːsk] vt. 掩饰,伪装
 n. 面具;面罩
~ed a. 隐形的,隐蔽的
imidazole[ˌimiˈdæzəul] n. 咪唑
asymmetric(al)[ˌæsiˈmetrik(əl)] a. 不对称的
~ synthetic reaction 不对称合成反应
tridentate[traiˈdenteit] a. 三齿的,三叉的
~ ligand 三齿配体
isolate[ˈaisəleit] vt. 分离,隔离,使脱离
nucleic[njuːˈkliːik] acid 核酸
purine[ˈpjuəriːn] n. 嘌呤
chlorophyll[ˈklɔːrəfil] n. 叶绿素
heme[hiːm] n. (亚铁)血红素
porphyrin[ˈpɔːfərin] n. 卟啉
diet[ˈdaiət] n. 饮食,食物
ingredient[inˈgriːdiənt] n. 配料,成分
thiamin (vitamin B1)[ˈθaiəmiːn] n. 硫胺
riboflavin (vitamin B2)[ˌraibəuˈfleivin] n. 核黄素
pyridoxol (vitamin B6)[ˌpiriˈdɔksɔl] n. 吡哆醇
nicotinamide (vitamin B3)[ˌnikəˈtinəmaid] n. 烟碱
ascorbic acid (vitamin C)[əsˈkɔːbik ˈæsid] 抗坏血酸,维生素C

origin ['ɔːrədʒin] n. 根源,起源；起因
natural product 天然产物
vat dye [væt dai] 瓮染料,还原染料
indigo ['indigəu] n. 靛蓝,靛青；靛蓝类染料
antibacterial ['æntibæk'tiəriəl] agent 抗(细)菌剂
lactam ['læktæm] n. 内酰胺
penicillin. [ˌpeni'silin] n. 青霉素,盘尼西林
biology [bai'ɔlədʒi] n. 生物学,生物
DNA deoxyribonucleic acid [diːˈɒksiˌraibəuˈnjuːkliːikˈæsid] 脱氧核糖核酸
RNA ribonucleic acid [raibəunjuːˈkliːikˈæsid] 核糖核酸
uracil ['juərəsil] n. 尿嘧啶
thymine ['θaimiːn] n. 胸腺嘧啶
cytosine ['saitəsiːn] n. 胞核嘧啶,氧氨嘧啶
biochemistry ['baiəu'kemistri] n. 生物化学
unbonded ['ʌn'bɔndid] a. 未键合的,自由的
ATP adenosine [ə'denəsiːn] triphosphate 三磷酸腺苷

词　组

a wide range of …　各种各样的,多种的
fit into …　适应于,符合于
change … for …　拿…换…,用…换…
armed with …　装备有…的,掌握有…的
carry through …　完成…,将…进行到底
be of …　具有…的
lie in …　在于,处于
deal with …　讨论,研究
go into …　深入研究,探究

前缀和后缀

hetero- [hetərə-] 异,杂,多； heterogeneous, heterocycle, heterochain, heteroatom
-ity [-iti] 度,性（名词词尾,状态,特性） acidity, basicity, purity, susceptibility, opticity

Exercises

1. Fill in the blanks with proper words or phrases.

___ heterocyclic chemical compounds may be inorganic compounds or organic ___, most contain at least one ___. While atoms that are neither carbon ___ hydrogen are normally referred to in organic chemistry as ___, this is usually in comparison to the all-carbon backbone. But this does not ___ a compound such as borazine (which has no carbon atoms) from being "heterocyclic". IUPAC ___ the Hantzsch-Widman nomenclature for naming heterocyclic compounds.

　　A. recommends　　B. Although　　C. labelled　　D. heteroatoms
　　E. compounds　　F. prevent　　G. nor　　H. carbon

2. Match the following explanation with the proper word.

　　___ diversity, ___ pigment, ___ mimic, ___ ingredient, ___ origin

　A. one of the foods that you use to make a particular food or dish

　B. the fact of including many different types of people or things

　C. the place or situation in which something begins to exist

　D. a natural substance that makes skin, hair, plants etc. a particular color

　E. to copy the way someone speaks or behaves

Lesson 22
POLYMERS

Introduction

The term macromolecule, or polymer, is applied to substances of high molecular weight that are composed of a large number (usually at least 100) of units of low molecular weight joined by covalent bonds. If the low molecular weight units making up the macromolecule are bonded end-to-end in a long chain and no covalent chemical bonds exist between the chains, the macromolecules are called linear polymers. Such polymers, unless of extremely high molecular weight (1 000 000), can usually be dissolved and, when heated, they soften or melt so that they can be extruded into fibers or molded into desired shapes. These polymers are said to be thermoplastic. On the other hand, if the polymer chains are linked together at numerous points, the polymer is one large three-dimensional molecule, infusible and insoluble. Such polymers are called cross-linked polymers, and the bonds connecting the chains are cross-links. Certain linear polymers, referred to as thermosetting, contain groups which, when heated, react to give cross-linked polymers。

The process by which small molecules undergo multiple combination to form macromolecules is polymerization. Small molecules from which a macromolecule or polymer can be made are called monomers. Two types of polymerization are recognized: (1) condensation polymerization and (2) addition polymerization. A polymer-forming reaction involving elimination of a small molecule such as water or alcohol between monomer units is described as condensation polymerization. In addition polymerization, unsaturated or cyclic molecules add to each other without elimination of any portion of the monomer molecule. The empirical formula of the polymer is then, of course, the same as that of the monomer.

Reactions capable of forming macromolecules by either addition or condensation polymerization must be functionally capable of proceeding indefinitely. Whenever two monomer molecules react, the product must contain a functional group capable of reacting with another molecule of monomer. In condensation polymerization, each monomer unit must have at least two functional groups. In addition polymerization, the monomer need have only one functional group ——the presence of two or more functional groups usually leads to the production of cross-linked addition polymers. Examples of some of the more important synthetic polymers formed by condensation polymerization are listed in Table 15.

1.Condensation Polymers

Typical of condensation polymers are the polyamides or "nylons" formed by the condensation reaction of a diacid with a diamine. Nylon 66, poly(hexamethyleneadipa-mide), is formed by heating an equimolar mixture of adipic acid and hexamethylenediamine at a temperature of 215℃ for several hours and then at 270℃ under vacuum for about one hour.

Table 15. Condensation Polymers

Name	Monomers	Repeating Unit	Principal Use
Dacron Mylar Terylene	Dimethyl terephthalate and ethylene glycol	—C(=O)—C₆H₄—COCH₂CH₂O—	Fiber, film
Nylon 66	Adipic acid and hexamethylene diamine	—C(=O)(CH₂)₄C(=O)NH(CH₂)₆NH—	Fiber, molded articles
Epoxy resins	Bisphenol-A and epichlorohydrin	—C₆H₄—C(CH₃)₂—C₆H₄—OCH₂CHOHCH₂O—	Protective coatings, adhesives, potting resins
Urethans	Toluene diisocyanate and poly(propylene glycol)	—C(=O)—NH—C₆H₃(Me)—NH—C(=O)—O—[CH₂—CH(CH₃)—O]ₓ—CH₂—CH(CH₃)—O—	Rigid and elastic foams
Thiokols (poly sulfide rubbers)	Ethylene chloride and sodium polysulfide	—CH₂CH₂S—S— (with S=S)	Oil-resistant rubber
Silicones	Methyl dichlorosilanes and / or trichorosilanes	—O—Si(CH₃)₂—	Oils, rubbers coatings, resins
phenol-formaldehyde	phenol and formaldehyde	—C₆H₃(OH)—CH₂— Ortho or para linkages	Molded articles, adhesives, laminating resins

In the laboratory, nylons are more conveniently prepared by a polymer-forming reaction called interfacial polymerization. The reaction is between a diacid chloride dissolved in a water-immiscible organic solvent and a water solution of a diamine. Reaction apparently occurs at the interface of two solutions.

Experimental (Preparation of Nylon 6—10)

Place a solution of 2.0mL of sebacoyl chloride in 100mL of carbon tetrachloride in a 200-mL tall-form beaker. Over the acid chloride solution carefully pour a solution of 2.2g of hexamethylenediamine and 1.5g of sodium hydroxide in 50mL of water. The addition of the diamine-sodium hydroxide solution is best done by pouring the solution through a funnel placed so that its outlet is just over the surface of the diacid solution. Grasp the polymeric film which forms at the interface of the two solutions with forceps and raise it from the beaker as continuously forming rope. If a mechanical windup device is placed above the beaker, the polymer may be wound up continuously until one of the reactants is exhausted.

Wash the polymer thoroughly with water and finally with a 50% acetone solution. Allow the washed polymer to air dry. Place about 0.1g of the dried polymer on a metal spatula or spoon and melt it carefully over a low flame care being taken not to char the polymer[1]. Touch a glass rod or a matchstick to the molten polymer and pull it slowly away to form a fiber.

2. Addition Polymers

Introduction

Addition polymerization usually must be catalyzed by a base, by an acid, or by free radicals. Three stages are involved in all addition polymerization reactions, no matter what the catalyst: these are initiation, propagation, and termination. In the initiation step, the catalyst molecules attack the monomers to give intermediates which, during the propagation stage, are capable of attacking other molecules of monomer with lengthening of the chain. In the termination step, chain growth is stopped by elimination of a group from the reacting end of the chain or by addition of a group to the end of the chain to form a molecule which is no longer a chain carrier.

Free radical-initiated polymerizations follow a similar course but with radical, rather

than ionic, intermediates. If two or more monomers undergo addition polymerization together, the process is called copolymerization, and the product is a copolymer. Should one of the monomers(even though present in only minor amounts) from which a copolymer is formed contain two or more groups capable of undergoing addition polymerization, the copolymer will be insoluble as a result of cross-linking[2].

Examples of some of the more important synthetic polymers formed by addition polymerization are listed in Table 16.

Table 16. Addition polymers

Name	Monomer	Repeating Unit	Principal Use
Polyethylene	$CH_2=CH_2$	$-CH_2-CH_2-$	Film, molded articles
Poly(vinyl chloride)	$CH_2=CHCl$	$-CH_2-CH(Cl)-$	Coatings, sheet
Poly(vinyl acetate)	$CH_2=CHOCOCH_3$	$-CH_2-CH(OCOCH_3)-$	Coatings, adhesives

Name	Monomer	Repeating Unit	Principal Use
Polyacrylonitrile, Orlon, Acrilan	$CH_2=CHC\equiv N$	$-CH_2-CH(C\equiv N)-$	Fiber
Polystyrene, Lustron, Styron	$CH_2=CHC_6H_5$	$-CH_2-CH(C_6H_5)-$	Molded articles, rigid foams
Poly(methyl methacrylate) Lucite, Plexiglass	$CH_2=C(CH_3)-COOCH_3$	$-CH_2-C(CH_3)(COOCH_3)-$	Coatings, molded articles, sheet
Polyformaldehyde, Delrin	$H_2C=O$	$-CH_2-O-$	Molded articles
Polyethylene oxide, Polyethyleneglycol, Carbowax	$\underset{O}{CH_2-CH_2}$ (epoxide)	$-CH_2-CH_2-O-$	Waxes
Polycaprolactam, Nylon 6	caprolactam (O=C-NH-(CH_2)_5 ring)	$-CO(CH_2)_5NH-$	Fiber

Preparation of Polystyrene

(1) Polymerization of Styrene with Benzoyl Peroxide. To a large test tube add 20mL of toluene and 5mL of styrene. Then add 0.3g of benzoyl peroxide and place the test tube in a beaker of water which is maintained at a temperature of 90~95℃. After 60 minutes, remove the test tube, allow the contents to cool for 5 minutes, and note the viscosity of the solution. Pour the solution into 200mL of methyl alcohol contained in a 400mL beaker. Collect the white precipitate of polystyrene by filtration, using a Büchner funnel, and wash the precipitate on the funnel with 50mL of methyl alcohol. Remove the precipitate from the funnel and spread it out to dry on a large, clean sheet of filter paper.

Place 3mL. of acetone in a clean test tube, add 0.2g of the dried polymer, and stir the mixture for several minutes. Is the polymer soluble? Use this same procedure to determine the solubility of polystyrene in water, ethyl alcohol, benzene, carbon tetrachloride, and petroleum ether. Place approximately 0.1g of the polymer on a metal spatula or spoon and warm it gently over a flame until the polymer melts. To the molten polymer touch a glass stirring rod or a matchstick and pull away gently to draw out a fiber. How would you describe the properties of the fiber as to brittleness, color, and strength? Compare the properties of this polystyrene fiber with those of the Nylon fiber. Allow the molten polymer on the spatula to cool. Describe the appearance of the cooled polymer. Scrape this material from the spatula, place it in a test tube, and determine its solubility in acetone. Did melting change the solubility of the polystyrene?

New words and Expressions

macromolecule[ˌmækrə'mɔlikjuːl] n. 大分子
polymer['pɔlimə] n. 聚合物,聚合体
linear['liniə] polymer 线形聚合物
soften['sɔfn] v. 变软,软化
extrud[e(i)ks'truːd] v. 使…喷出,挤压成
fiber['faibə] n. 纤维
mold[məuld] vt. 铸塑(into)
thermoplastic['θəːməu'plæstik] a. 热塑的
infusible[in'fjuːzəbl] a. 不熔的
cross-linked polymer[krɔs liŋkt 'pɔlimə] a.n. 交联聚合物
cross-link 交联;交联键
thermosetting['θəməuˌsetiŋ] a.n. 热固(的)
multiple['mʌltipl] a. 多重的,重复的,多次的
monomer['mɔnəmə] n. 单体
condensation polymerization[ˌkɔnden'seiʃənˌpɔlimərai'zeiʃən] n. 缩聚,缩合聚合
addition[ə'diʃən] polymerization 加聚,加成聚合
polymer-forming 形成聚合物
empirical formula[em'pirikəl 'fɔːmjulə] 实验式
indefinite[in'definit] a. 无限制的,不确定的
dacron['deikrɔn] n. (商品名)的确良,涤纶,聚对苯二甲酸乙二醇酯
mylar['mailə] n. (商品名)涤纶,聚酯薄膜,聚对苯二甲酸乙二醇酯
terylene['teriliːn] n. (商品名)涤纶,特丽纶,聚对苯二甲酸乙二醇酯
nylon['nailɔn] n.a. 尼龙,锦纶,聚酰胺
epoxy[e'pɔksi] a. 环氧的 n. 环氧(树脂)
epoxy resin['rezin] 环氧树脂

urethans(es)['jurəθənz] n 聚氨酯(橡胶)
thiokol['θaiəkɔl] n. = polysulfide rubber 聚硫橡胶
silicone(s)['silikəun] n. 聚硅氧烷,(聚)硅酮
dimethyl terephthalate[dai'meθilˌteref'θæleit] 对苯二甲酸二甲酯
hexamethylenediamine[heksəˌmeθiliː ndai'æmiːn] n. 己二胺
bisphenol-A[bis'fiːnɔl-ei] n. 双酚A
epichlorohydrin[epə'klɔːrə'haidrin] n. 3-氯-1,2-环氧丙烷,环氧氯丙烷
toluene diisocyanate['tɔljuiːn daiˌaisə' saiəneit] 甲苯二异氰酸酯
ethylene chloride 1,2-二氯乙烷,氯化乙烯
sodium polysulfide[ˌpɔli'sʌlfaid] 多硫化钠
silane['silein] n. 硅烷
molded article[məuldid'ɑːtikl] 模制品
coating['kəutiŋ] n. 涂料,涂层
adhesive[əd'hiːsiv] n. 粘合剂,胶粘剂
potting['pɔtiŋ] a. 陶器制造的
elastic[i'læstik] a. 有弹性的,弹性的
foam[fəum] n. 泡沫;v. 起泡沫,发泡沫
oil-resistant rubber 耐油橡胶
laminating resin['læmineitiŋ 'rezin] 层压树脂
polyamide[ˌpɔli'æmaid] n. 聚酰胺
diamine[dai'æmiːn] n. 二胺
poly(hexamethyleneadipamide)[-'meθiliː nˌædi'pæmaid] n. 聚亚己基己二酰胺,聚己撑己二酰胺
vacuum['vækjuəm] n. 真空
interfacial[ˌitə'feiʃəl] polymerization 界面聚

合(作用)
immiscible[i'misəbl]a.不混溶的,难溶的
interface['intəfeis]n.界面,分界面
sebacoyl[si'bæsəuil]n.癸二酰
polyhexamethylenesebacamide[sibə'kæmaid]n.聚亚己基癸二酰胺,聚己二撑癸二酰胺;尼龙66
tall-form beaker[tɔ:l fɔ:m 'bi:kə]高形烧杯
outlet['autlet]n.出口,排水口
polymeric['pɔli'mɛərik]a.聚合的,聚合体的
forceps['fɔ:seps]n.镊子
windup['waindʌp]a.装有发条的
device[di'vais]n.装置,设备,器
wind['waind](wound[waund])v.缠绕,盘绕,卷
exhaust[ig'zɔ:st]vt.耗尽
propagation[ˌprɔpə'geiʃən]n.传播,传递
lengthen['leŋθən]v.加长,延长,变长
carrier['kæriə]n.传递者,载体
copolymerization[kəu'pɔliməraiˈzeiʃən]n.共聚合(作用)
copolymer[kəu'pɔlimə]n.共聚物,共聚体
polyethylene[ˌpɔli'eθili:n]n.聚乙烯
polyvinyl chloride[ˌpɔli'vainil 'klɔ:raid]聚氯乙烯
polyvinyl acetate['æsitit]聚醋酸乙烯酯
polyacrylonitrile[ˌpɔli'ækrilə'naitril]n.聚丙烯腈
orlon['ɔ:lən]n.(商品名)腈纶,聚丙烯腈纤维
acrilan['ækrilən]n.(商品名)腈纶,聚丙烯腈纤维
polystyrene[ˌpɔli'staiəri:n]n.聚苯乙烯
lustron['lʌstrɔn]n.(商品名)聚苯乙烯(塑料)
styron['stairɔn]n.(商品名)肉桂塑料,聚苯乙烯塑料
polymethyl methacrylate[meθə'krileit]聚甲基丙烯酸甲酯
lucite['lu:sait]n.(商品名)有机玻璃,聚甲基丙烯酸甲酯

plexiglass['pleksiglɑ:s]n.(商品名)有机玻璃,聚甲基丙烯酸甲酯
delrin['delrin]n.(商品名)聚甲醛塑料
ethylene oxide['eθili:n 'ɔksaid]环氧乙烷
polyethylene glycol[ˌpɔli'eθili:n 'glaikɔl]聚乙二醇
Carbowax[ˌkɑ:bə'wæks]n.(商品名)聚乙二醇
polycaprolactam[ˌpɔli'kæprəu'læktæm]n.聚己内酰胺
rigid foam['ridʒid fəum]硬泡沫(塑料)
benzoyl peroxide['benzəuil pə'rɔksaid]过氧化苯甲酰、苯甲酰过氧化物
ethyl alcohol['eθil 'ælkəhɔl]乙醇
petroleum ether[pi'trəuljəm 'i:θə]石油醚
brittleness['britlnis]n.脆性,易脆性
scrape[skreip]vt.刮落,擦,掏

前缀

macro-[mækrou(ə)]大,常量,宏观
　　marcromolecule,macrocyclic;
　　macroanalysis(常量分析);
　　macroscopic,macrostructure
poly-[ˌpɔli]聚,多
　　polymer,polyamide;polyatomic,polycyclic
thermo-[θə:məu]热
　　thermoplastic,thermometer(温度计),
　　thermostatic(恒温的).
ep(i)-[ep(ə)-]环
　　epoxide,epichlorohy-drin.

词组

end-to-end　头尾连接(相接)
wind up　缠绕,卷起
low flame[ləu fleim]　低温火焰
no matter what…　无论什么,不管是什么
no longer　不再
spread…out…　把……摊开(铺开)……
draw out…　抽出,拉长

Notes

1. Place about 0.1g.of the dried polymer on a metal spatula or spoon and melt it carefully over a low flame care being taken not to char the polymer. 该句是由 and 连接的并列祈使句,第二个祈使句中分词独立结构""care being…the polymer"作状语,表示伴随。**译文**:"把约0.1克的干燥聚合物放在一个金属刮刀或小勺上,在低温火焰上小心地熔化,当心不要使聚合物烧焦"。

2. "Should one of the monomers……as a result of cross-linking."是主从复合句,"Should one of…addition polymerization"是虚拟条件从句,其中包括了"from which…polymerization"这个定语从句,修饰 monomers. 主句是"copolymer…of cross-linking",主句中谓语不是虚拟语气。这种从句中的谓语用 should 加原形动词构成,而主句不用虚拟语气时,表示该事物实现的可能性虽较小,但仍然是可能的。

Exercises

1. Fill in the blacks with proper words or phrases.

   There are three main ___ of biopolymers: polysaccharides, polypeptides, and polynucleotides. In living ___, they may be synthesized by enzyme-mediated processes, ___ as the formation of catalyzed by DNA polymerase. The synthesis of proteins involves multiple enzyme-mediated processes to transcribe genetic ___ from the DNA to RNA and subsequently ___ that information to synthesize the specified protein from ___ acids. The protein may be modified further following translation in order to provide ___ structure and functioning.

   A. DNA   B. translate   C. cells   D. amino
   E. such   F. appropriate   G. classes   H. information

2. Match the following explanation with the proper word.

   ___ extrude, ___ coating, ___ interface, ___ device, ___ exhaust

   A. a thin layer of something that covers a surface
   B. to make someone feel extremely tired
   C. a surface forming a common boundary between two things
   D. to push or force something out through a hole
   E. a machine or tool that does a special job

Lesson 23
IONIC LIQUID

An ionic liquid is a liquid that contains essentially only ions. Some ionic liquids, such as ethylammonium nitrate, are in a dynamic equilibrium where at any time more than 99.99% of the liquid is made up of ionic rather than molecular species. In the broad sense, the term includes all molten salts, for instance, sodium chloride at temperatures higher than 800 °C. Today, however, the term "ionic liquid" is commonly used for salts whose melting point is relatively low (below 100℃). In particular, the salts that are liquid at room temperature are called room-temperature ionic liquids, or RTILs.

ethylammonium nitrate   1-butyl-3-methylimidazolium salts or bmim

History

Whereas the date of discovery, as well as discoverer, of the "first" ionic liquid is disputed, one of the earlier known ionic liquids was $[EtNH_3]^+ [NO_3]^-$ (m.p. 12℃), the synthesis of which was published in 1914. Much later, series of ionic liquids based on mixtures of 1,3-dialkylimidazolium or 1-alkylpyridinium halides and trihalogenoaluminates, initially developed for use as electrolytes, were to follow. An important property of the imidazolium halogenoaluminate salts was that they were tuneable — viscosity, melting point and the acidity of the melt could be adjusted by changing the alkyl substituents and the ratio of imidazolium1 or pyridinium halide to halogenoaluminate.

A major drawback was their moisture sensitivity and, though to a somewhat lesser extent, their acidity/basicity, the latter which can sometimes be used to an advantage. In 1992, Wilkes and Zawarotko reported the preparation of ionic liquids with alternative, 'neutral', weakly coordinating anions[2] such as hexafluorophosphate ($[PF_6]^-$) and tetrafluoroborate ($[BF_4]^-$), allowing a much wider range of applications for ionic liquids. It was not until recently that a class of new, air and moisture stable, neutral ionic liquids, was available that the field attracted significant interest from the wider scientific community.

More recently, people have been moving away from $[PF_6]^-$ and $[BF_4]^-$ since they are highly toxic, and towards new anions such as bistriflimide $[(CF_3SO_2)_2N]^-$ or even away from halogenated compounds completely. Moves towards less toxic cations have also been growing, with compounds like ammonium salts (such as choline) being just as flexible a scaffold as imidazole.

Characteristics

Ionic liquids are electrically conductive and have extremely low vapor pressure. (Their noticeable odours are likely due to impurities.) Their other properties are diverse. Many have low combustibility, excellent thermal stability, a wide liquid range, and favorable solvating properties for diverse compounds. Many classes of chemical reactions, such as Diels - Alder reactions and Friedel - Crafts reactions, can be performed using ionic liquids as solvents. Recent work has shown that ionic liquids can serve as solvents for biocatalysis. The miscibility of ionic liquids with water or organic solvents varies with side chain lengths on the cation and with choice of anion. They can be functionalized to act as acids, bases or ligands, and have been used as precursor salts in the preparation of stable carbenes. Because of their distinctive properties, ionic liquids are attracting increasing attention in many fields, including organic chemistry, electrochemistry, catalysis, physical chemistry, and engineering; see for instance magnetic ionic liquid.[3]

Despite their extremely low vapor pressures, some ionic liquids can be distilled under vacuum conditions at temperatures near 300 ℃. Some ionic liquids (such as 1 - butyl - 3 - methylimidazolium nitrate) generate flammable gases on thermal decomposition. Thermal stability and melting point depend on the components of the liquid.

The solubility of different species in imidazolium ionic liquids depends mainly on polarity and hydrogen bonding ability. Simple aliphatic compounds are generally only sparingly soluble in ionic liquids, whereas olefins show somewhat greater solubility, and aldehydes can be completely miscible. This can be exploited in biphasic catalysis, such as hydrogenation and hydrocarbonylation processes, allowing for relatively easy separation of products and/or unreacted substrate(s). Gas solubility follows the same trend, with carbon dioxide gas showing exceptional solubility in many ionic liquids, carbon monoxide being less soluble in ionic liquids than in many popular organic solvents, and hydrogen being only slightly soluble (similar to the solubility in water) and probably varying relatively little between the more popular ionic liquids. (Different analytical techniques have yielded somewhat different absolute solubility values).

Room temperature ionic liquids

Room temperature ionic liquids consist of bulky and asymmetric organic cations such as 1-alkyl-3-methylimidazolium, 1-alkylpyridinium, N-methyl-N-alkylpyrrolidinium and ammonium ions. A wide range of anions is employed, from simple halides, which generally inflect high melting points, to inorganic anions such as tetrafluoroborate and hexafluorophosphate and to large organic anions like bis-trifluorosulfonimide, triflate or tosylate. There are also many interesting examples of uses of ionic liquids with simple non-halogenated organic anions such as formate, alkylsulfate, alkylphosphate or glycolate. As an example, the melting point of 1-butyl-3-methylimidazolium tetrafluoroborate or [bmim][BF4] with an imidazole skeleton is about -80℃, and it is a colorless liquid with high viscosity at room temperature.

It has been pointed out that in many synthetic processes using transition metal catalyst, metal nanoparticles play an important role as the actual catalyst or as a catalyst reservoir. It also been shown that ionic liquids (ILs) are an appealing medium for the formation and stabilization of catalytically active transition metal nanoparticles. More importantly, ILs can be made that incorporate co-ordinating groups, for example, with nitrile groups on either the cation or anion (CN-IL). In various C-C coupling reactions catalyzed by palladium catalyst, it has been found the palladium nanoparticles are better stabilized in CN-IL compared to non-functionalized ionic liquids; thus enhanced catalytic activity and recyclability are realized.

Low temperature ionic liquids

Low temperature ionic liquids (below 130 degrees Kelvin) have been proposed as the fluid base for an extremely large diameter spinning liquid mirror4 telescope to be based on the earth's moon. Low temperature is advantageous in imaging long wave infrared light which is the form of light (extremely red-shifted) that arrives from the most distant parts of the visible universe. Such a liquid base would be covered by a thin metalic film that forms the reflective surface. A low volatility is important for use in the vacuum conditions present on the moon.

Food science

The application range of ionic liquid also extends to food science. [bmim]Cl for instance is able to completely dissolve freeze dried[5] banana pulp and the solution with an additional 15% DMSO lends itself to Carbon-13 NMR analysis. In this way the entire banana compositional makeup of starch, sucrose, glucose, and fructose can be monitored as a function of banana ripening.

Safety

Due to their non-volatility, effectively eliminating a major pathway for environmental release and contamination, ionic liquids have been considered as having a low impact on the environment and human health, and thus recognized as solvents for green chemistry. However, this is distinct from toxicity, and it remains to be seen how 'environmentally-friendly' ILs will be regarded once widely used by industry. Research into IL aquatic toxicity has shown them to be as toxic as or more so than many current solvents already in use. A new review paper on this aspect has just appeared. Available research also shows that mortality isn't necessarily the most important metric for measuring their impacts in aquatic environments, as sub-lethal concentrations have been shown to change organisms' life histories in meaningful ways. According to these researchers balancing between zero VOC emissions, and avoiding spills into waterways (via waste ponds/streams, etc.) should become a top priority. However, with the enormous diversity of substituents available to make useful ILs, it should be possible to design them with useful physical properties and less toxic chemical properties.

With regard to the safe disposal of ionic liquids, a 2007 paper has reported the use of ultrasound to degrade solutions of imidazolium-based ionic liquids with hydrogen peroxide and acetic acid to relatively innocuous compounds.

Despite their low vapor pressure many ionic liquids have also found to be combustible and therefore require careful handling. Brief exposure (5 to 7 seconds) to a flame torch will ignite these IL's and some of them are even completely consumed by combustion.

New words and Expressions

ionic liquid (IL) 离子液
ethylammonium ['eθilə'məunjəm] nitrate 硝酸乙基铵
room-temperature 室温,常温
~ ionic liquid (RTIL) 室温离子液
1,3-dialkylimidazolium [dai'ælkil,imidæ'zəuliəm] halide 1,3-二烷基咪唑卤盐,1,3-二烷基咪唑卤化盐
1-alkylpyridinium ['ælkil,piri'diniəm] halide 烷基吡啶卤盐
trihalogenoaluminate [trai'hælədʒənəuə'lju:mineit] n. 三卤代(化)铝酸盐
imidazolium halogenoaluminate salt 咪唑卤代铝酸盐
tuneable ['tju:nəbl] (= tunable) a. 可调整的,可协调的
bistriflimide [,bistrai'flimaid] = bis-(trifluoromethane) sulfonimide n. 二(三氟甲基)磺酰亚胺阴离子 [(CF₃SO₂)₂N]⁻
halogenate ['hælədʒəneit] v. 卤化
choline ['kəuli:n] n. 胆碱
scaffold ['skæfəld] n. 脚手架
biocatalysis [,baiəukə'tælisis] n. 生物催化(作用)

1-butyl-3-methylimidazolium (bmim) 1-丁基-3-甲基咪唑阳离子
~ nitrate 1-丁基-3-甲基咪唑硝酸盐
miscible ['misibl] a. 易混合的,可混溶的
biphasic [bai'feizik] a. [化]两相的,[植]两阶段的
hydrocarbonylation [,haidrə,ka:bəni'leiʃən] n. 氢甲酰化
tetrafluoroborate [,tetrə'fluərə'bɔ:reit] n. 四氟硼酸盐
hexafluorophosphate n. 六氟磷酸盐
bis-trifluorosulfonimide = bis(trifluoromethane)sulfonimide
triflate ['traifleit] n. 三氟甲基磺酸盐(酯)
tosylate ['tɔsileit] n. 对甲苯磺酸盐(酯)
nanoparticle [,nænə'pa:tikl] n. 纳米粒子
reservoir ['rezəvwa:] n. 贮存器,水库,蓄水池
coupling reaction 偶联反应
recyclability [ri:,saiklə'biliti] n. 可再循环性
Kelvin ['kelvin] n. 绝对温度
spinning ['spiniŋ] a. 旋转的;自旋的
liquid mirror ['mirə] 液体镜
red-shift ['red,ʃift] n. 红移,红向移动
freeze dry 冷冻干燥,冷冻脱水
DMSO = dimethyl sulfoxide 二甲基亚砜

makeup [ˈmeikˌʌp] n. 化妆品
　compositional ~ 组合化妆品, 结构化妆品
starch [stɑːtʃ] n. 淀粉
sucrose [ˈsuːkrəus; ˈsuːkrəuz] n. 蔗糖
fructose [ˈfrʌktəuz; ˈfrʌktəus] n. 果糖
green chemistry 绿色化学
aquatic [əˈkwætik] n. 水生动植物;
　a. 水生的
mortality [mɔːˈtæliti] n. 死亡率
sublethal [sʌbˈliːθəl] a. 亚致死(量)的
meaningful [ˈmiːniŋful] a. 意味深长的,
　重要的
VOC = volatile organic compound 易挥发有机化合物
emission [iˈmiʃən] n. (光、热等的)散发, 发射
spill [spil] n.v. 溢出, 溅出, 流出
waterway [ˈwɔːtəwei] n. 水路, 排水沟
ultrasound [ˈʌltrəˌsaund] n. 超声波
innocuous [iˈnɔkjuəs] a. 无害的, 无毒的

词　组

in the broad sense 在广义上

in particular 特别是
as well as 以及
(move) away from … 离开…, 远离…
move towards 接近, 走向
act as … 起 … 的作用, 作为
(be) able to (+ inf.) 能…, 会…
lend itself to … 有助于…, 适用于…
(in) this way 这样, 因此, 用这种方法
it remains to be seen … 还需观察…, 有待看…
research into 调查, 探究
with regard to … 关于, 对于

前缀和后缀

-ium [iəm] 金属元素或有机阳离子、鎓化合物词尾, 如: aluminium, calcium, pyridinium, imidazolium
-ility [-iliti:] 名词词尾, 从形容词词尾 -able, -ble, -ile, il 等转化而来, 如: recyclability, volatility, availability
bio- [ˌbaiəu-] 生物
　biocatalysis, biochemistry

Notes

1. 在英语中, 如果一个杂环有机阳离子的正电荷是通过在母体化合物的杂原子上固定一个质子而得, 而母体化合物的的名称又不是"amine", 则该有机阳离子的名称是通过去掉原名称中的词尾"e", 加上"-ium"形成, 如 imidazole →imidazolium, pyridine → pyridinium, proline → prolinium, 1,4-dioxane →1,4-dioxanium.

2. "weakly coordinating anions": Anions that interact weakly with cations are optimistically termed non-coordinating anions, although a better term is 'weakly coordinating anion'. Non-coordinating anions are useful in studying the reactivity of electrophilic cations. They are commonly found as counterions(抗衡离子) for cationic metal complexes with an unsaturated coordination sphere.

3. "magnetic ionic liquid": A magnetic ionic liquid was identified by Satoshi Hayashi and Hiro-o Hamaguchi of the University of Tokyo in 2004 as an ionic liquid based on the imidazole 1-butyl-3-methylimidazolium chloride and ferric chloride ([bmim]$FeCl_4$). Due to the presence of high spin $FeCl_4$, the liquid is paramagnetic and a magnetic susceptibility of 40.6×10^{-6} emu·g^{-1} is reported.

A simple magnet suffices to attract the liquid in a test tube.

4. Liquid mirrors are mirrors made with reflective liquids. The most common liquid used is mercury, but other liquids will work as well (for example, gallium alloys). Liquid mirrors can be a low cost alternative to conventional large telescopes.

5. Freeze drying (also known as lyophilization(冻干法)) is a dehydration process typically used to preserve a perishable material or make the material more convenient for transport. Freeze drying works by freezing the material and then reducing the surrounding pressure and adding enough heat to allow the frozen water in the material to sublime directly from the solid phase to gas.

Exercises

1. Fill in the blacks with proper words or phrases.

The discovery ____ of the "first" ionic liquid is disputed, along with the ____ of its discoverer. Ethanolammonium nitrate (m.p. 52–55 °C) was ____ in 1888 by S. Gabriel and J. Weiner. One of the ____ truly room ____ ionic liquids was ethylammonium nitrate $(C_2H_5)NH_3^+ \cdot NO_3^-$ (m.p. 12°C), reported in 1914 by Paul Walden. In the 1970s and 1980s, ionic liquids _____ ____ on alkyl-substituted imidazolium and pyridinium ____, with halide or tetrahalogenoaluminate anions, were ____ as potential electrolytes in batteries.

  A. temperature   B. date   C. cations   D. earliest

  E. developed   F. reported   G. based   H. identity

2. Match the following explanation with the proper word.

  ____ miscible, ____ biocatalysis, ____ biphasic, ____ reservoir, ____ spill

  A. the use of natural catalysts, such as protein enzymes, to perform chemical transformations on organic compounds

  B. a liquid it accidentally flows over the edge of a container

  C. a lake, especially an artificial one, where water is stored before it is supplied to people's houses

  D. capable of being mixed

  E. a system is one which has two phases.

Lesson 24
VOLUMETRIC ANALYSIS

General principles

Chemical analyses can be made by determining how much of a solution of known concentration is needed to react fully with an unknown test sample[1]. The method is generally referred to as volumetric analysis and consists of titrating the unknown solution with the one[2] of known concentration (a standard solution). By titration, you can determine exactly how much of a reagent is required to bring about complete reaction of the test solution. Usually, completion of the reaction is indicated by a sudden, visible change in the reaction system that coincides with the stoichiometric relationship between moles or equivalents of the reagent solution and the reactant in the test solution. A drop or two of an appropriate indicator solution produces a color change at the point where the reaction is complete-referred to as the endpoint.

Molarity is the number of moles (gram-molecular weights) of substance per liter of solution. The mole weight of sulfuric acid is 98.08 g, and therefore, 1 mole of H_2SO_4 contains 98.08 g. If 49.04 g are diluted to 1 liter then the concentration is 0.49 or 0.5M. In the case of hydrochloric acid, HCl, a 1 M solution is prepared by taking 36.465 g of HCl and diluting to 1 liter. The procedure is the same for bases.

Normality is the number of equivalent weights of substance per liter of solution. The equivalent weight of an acid is the weight of that acid capable of furnishing 1 mole of protons (H^+), and the equivalent weight of a base is the weight of base capable of receiving 1 mole of protons. The equivalent weight of H_2SO_4 is 98.08g/2 or 49.04 g. Therefore, a normal solution (N) of H_2SO_4 contains 49.04g per liter.

The normality of an acid or base of unknown concentration may be determined by titration. The advantage of using normality rather than molarity is that equal volumes of solutions of equal normalities have identical capacities for neutralization, because they contain the same number of equivalent weights.

In a titration, we compare equivalent weights of acid and base. The number of equivalents of acid is equal to the product of the volume of the acid solution and its normality:

$$V_a \times N_a = \text{equivalents of acid}$$

The number of equivalents of base is the product of the volume of the base solution and its normality:

$$V_b \times N_b = \text{equivalents of base}$$

That's true because:

$$(\text{volume})(\text{normality}) = (\text{liters})\left(\frac{\text{equivalents}}{\text{liter}}\right) = \text{equivalents}$$

Neutralization has taken place when the number of equivalents of acid is equal to the number of equivalents of base:

$$V_a \times N_a = V_b \times N_b$$

Procedure

Care must be exercised throughout the titration procedure[3]. The burette should be thoroughly cleaned with soap and water, rinsed with tap water, and finally, rinsed with distilled water. Just before use, the burette should be rinsed with two 5-ml portions of the solution to be used in the burette. This is done by holding the burette in a semi-horizontal position and rolling the solution around the entire inner surface. Allow the final rinsing to drain through the tip.

Fill the burette to a point above the top marking and allow the solution to run out until the bottom of the meniscus is just at the top marking of the burette. The burette tip must be completely filled to deliver the volume measured.

In addition, the burette must be cleaned thoroughly after use because sodium hydroxide and other types of solutions will eventually frost the glass and render an expensive piece of equipment useless.

1. Titration of Vinegar

Measure 50 ml of vinegar with a pipette and pour into a 250-ml beaker. Add 2 drops of phenolphthalein indicator. Fill a burette with a 1 N solution of sodium hydroxide (NaOH) and draw out the excess as described above. From the burette add NaOH to the beaker of vinegar untill 1 drop of NaOH produces a pale pink color in the solution. Maintain constant stirring. The appearance of pink tells you that the acid has been neutralized by the base and there is now 1 drop of excess base which has turned the indicator. Read the burette and record this reading as the volume of base used to neutralize the acid.

According to the equation:

$$NaOH + CH_3COOH \rightarrow Na^+ + CH_3COO^- + H_2O$$

One molecule of NaOH neutralizes one molecule of acetic acid, or one gram-molecular weight of NaOH neutralizes one gram-molecular weight of acetic acid. Calculate the amount of acetic acid present in the vinegar. Report this amount as the percentage of acetic acid.

2. Standard Titration Curve

If a pH meter is available, repeat the above process using a pH meter for constantly determining the pH. When the endpoint is reached, continue adding the base to expand the curve further. Make a graph for this titration.

3. Equivalents of Acid

Using the 1 N solution of NaOH, determine the number of equivalents in two samples of benzoic acid.Carry out the procedure for the two determinations simultaneously.From this value calculate the equivalent weight of the acid.The solid should be weighed in a beaker and should be dissolved in about 25 ml of ethyl alcohol before titration with the base.Between 2.0 and 2.2 g of the solid provide the best results. Record all data and make all calculations necessary to determine the equivalent weight of the solid acid. Compare your experimental value with the equivalent weight of benzoic acid（calculated from the formula）and determine the percentage of error of your work.

Questions to Consider

(1)Calculate the percentage of acetic acid in vinegar.

(2)What is the pH range for phenolphthalein?

(3)Why should the solution in the flask constantly be stirred?

(4)What determines the pH of a solution at the end of a titration of an acid with a base?

(5)Consider a hypothetical experiment in which you weighed out 2.0 g of oxalic acid and titrated it with 43 ml of 1.5N NaOH.What is the equivalent weight of oxalic acid?

(6)When all of the acetic acid was neutralized by the sodium hydroxide, was the pH=7? Explain.

New words and Expressions

volumetric[ˌvɔljuˈmetrik]a.体积的,容量的
unknown[ˈʌnˈnəun]a.未知的;n.未知物
 unknown test sample 未知试样
coincide[ˌkəuinˈsaid]vi.一致,符合(with)
stoichiometric[ˌstɔikiːəˈmetrik]a.化学计量的
molarity[məuˈlæriti]n.摩尔浓度
gram molecular weight 克分子量
mole weight (克)分子量
normality[nɔːˈmæliti]n.当量浓度
equivalent weight 当量
curve[kəːv]n.曲线
graphic[ˈɡræfik]a.图表的
representation[ˌreprizenˈteiʃən]n.代表,象征
burette[bjuəˈret]n.滴定管
rinse[rins]vt.清洗,冲洗
tap[tæp]water 自来水
drain[drein]n.徐徐流出,渐次排出
meniscus[məˈniskəs]n.液面,弯月面

marking[ˈmɑːkiŋ]n.识别标志,条纹,刻度
deliver[diˈlivə]vt.供应,放出,出产
frost[frɔst]n.霜;vt.霜化
render[ˈrendə]vt.致使、使成,使变为
vinegar[ˈviniɡə]n.醋
methyl red 甲基红
record[riˈkɔːd]v.记录;
  [ˈrekɔːd]n.记录,报告
reading[ˈriːdiŋ]n.(仪器的)读数
pH meter[piːeitʃˈmiːtə]pH计(仪)
graph[ɡræf]n.曲线图
simultaneously[ˌsiməlˈteinjəsli]ad.同时地
weigh[wei]vt.称(重)
error[ˈerə]n.错误,误差

前 缀

mol(e)- [məul-]摩尔,分子
 mole, molarity, molecule, molecular

nor-[nɔ:-](1)正,正常
  normal,normalize,nomality(当量浓度)
  (2)降(少一个CH_2^-),去甲
  norbornene(降冰片烯),noradrenaline
  (去甲肾上腺素)

词　组

bring about… 导致,引起
in the case of… 在…情况下,要是
run out 溢出,放出
question to consider 思考题

Notes

1. Chemical analyses can be made by determining how much of a solution of known concentration is needed to react fully with an unknown test sample. 该句中"how much…sample"是动名词 determining 的宾语从句。
2. one: solution
3. Care must be exercised throughout the titration procedure. **译文**:"所有滴定步骤都必须谨慎(操作)"。

Exercises

1. Fill in the blanks with proper words or phrases.

    Titration, also ___ as titrimetry, is a common ___ method of ___ chemical analysis that is used to determine the ___ of an identified analyte. Since volume measurements play a ___ role in titration, it is also known as volumetric analysis. A reagent, called the titrant or titrator is ___ as a standard solution. A known concentration and ___ of titrant reacts with a solution of analyte or titrand to determine concentration. The volume of titrant ___ is called titration volume.

    A. quantitative　B. known　　　C. prepared　　D. volume
    E. reacted　　　F. concentration　　G. key　　　H. laboratory

2. Match the following explanation with the proper word.

    ___ coincide, ___ curve, ___ rinse, ___ deliver, ___ frost

    A. to wash clothes, dishes, vegetables etc. quickly with water, especially running water, and without soap

    B. be the same

    C. provide with a rough or speckled surface or appearance

    D. to take goods, letters, packages etc. to a particular place or person

    E. a line that gradually bends like part of a circle

Lesson 25
COMBINATORIAL CHEMISTRY

Combinatorial chemistry is one of the important new methodologies developed by academics and researchers in the pharmaceutical, agrochemical, and biotechnology industries to reduce the time and costs associated with producing effective, marketable, and competitive new drugs. Simply put, scientists use combinatorial chemistry to create large populations of molecules, or libraries, that can be screened efficiently en masse. By producing larger[1], more diverse compound libraries, companies increase the probability that they will find novel compounds of significant therapeutic and commercial value. The field represents a convergence of chemistry and biology, made possible by fundamental advances in miniaturization, robotics, and receptor development. And not surprisingly, it has also captured the attention of every major player in the pharmaceutical, biotechnology, and agrochemical arena.

While combinatorial chemistry can be explained simply, its application can take a variety of forms, each requiring a complex interplay of classical organic synthesis techniques, rational drug design strategies, robotics, and scientific information management. This article will provide a basic overview of existing approaches to combinatorial chemistry, and will outline some of the unique information management problems that it generates.

Approaches to Combinatorial Chemistry

As with traditional drug design, combinatorial chemistry relies on organic synthesis methodologies. The difference is the scope—instead of synthesizing a single compound, combinatorial chemistry exploits automation and miniaturization to synthesize large libraries of compounds. But because large libraries do not produce active compounds independently, scientists also need a straightforward way to find the active components within these enormous populations. Thus, combinatorial organic synthesis (COS) is not random, but systematic and repetitive, using sets of chemical "building blocks" to form a diverse set of molecular entities. Scientists have developed several different COS strategies, each with the same basic philosophy—stop shooting in the dark and instead, find ways to determine active compounds within populations, either spatially, through chemical encoding, or by systematic, successive synthesis and biological evaluation (deconvolution)[2].

There are three common approaches to COS. During arrayed, spatially addressable synthesis, building blocks are reacted systematically in individual reaction wells or positions to form separated "discrete molecules[3]." Active compounds are identified by their location on the grid. This method has been applied in scale (as in the Parke - Davis Pharmaceutical **DIVERSOMER** technique), as well as in miniature (as in the Affymax VLSIPS technique). The second technique, known as encoded mixture synthesis, uses nucleotide, peptide, or other types of more inert chemical tags to identify each compound.

During deconvolution, the third approach, a series of compound mixtures is synthesized combinatorially, each time fixing some specific structural feature. Each mixture is assayed as a mixture and the most active combination is pursued. Further rounds systematically fix other structural features until a manageable number of discrete structures can be synthesized and screened. Scientists working with peptides, for example, can use deconvolution to optimize, or locate, the most active peptide sequence from millions of possibilities. You could say that combinatorial chemistry is a technologically advanced way of finding a needle in a haystack. The whole idea is to remove the guesswork and instead, to create and test as many compounds or mixtures as possible—logically and systematically—to obtain a viable set of active leads.

Managing Combinatorial Chemistry Libraries

As with traditional drug design, the ability to integrate different types of chemical, biological, and corporate information is crucial to combinatorial chemistry techniques. But combinatorial chemistry also generates an enormous amount of information which present day information systems have a hard time managing. Combinatorial chemists also ask different questions in different ways, and their information systems need to adapt to find these answers quickly.

For example, chemists planning a traditional synthesis typically conduct a retrosynthetic analysis to determine the best, and perhaps cheapest, way to obtain the target. And while combinatorial chemists also look at retrosynthetic trees to build combinatorial libraries, their priorities differ. "By which modes of forward synthesis are the most building blocks available or obtainable?" they might ask. "And if I allow the synthesis to proceed by this course, what is the scope and reliability of the necessary reactions?" Combinatorial chemists need a way to access this type of reaction information efficiently. In addition, one of the largest bottlenecks in the construction of combinatorial libraries is in obtaining the basic building blocks necessary to run each reaction. Chemical information systems that can quickly access updated databases of inventory and commercially available reagents are invaluable tools in reagent acquisition (see Figure A).

An Archival Revolution

Once built, combinatorial libraries produce unprecedented amounts of information. Reaction histories for each compound must be archived. Robots and other laboratory instruments need to be controlled, and the data they acquire archived for future reference. Scientists need to integrate screening results and biological data with structural information. As in single - molecule archival systems, the archival of combinatorial libraries and their corresponding data is essential to cost-effective research and development. Basic archival and reporting capabilities can provide managers with the facts they need to justify the costs associated with combinatorial chemistry. And researchers can use scientific information management systems to avoid past mistakes, learn from previous successes, and answer critical questions, such as, "How much of this library overlaps with other corporate libraries?" or "Which building blocks have proven most successful?" or "What is the difference between the biological performance of a molecule produced in a mixture rather than in a discrete format?"

Combinatorial chemistry is a promising new field that stands to revolutionize the chemical industry, and demands completely new scientific information management solutions. Combinatorial chemists will be able to meet their goals if they can find ways to plan libraries quickly, produce libraries that better interrogate biological assays, and learn from past screening results. Using software that can orchestrate the planning, building, screening, and interpretation of synthesized libraries, combinatorial chemistry programs will begin to realize their promise of minimizing the time and cost associated with bringing new molecular entities to market.

Figure A

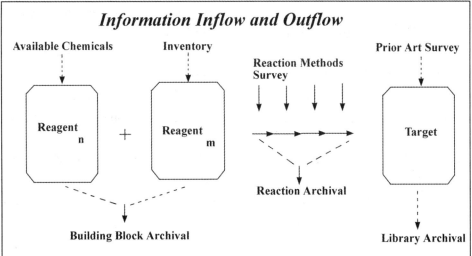

Information sources useful throughout the planning of a combinatorial library. Databases of reaction methodology, reagent availability, and relevant prior-art[4] are extremely useful in combinatorial synthesis. Also portrayed is data capture, discussed in the article on Project Library.

New words and Expressions

combinatorial [ˌkɔmbinəˈtɔːriəl] a. 组合的
　~ chemistry 组合化学
methodology [ˌmeθəˈdɔlədʒi] n. 方法学，
　方法论
academics and researchers　学者
biotechnology [ˌbaiəutekˈnɔlədʒi] n. 生物工程，生物工艺学
marketable [ˈmɑːkitəbl] a. 适销的，可销售的
create [kriˈeit] vt. 创造，创立，建起，造成
population [ˌpɔpjuˈleiʃən] n. 群体；族，组
library [ˈlaibrəri] n. 图书馆，(数据)库
screen [skriːn] vt. 筛选 n. 筛子，掩蔽物，屏风
en masse [ɑŋˈmæs] ad. 整体地，全体地，一同地
therapeutic [ˌθerəˈpjuːtik] a. 治疗(学)的；n. 治疗剂，治疗学家
convergence [kənˈvəːdʒəns] n. 融合，集中，收敛
miniaturize [ˈminiətʃəˌraiz] vt. 使微型化；
　-ation n.
robotics [rəuˈbɔtiks] n. 机器人技术，计算机技术
capture [ˈkæptʃə] vt.n. 引起(注意等)，俘获，捕获
player [ˈpleiə] n. 从业者，演员，表演者
strategy [ˈstrætidʒi] n. 策略，方案
exploit [iksˈplɔit] vt. 利用，开拓，开发，剥削
automation [ˌɔːtəˈmeiʃən] n. 自动控制，自动操作
straightforward [streitˈfɔːwəd] a. 直接的，简单的，易懂的
random [ˈrændəm] n.a. 随意，任意
chemical 'building blocks' 化学模块，化学建筑块
entity [ˈentəti] n. 实体
encode [inˈkəud] vt. 编码，译码
deconvolution [ˌdiːkɔnvəˈluːʃən] n. 去卷积，反褶积
discrete [disˈkriːt] a. 不连续的，离散的；
　~ molecule 离散分子
nucleotide [ˈnjuːkliətaid] n. 核苷酸
peptide [ˈpeptaid] n. 肽，缩氨酸
tag [tæg] n. 标签，标记符，名称
assay [əˈsei] n.v. 化验，鉴定，测试
combination [ˌkɔmbiˈneiʃən] n. 化合，化合物，结合，组合
pursue [pəˈsjuː] vt. 追踪，跟踪
locate [ləuˈkeit] v. 定位，位于，确定…的位置
needle [ˈniːdl] n. 针，指针
lead [liːd] n. 先导(化合)物，领导，领先
guesswork [ˈgeswəːk] n. 臆测，猜测
crucial [ˈkruːʃiəl, ˈkruːʃəl] a. 至关紧要的
priority [praiˈɔriti] n. 优先(顺序)，优先权
access [ˈækses] vt.n. 存取，接近，通路，访问，入门
reliability [riˌlaiəˈbiliti] n. 可靠性
archival [ɑːˈkaivəl] a. 档案的
cost-effective a. 低成本的，高效益的
research and development (产品等的)研究与开发
critical [ˈkritikəl] a. 临界的，评论的，没有定论的
prior-art [ˈpraiə ɑːt] 优先技巧，优先策略

词组

(be) associate with …　涉及，与…相联系
simply put　简单地说，简言之
make possible　使…变成可能
as with …　正如…的情况一样
rely on (upon)　依赖，依靠
a set of …　一组，一套
each time　每次
work with …　从事…工作(研究)，对…

起作用

provide … with 给 … 提供

learn from … 向 … 学习

Notes

1. "by producing larger"意思是"(采用)大规模生产",这里 producing 是名词,larger 是副词比较级。
2. deconvolution: In mathematics, deconvolution is a process used to reverse the effects of convolution on recorded data. The concept of deconvolution is widely used in the techniques of signal processing and image processing. Since these techniques are in turn widely used in many scientific and engineering disciplines, deconvolution finds many applications.
3. discrete molecule: Basically a "discrete molecule" is a covalent molecule in which the intermolecular forces are really weak, hence the low melting and boiling points of these molecules.
4. prior art (also known as or state of the art, which also has other meanings): In in most systems of patent law constitutes all information that has been made available to the public in any form before a given date that might be relevant to a patent's claims of originality. If an invention has been described in prior art, a patent on that invention is not valid.

Exercises

1. Fill in the blacks with proper words or phrases.

    Combinatorial chemistry ____ chemical synthetic methods that make it possible to ____ a large number of compounds in a ____ process. These compound libraries can be made as ____, sets of individual compounds or chemical structures ____ by computer software. Combinatorial chemistry can be used for the ____ of small molecules and for peptides.

    Strategies that ____ identification of useful components of the libraries are also part of combinatorial chemistry. The methods used in combinatorial chemistry are ____ outside chemistry, too.

    A. mixtures   B. comprises   C. allow   D. prepare
    E. generated   F. synthesis   G. single   H. applied

2. Match the following explanation with the proper word.

    ____ population, ____therapeutic, ____strategy, ____random, ____locate

    A. a planned series of actions for achieving something

    B. the number of people living in a particular area, country etc.

    C. to find the exact position of something

    D. happening or chosen without any definite plan, aim, or pattern

    E. relating to the treatment or cure of an illness

Lesson 26
VAPOR – PHASE CHROMATOGRAPHY

Vapor - phase chromatography (V. P. C.), also known as gas chromatography or gas - liquid partition chromatography (G. L. P. C.), is used extensively in the separation and identification of mixtures of volatile compounds. It is also one of the most useful techniques for the quantitative determination of the components of a mixture.

In gas chromatography each component of a volatile mixture of compounds is partitioned between vapor and liquid phases as the mixture is passed through a column containing a suitable, relatively non - volatile liquid (the adsorbent) impregnated in a finely divided, inert, porous solid material (the solid support). The various components of the mixture pass through the column at different rates, and each component can be detected and, if desired, collected as it emerges from the column.

In actual practice, a very small sample of the mixture is injected into a heated chamber where it is vaporized and swept through the column (which is encased in an oven which also is usually heated) with the aid of an inert carrier gas, most commonly helium. After the components of the mixture have been separated as they!pass through the column, they pass through a detector located at the end of the column and are recorded as peaks on a mechanical recorder. A schematic diagram of a gas chromatograph is shown in Figure 2.

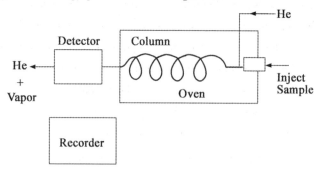

Figure 2. Schematic diagram of a gas chromatograph.

The time required for a compound to pass through a column is known as its retention time, which depends on many variables, including the following:

1. The nature of the adsorbent.
2. The concentration of adsorbent in the inert solid support.
3. The degree to which the compound being analyzed is adsorbed by adsorbent.

4. The volatility of the compound.
5. The column temperature.
6. The rate of flow of carrier gas.
7. The nature of the carrier gas.
8. The dimensions of the column.

By suitable adjustments of these variables it is usually possible to effect a clean separation of all of the components of a mixture of compounds. Some of the liquid adsorbents in common use are silicone rubbers, apiezon greases, waxes, oils, carbowaxes and dialkyl phthalates. Some of the inert, porous solid supports for the adsorbents are firebrick, Celite and Teflon. Nitrogen or argon is sometimes used as the carrier gas in place of helium.

The detector must be capable of measuring some difference between the carrier gas alone and the carrier gas mixed with an effluent compound. Differences in thermal conductivity are easy to measure and provide a high sensitivity to detection. Thus, this type of detector is the one most frequently used. Changes in the thermal conductivity of the effluent gases are sensed by the detector and relayed to a recorder by electronic means. The recorder provides a display of peaks as a function of time. Each peak represents a new compound that has passed through the detector. Furthermore, the area under each peak, when suitably calibrated, represents a measure of the amount of the compound present in the original mixture.

As mentioned above, the retention time of a compound under a specified set of conditions is a physical constant that can be used in qualitative analysis. For example, you might wish to use p-xylene as a reagent, but might suspect that a commercial sample of the compound is contaminated with small amounts of o-xylene and m-xylene. If your suspicion is correct, you will find that a vapor-phase chromatogram of the reagent shows one major peak and two minor peaks. You can then confirm your suspicion as to the nature of one of the contaminants by deliberately adding some o-xylene to the reagent and obtaining a new vapor-phase chromatogram[1]. If the contaminant in the reagent is indeed o-xylene, one of the minor peaks will be larger in the second chromatogram than in the first. The same type of control experiment can then be carried out with addition of m-xylene to the reagent. Alternatively, the retention times of pure o-xylene and m-xylene can be obtained under the same conditions as used for the reagent, p-xylene. The retenion times of the two impurities present in the p-xylene should correspond to those of pure o-xylene and m-xylene, respectively, if these are the impurities. Reasonable care must be exercised in the interpretation of vapor-phase chromatograms for purposes of qualitative analysis. Just as two different solid compounds may possess the same or very nearly the same melting point, two different volatile compounds may exhibit identical retention times under a given set of conditions. Thus, it is usually wise to confirm a tentative identification of a compound through its retention time by some other means[2]. One of the best ways to do this is to condense the compound as it leaves the detector unit of the gas chromatogragh, take the infrared spectrum of the condensate and then compare it with the spectrum of the compound whose presence is suspected.

The best use of vapor-phase chromatography is in the quantitative analysis of mixtures of volatile compounds. As a first approximation, it can be assumed that the peak area of a given compound in its vapor-phase chromatogram will be directly proportional to its weight per cent in the original mixture. However, this is not strictly true, and suitable calibration curves must be constructed to derive accurate values. A variety of methods can be used to measure the peak areas of the components of a mixture. One sophisticated method is to use an electronic integrator connected to the recorder. However, while[3] this approach is routine in nuclear magnetic resonance spectral determinations there are relatively few commercial vapor-phase fractometers that are equipped with an integrator. Since the recorder paper used in vapor-phase chromatography has a very uniform thickness and density, a simple way to calculate relative areas is to cut out the peaks and weigh the pieces of paper on an analytical grade balance. Alternatively, you may use a planimeter to measure the area under each peak. A fourth method is to calculate the area under each peak by use of the relationship:

$$\text{Area} = \text{Peak Height} \times \text{Width at Half Height}$$

It should be kept in mind that this method can be used satisfactorily only when the peaks are symmetrical and well separated from one another.

New words and Expressions

vapor-phase chromatography ['veipə'feiz ˌkrəuməˌtɔgrəfi] 气相色谱(法)
chromatography [ˌkrəumə'tɔgrəfi] n. 层析法,色谱法
gas chromatography 气体色谱(法)
partition [pɑː'tiʃən] n.vt. 分离,分开,分配
gas-liquid partition chromatography 气—液分配色谱法
determination [diˌtəːmi'neiʃən] n. 测定
identification [aiˌdentifi'keiʃən] n. 鉴定,鉴别
volatile ['vɔlətail] a. 易挥发的,挥发性的
adsorbent [æd'sɔːbənt] n. 吸附剂
impregnate ['impregneit] vt. 浸渍
porous ['pɔːrəs] a. 多孔的
collect [kə'lekt] vt. 收集
emerge [i'məːdʒ] vi. 出现,形成
practice ['præktis] n. 实践,操作
sample ['sɑːmpl] n. 试样
inject [in'dʒekt] vt. 注射 (into)

chamber ['tʃeimbə] n. 室,房间
vaporize ['veipəraiz] v. 汽化
encase [in'keis] vt. 把…放入套内
oven ['ʌvn] n. 炉,烘箱
carrier gas ['kæriə gæs] 载气
detector [di'tektə] n. 检测器
recorder [ri'kɔːdə] n. 记录器(仪)
schematic diagram [ski'mætik 'daiəgræm] 示意图
retention time [ri'tenʃən taim] 保留时间
degree [di'griː] n. 程度,度
analyze ['ænəlaiz] vt. 分析
adsorb [æd'sɔːb] vt. 吸附
volatility [ˌvɔlə'tiliti] n. 挥发性,挥发度
dimension [di'menʃən] n. 尺寸
effect [i'fekt] vt. 实现; n. 效应,影响,效果
clean [kliːn] a. 彻底的,完全的
silicone rubber ['silikən 'rʌbə] 硅橡胶
apiezon [ə'piːzɔn] n. 阿匹松,饱和烃

phthalate[fˈθæleit,ˈθæleit]n.邻苯二甲酸盐（酯）
dialkyl[daiˈælkil]phthalate 邻苯二甲酸二烷基酯
firebrick[ˈfaiəˈbrik]n.耐火砖
celite[siˈ:lait]n.硅藻土
Teflon[ˈteflɔn]n.(商品名)特氟隆,聚四氟乙烯
effluent[ˈefluənt]n.流出物；a.流出的
sensitivity[ˌsensiˈtiviti]n.敏感度,灵敏度
sense[sens]vt.(自动)检测
relay[ˈri:lei]vt.传递
display[disˈplei]vt.n.显示,表现
calibrate[ˈkælibreit]vt 校准
detector signal[signl]检测器信号
methyl ester　甲酯
fatty acid[ˈfætiˈæsid]脂肪酸
constant[ˈkɔnstənt]n.常数
suspect[səsˈpekt]v.猜想,怀疑
suspicion[səsˈpiʃən]n.猜想,怀疑
chromatogram[ˈkrəumətəgræm]n.色谱图
minor[ˈmainə]a.较小的,较次要的
contaminant[kənˈtæminənt]n.沾污物,污染物
deliberately[diˈlibəritli]adv.故意地,审慎地
control[kənˈtrəul]n.对照(物)；控制
wise[waiz]a.明智的,考虑周到的,慎重的
tentative[ˈtentətiv]a.不明确的
condense[kənˈdens]v.冷凝,缩合
infrared[ˈinfrəˈred]a.红外线的
spectrum[ˈspektrəm]n.谱,波谱,光谱
assume[əˈsju:m]vt.假定,设想
construct[kənˈstrʌkt]vt.作(图)
sophisticate[səˈfistikeit]vt.使复杂
integrator[ˈintigreitə]n.积分仪,求积仪
routine[ru:ˈti:n]n.a.常规,例行(手续)
nuclear magnetic resonance spectra 核磁共振谱
fractometer[ˈfræktəuˈmi:tə]n.色层分离仪

planimeter[plæˈnimitə]n.面积仪

前　缀

chrom(o)-[ˈkrəum(ə)-]颜色,铬
　chromatography,chromometer(比色仪),
　chromatron(彩色显像管)；chromic
phthal-[fθæl-](邻)苯二甲,与邻苯二甲酸有关的化合物
　phthalaldehyde,phthalate,phthalonitrile
sens-[sens-]感觉,灵敏
　sensitivity,photosensitive,sensitizer

后　缀

-gram[græm](名词词尾)
　(1)记录,图像,图表
　chromatogram,telegram,program(说明书),diagram
　(2)重量单位(克)
　kilogram,microgram
-meter[mitə](名词词尾)
　(1)仪,计,表
　planimeter,thermometer,spectrometer
　(2)计量
　kilometer diameter,parameter(参变数)

词　组

pass through　通过
sweep through　冲过,带过,掠过
with the aid of …　借助于……,用…
in common use …　被普遍使用,通用
in place of…　代替……
as to …　关于……；就……而论
for purposes of…　为……起见,对……来说
just as …　正如……一样
separate from…　和……分开,从……分离出来

Notes

1. You can then confirm your suspicion as to the nature of one of the contaminants by deliberately adding some o-xylene to the reagent and obtaining a new vapor-phase chromatogram. 该句为简单句,"as to the… contaminants"为suspicion的定语,"by deliberately…chromatogram"为方式状语。**译文**:"然后你可以通过有意加入一些邻二甲苯到试剂中和得到一个新的气相色谱图而证实你的关于污染物之一的种类的设想。"

2. Thus, it is usually wise to confirm a tentative identification of a compound through its retention time by some other means. 该句中it是形式主语,真正的主语是"to confirm…other means"。
译文:"通常利用一些其它方法来证实通过化合物的保留时间的不明确鉴别是明智的。"

3. while: although, 引导让步状语从句。

Exercises

1. Fill in the blanks with proper words or phrases.

A gas chromatograph is a chemical analysis ____ for separating chemicals in a complex sample. A gas chromatograph uses a flow-through narrow tube known as the ____, through which different chemical constituents of a sample

____ in a gas stream at different rates depending on their various chemical and physical and their interaction with a specific column ____, called the stationary phase. As the chemicals exit the end of the column, they are detected and ____ electronically. The function of the stationary phase in the column is to separate different components, causing each one to exit the column at a different __ time. Other parameters that can be used to ____ the order or time of retention are the carrier gas flow rate, column length and the temperature.

   A. pass   B. instrument   C. filling   D. alter,

   E. column   F. retention   G. identified   H. properties

2. Match the following explanation with the proper word.

   ____ partition, ____ porous, ____ vaporize, ____ calibrate, ____ routine

   A. full of pores or vessels or holes

   B. to check or slightly change an instrument or tool, so that it does something correctly

   C. turn into gas

   D. the usual order in which you do things, or the things you regularly do

   E. separation by the creation of a boundary that divides or keeps apart

Lesson 27
INFRARED SPECTROSCOPY

That part of the electromagnetic spectrum of longer wavelength than the visible region and shorter wavelength than the microwave region is known as the infrared region. However, the decidedly limited portion of the infrared region which lies between wavelengths of 2.5 μ and 15 μ (1μ=1 micron= 10^{-4}cm =10,000Å) or between frequencies of 4000 cm^{-1} and 660 cm^{-1} (1cm^{-1}= 10,000/μ) is of greatest practical use to the organic chemist. Radiation energy in this region may be absorbed by an organic molecule and converted into molecular vibrational energy. Organic molecules undergo continual stretching, bending, and rotational motions, and, when the frequency of the infrared light passing through an organic compound corresponds to the frequency of one of the molecular motions cited above, the light is absorbed. Since rotational frequencies fall outside the 4000~660 cm^{-1} region, a plot of intensities (either as per cent transmittance or as absorbance, usually the former) versus wavelength (microns) or frequency (wave numbers) represents a record of the relative amounts of stretching and bending motions of various atoms in the molecule under consideration. The energy absorption pattern thus obtained is known as the infrared spectrum of the compound. One restriction of the above discussion is worthy of mention, namely, the stretching or bending motion of atoms within a molecule must cause a change in the instantaneous dipole moment in order to produce an absorption peak in the infrared spectrum.

The organic chemist has learned to interpret infrared spectra on an empirical basis. He is able to take advantage of the fact that even a simple molecule can have a complex infrared spectrum which is of value when he compares the spectrum of an unknown compound with that of an authentic sample of the compound. When the two spectra are exactly superposable, this fact constitutes excellent evidence for the identity of the unknown.

Of even greater usefulness to the organic chemist is the fact that certain groups of atoms (certain functional groups) give rise to stretching or bending vibrational absorption bands at or near the same wavelength, regardless of the structure of the remainder of the molecule[1]. Thus, it is frequently possible for an experienced chemist to examine the infrared spectrum of an unknown compound and decide what functional groups are present in the molecule, even though he might not have any other information about the compound whatsoever[2]. One can readily appreciate the value of this approach when he realizes that a natural products chemist, for example, after weeks of work in the isolation of a new alkaloid or steroid, might end up with only a few miligrams of the pure compound in his possession. Furthermore, after he has obtained knowledge of the functional groups present by examination of the infrared spectrum of the compound, he still has the entire sample available for other work. In other words, he has not consumed a portion of his compound by carrying out qualitative tests for functional groups.

Infrared spectroscopy is also used extensively for quantitative analysis of mixtures of organic compounds. Within an accuracy of ±1~5 per cent, depending on the precautions employed in the determination, it is frequently possible to determine the percentage of each component in a mixture of two or more compounds. For example, one can readily determine the percentages of *ortho*, *meta*, and *para* isomers obtained by bromination of toluene in the presence of ferric bromide catalyst by the use of infrared spectroscopy. The chemical literature contains this and thousands of additional applications of the use of infrared spectroscopy in quantitative analysis.

Although techniques are available for taking the infrared spectra of gaseous, solid, and neat liquid samples of organic compounds, most infrared spectra are taken of compounds in an appropriate solution. The solvents most commonly used for such determinations are chloroform, carbon tetrachloride, and carbon disulfide. The ideal solvent would be one which is transparent over the entire wavelength range of 2.5~15μ, undergoes no interaction with the solute, exerts excellent solvent action on all organic compounds, is nonflammable, nontoxic, inexpensive, and completely noncorrosive. Unfortunately, no such ideal solvent is known. However, the three solvents cited above give reasonably good results when properly used. Since the most commonly used cells for holding solutions in the infrared spectrophotometer contain windows constructed of sodium chloride, it is important that the solutions always be free of water. A sample containing water dissolves a portion of the windows and thus ruins a set of expensive cells. Furthermore, water and other solvents containing hydroxyl groups enter into hydrogen bonding with many solutes, and this causes unpredictable shifts in the characteristic absorption peaks of some functional groups.

Certain generalizations about structure and frequency of absorption are of value. The characteristic absorption peaks arising from vibrations about a single bond generally fall in the region between 660 cm^{-1} and 1600 cm^{-1}, those from[3] a double bond in the 1600~2000 cm^{-1} region, and those from a triple bond in the 2000~2500 cm^{-1} region. However, the stretching frequencies of the bonds between hydrogen and oxygen, hydrogen and sulfur, and hydrogen and nitrogen are found in the 2500~3700 cm^{-1} region of the infrared spectrum. On the basis of the theory of resonance or the simple molecular orbital theory, the student could predict that conjugation of multiple bonds decreases their double bond character. Thus, the effect of conjugation on the position of the absorption peaks attributable to the multiple bonds is to decrease their frequencies (increase their wavelengths). For example, simple saturated aldehydes show absorption owing to the presence of the carbon-oxygen double bond at about 1725 cm^{-1}. Conjugation with a carbon-carbon double bond lowers this value to about 1700 cm^{-1}.

$$-\overset{}{C}=\overset{\overset{H}{|}}{C}-\overset{}{C}=O \longleftrightarrow -\overset{}{\underset{+}{C}}-\overset{\overset{H}{|}}{C}=\overset{}{C}-O^{-}$$

$$\text{I} \qquad\qquad\qquad \text{II}$$

(The carbonyl group has increased single bond character owing the contribution of structure II.)

The literature on the subject of the correlation of structure with infrared absorption spectra is vast, and no attempt will be made here to give complete details[4].

New words and Expressions

electromagnetic [iˈlektrəumægˈnetik] a. 电磁的
wavelength [ˈweivleŋθ] n. 波长
visible [ˈvizəbl] a. 可见的，显而易见的
microwave [ˈmaikrəweiv] n. 微波
limited [ˈlimitid] a. 有限制的，狭窄的
frequency [ˈfriːkwənsi] n. 频率
absorb [əbˈsɔːb] vt. 吸收
vibrational [vaiˈbreiʃnl] a. 振动的，摆动的
bend [bend] vt. 弯曲，变角
rotational [rəuˈteiʃənəl] a. 转动的，旋光的
motion [ˈməuʃən] n. 运动
transmittance [trænzˈmitəns] n. 透过率，透射比
absorbance [əbˈsɔːbəns] n. 吸收率
former [ˈfɔːmə] n. 前者
versus [ˈvəːsəs] prep. …对…，比较
micron [ˈmaikrɔn] n. 微米(μ)，百万分之一米
wave mumber 波数
restriction [risˈtrikʃən] n. 限制
instantaneous [ˌinstənˈteinjəs] a. 瞬间的，瞬时的
interpret [inˈtəːprit] vt. 解释
authentic [ɔːˈθentik] a. 可信的，可靠的，真正的
superposable [ˈsjuːpəˈpəuzəbl] a. 能重叠的
experienced [iksˈpiəriənst] a. 已有经验的，熟练的
appreciate [əˈpriːʃieit] vt. 重视，赏识
alkaloid [ˈælkəˈlɔid] n. 生物碱
steroid [ˈsterɔid] n.a. 甾族化合物，类固醇
possession [pəˈzeʃən] n. 持有，财产
consume [kənˈsjuːm] vt. 消耗，耗尽，浪费
accuracy [ˈækjurəsi] n. 准确度
precaution [priˈkɔːʃən] n. 事先之准备，细心，小心
bromination [ˌbrəumiˈneiʃən] n. 溴化

carbon disulfide [ˈkɑːbən daiˈsʌlfaid] 二硫化碳
transparent [trænsˈpɛərənt] a. 透明的，透过的
interaction [ˌintəˈrækʃən] n. 相互作用，相互影响
nonflammable [nɔnˈflæməbl] a. 不燃的，难燃的
nontoxic [nɔnˈtɔksik] a. 无毒的
noncorrosive [nɔnkəˈrəusiv] a. 不腐蚀的
properly [ˈprɔpəli] ad. 正当地，精确地
cell [sel] n. 小室，(电)池
ruin [ruin] vt.n. 破坏，毁灭
solute [ˈsɔljuːt] n. 溶质
unpredictable [ˌʌnpriˈdiktəbl] a. 不可预测的
shift [ʃift] n.vt. 移动，位移，变动
generalization [ˌdʒənərəlaiˈzeiʃən] n. 概括，一般法则
simple molecular orbital theory 简单分子轨道理论
conjugation [ˌkɔndʒuˈgeiʃən] n. 共轭
correlation [ˌkɔriˈleiʃən] n. 相互关系

前 缀

spectr(o)- [ˌspektr(ə)-] 影像，视觉，谱
　spectrum, spectroscopy, spectrophotometer, spectropolarimeter (旋光仪)
electr(o)- [iˈlektr(əu)-] 电，电的
　electron, electromagnetic, electrochemical, electrolyte

后 缀

-scopy [-skəpi] 观察，检验方法（名词词尾）
　spectroscopy, microscopy
-oid [-ɔid] (n. 或 a. 词尾) 似，像
　alkaloid, colloid, metaloid

词 组

under consideration　在研究中，在考虑中

take advantage of… 利用,运用	end(up)with… 以……结束
worthy of … 值得(……)的	attributable to … 归因于
of value 有用的,有价值的	owing to … 由于,因为
regardless of … 不管,不顾	on the subject of … 关于,涉及

Notes

1. Of even greater usefulness to the organic chemist is the fact that certain groups of atoms (certain functional groups) give rise to stretching or bending vibrational absorption bands at or near the same wave length, regardless of the structure of the remainder of the molecule. 该句主句是"Of even…the fact",为倒装句,主语是 the fact,"that certain…of the molecule"是修饰主句中主语的定语从句。**译文**:"一些原子团(官能团)不管分子其余部分结构如何,在相同波长产生伸缩和变形振动吸收带,这一类对有机化学家有更大的用处。"
2. whatsoever: at all.
3. those from: the characteristic absorption peaks arising from vibrations about.
4. The literature on the subject of the correlation of structure with infrared absorption spectra is vast, and no attempt will be made here to give complete details. 该句是并列句,第二个分句是"and no …details",此分句中 attempt 是主语,"to give complete details"是主语补足语。**译文**:"涉及结构与红外吸收谱关系的文献很多,这里不试图给出全部细节。"

Exercises

1. Fill in the blanks with proper words or phrases.

   Infrared radiation (IR) is electromagnetic ____ with longer ____ than those of visible light, and is therefore generally ____ to the human eye, although IR at wavelengths up to 1050 ____ from specially pulsed lasers can be seen by humans ____ certain conditions. IR wavelengths extend from the nominal red ____ of the visible spectrum at 700 nanometers, to 1 millimeter. Most of the thermal radiation emitted by objects near room ____ is infrared. As with all EMR, IR ____ radiant energy and behaves both like a wave and like its quantum particle, the photon.

   A. nanometers   B. radiation   C. edge   D. invisible
   E. wavelengths F. under   G. carries H. temperature

2. Match the following explanation with the proper word.
   ____ vibration, ____ rotation, ____ former, ____ interpret, ____ transparent

   A. happening or existing before, but not now
   B. when something turns with a circular movement around a central point
   C. to explain the meaning of something
   D. a continuous slight shaking movement
   E. something that you can see through it

Lesson 28
NUCLEAR MAGNETIC RESONANCE(Ⅰ)

Almost unbelievably detailed information about the structure of a typical organic compound can be obtained by examination of its nuclear magnetic resonance spectrum. It is no exaggeration to state that the advent of NMR spectroscopy has caused a revolution in the practice of organic chemistry. Almost every article published in an organic chemical journal today reveals that NMR spectroscopy was used at some point in the solution of the problem under consideration. Unfortunately, NMR spectrometers are very expensive instruments to purchase and to maintain. Thus, this type of instrumentation is not available in every college laboratory.

Of the various isotopes of elements commonly found in organic compounds, only three have nuclei which may be said to spin and can give usable nmr signals. These are H^1, F^{19} and P^{31}. The spinning of these charged particles generates a magnetic moment along the axis of the spin; consequently, these nuclei act like small bar magnets. Other nuclei which can give usable nmr signals are B^{11}, N^{14}, Si^{29}, H^2, C^{13}, N^{15}, O^{17} and S^{33}. It is fortunate that the nuclei of C^{12} and O^{16} are nonmagnetic; otherwise, proton spectra would be even more complicated than they are.

When a proton is located in an external magnetic field, its magnetic moment can be aligned either with or against the external field. Since alignment with the field represents a more stable state than alignment against the field, energy must be supplied to flip the proton from the more stable to the less stable alignment. The larger the external field, the greater the energy difference between nuclei aligned with or against the field. When nuclei are irradiated with electromagnetic radiation of an energy corresponding to the difference in energy between the two states, a transition or flip from one state to the other takes place. This is the NMR signal.

To obtain some idea of the amount of energy needed to bring about the flip of the proton in a field of 14,092 gauss, a value well within the range of field strengths obtainable by use of the magnet that is part of a proton magnetic resonance (pmr) spectrometer, we can make use of the equation[1].

$$\nu = \frac{\gamma H_0}{2\pi}$$

where ν=frequency in cycles per second

H_0=strength of the magnetic field, in gauss

γ=the gyromagnetic ratio, a nuclear constant, 26,750 for the proton.

The energy required for the flip corresponds to electromagnetic radiation having a frequency of 60 Mcs (60 megacycles or 60 million cycles per second), and this represents radiation in the radio-frequency range. This is radiation of much lower energy (i.e., much longer wavelength) than that of the infrared region of the electromagnetic spectrum, and, of course, of very much lower energy than that of the visible and ultraviolet regions. In principle, then, the technique of nmr spectroscopy could be to place a concentrated solution of the compound being studied (or the pure compound of it is a liquid) in a strong, homogeneous magnetic field and to pass a steadily changing radio-frequency signal through it. The frequency at which radiation is absorbed would be recorded. In practice, however, it is more convenient to maintain the radiation frequency at a constant value and to vary the strength of the magnetic field. At that value of the field strength where the energy required to flip the proton matches the energy of the radiation, absorption occurs, and a signal is transmitted to a recorder by electronic means[2]. The intensity if absorption of the radio-frequency signal is plotted as a function of increasing magnetic field strength, and such a plot is termed an NMR spectrum.

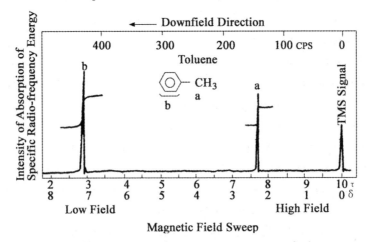

Figure 3. Nuclear magnetic resonance spectrum of toluene.

Let us now consider an actual nmr spectrum, that of toluene, shown in Fig.3. Two questions immediately come to mind: (1) Why are there two signals coming at different positions? (2) Why are the intensities of these signals different? Most of you, by intuition, would probably arrive at essentially correct answers to these questions. (1) There are two signals because there are two fundamentally different classes of hydrogen atoms in toluene, aromatic and benzylic. (2) The intensities are different because the numbers of the different types of hydrogen atoms are different; there are three aliphatic hydrogen atoms and five aromatic hydrogen atoms present in toluene.

In common with other forms of spectroscopy, the intensity of absorption of the radio-frequency signal in NMR spectroscopy is proportional, under proper experimental conditions, to the number of nuclei causing the absorption. Thus the spectrum reveals not only the number of different environments in which the protons are located, but also the relative numbers of protons in each equivalent set.[3]

As mentioned previously, NMR spectra are taken either on liquid samples or on solutions. The usual concentration required when solutions are used is 5~20%. Since the number of suitable solvents is limited, this concentration requirement can impose a limitation on the technique. The actual amount of compound required in order for its NMR spectrum to be taken is only 2.5~30 mg. Carbon tetrachloride is probably the best general solvent for NMR spectral determinations; many compounds, however, are not sufficiently soluble in this solvent, and therefore solvents such as deuteriochloroform ($CDCl_3$), carbon disulfide (CS_2), tetrachloroethylene (C_2Cl_4), CD_3CO_2D, CD_3SOCD_3, CD_3OD, CD_3COCD_3 and D_2O find wide use. For the examination of limited portions of the NMR spectrum, even some proton-containing solvents may be used.

Because of limited availability of a compound or, more commonly, limited solubility in any of the suitable solvents, less than 2.5mg. of a compound must sometimes be used in nmr spectral determinations. There are two ways in which an nmr spectrum can be obtained under these circumstances. One is to employ a more sensitive spectrometer. The second method is to scan the spectrum repeatedly and to analyze the results with the aid of a computer.

As shown in the NMR spectrum of toluene (Fig. 3), the intensity of absorption of radio-frequency energy of about 60 Mcps is plotted as the ordinate against increasing magnetic field strength as the abscissa. The magnetic field strength is most easily measured in frequency units (cps), and changes of less than one part in 60 million are significant in the spectrum in the strong magnetic field (about 14 000 gauss). Since it would be very difficult to measure to the required accuracy this large field strength in absolute terms, an indirect method is employed[4]. Tetramethylsilane (TMS), all of the protons of which absorb at the same magnetic field strength to produce a sharp, single peak which is upfield from where most protons of organic compounds absorb, is added to the solution or liquid compound.[5] Then, all observed field strengths where the protons absorb are measured in frequency units with reference to the field strength at which the TMS signal is observed. By definition, TMS absorbs at 0 cps. Note the signal attributable to the added TMS in the NMR spectrum of toluene (Fig.3).

New words and Expressions

journal['dʒə:nl]n. 杂志,
advent['ædvənt]n. 到来,出现
purchase['pə:tʃəs]n.vt. 购买
instrumentation[ˌinstrumen'teiʃən]n. 测试设备
spin[spin]n.v. 自旋,自转
generate['dʒenəreit]vt. 产生,造成
magnetic moment 磁矩
axis['æksis]n. 轴
bar[ba:]n. 棒条,巴(压力单位)

magnet['mægnit]n. 磁铁,磁石
usable['ju:zəbl]a. 可用的,有用的
external[eks'tə:nl]a. 外部的,外界的
align[ə'lain]vt. 排列,调整
flip[flip]v.n. 跃迁,猝然跳动
irradiate[i'reidieit]vt. 照射
electromagnetic radiation 电磁辐射
proton magnetic resonance 质子磁共振
cycle['saikl]n. 周波
magnetic field 磁场

gauss[gaus]n. 高斯
gyromagnetic ratio[ˌdʒaiərəumæg'netik 'reiʃiəu]磁旋比,回转磁比率
Mcs=megacycles per second 兆赫,兆周/秒
radio-frequency 射频,无线电频率
steadily['stedili]ad. 有规则地,均匀地
match[mætʃ]v. 相等,适合,一致
transmit[trænz'mit]vt. 传送,传导
downfield[ˌdaun'fi:ld]ad.n. 低磁场
specific radiofrequency energy 比射频能
intuition[ˌintju(:)'iʃən]n. 直观,直觉
fundamental[ˌfʌndə'mentl]a. 基本的,根本的
benzylic[ben'zilik]a. 苄基的,苯甲基的
impose[im'pəuz]vt. 强使,加于
deuteriochloroform[djuː'tiəriə'klɔrəfɔːm] n. 重氢化氯仿,氘化氯仿
D_2O = deuterium oxide=heavy water 重水
circumstance['səːkəmstəns]n. 情况,环境
spectrometer[spek'trɔmitə]n. 分光计,波谱仪
scan[skæn]vt.n. 扫描
ordinate['ɔːdinit]n. 纵座标
abscissa[æb'sisə]n. 横座标

tetramethylsilane[tetrə'meθil'silein]=TMS n. 四甲基硅烷
upfield['ʌp'fiːld]n.ad. 高磁场
definition[ˌdefi'niʃən]n. 定义,明晰度

前 缀

ex-[eks-,iks-] 由…出来,向外,超过
 external, export, extract, excess extra, exhaust(排气)
deutero-['djuːtərə-,'djuːtiəriə-]氘化
 deuterochloroform, deuteroxide(重水)

词 组

(the)sams as…… 像……一样,与……相同
in common with… 和……一样
proportional to… 与……成(正)比例
take on… 采取
in order for…to(inf.) 为了使……,以便让……
more commonly 更通俗地(说)
plot A against B 画出A对B的关系曲线
with(in)reference… 关于,根据
by definition 根据定义
divide A by B 用B除A

Notes

1. To obtain some ides of the amount of energy needed to bring about the flip of the proton in a field of 14 092 gauss, a value well within the range of field strengths obtainable by use of the magnet that is part of a proton magnetic resonance (pmr) spectrometer, we can make use of the equation.
该句为复合句,主句是"we can…equation"."to obtain…spectrometer"是目的状语,在此状语中"a value…the magnet"是"14 092 gauss"的同位语,"that is …spectrometer"是"the magnet"的定语从句。译文:"为得到在14 092高斯磁场中产生质子跃迁所需能量的某些概念,而14 092高斯恰好是使用的质子磁共振波谱仪的磁铁所达到的场强范围内的数值,我们可利用以下方程式 $\nu = \dfrac{\gamma H_0}{2\pi}$。"

2. "At that value of the field strength where the energy required to flip the proton matches the energy of the radiation, absorption occurs, and a signal is transmitted to a recorder by electronic means." 系并列复合句,主句是"absorption occurs"和"and a signal is …by electronic means"两个并列

句,"At that…of the radiation"是状语,此状语中"where the energy…of the radiation"是修饰 the field strength 的定语从句。**译文:**"在跃迁质子需要的能量与辐射能量一致时的场强值,产生吸收且通过电子方法把讯号送到记录器。"

3. set:environment

4. "Since it would be very difficult to measure to the required accuracy this large field strength in absolute term, an indirect method is employed."为主从复合句,"Since it would…in absolute terms"为原因状语从句,此从句中真正的主语是 to measure,"this large…terms"是 to measure 的宾语,"to the required accuracy"为状语,主句是"an indirect method is employed"。**译文:**"因为按照要求的精确度在绝对条件下测量这么大的场强将是非常困难的,所以利用了间接法。"

5. Tetramethylsilane(TMS), all of the protons of which absorb at the same magnetic field strength to produce a sharp, single peak which is upfield from where most protons of organic compounds absorb, is added to the solution or liquid compound. 该句系主从复合句,主句是 "Tetramethylsilane(TMS) is added to …compound"."all of the protons…of organic compounds absorb"为修饰 TMS 的非限制性定语从句,其中"which is upfieled…absorb"是修饰 peak 的定语从句,此从句中又包含了修饰 upfield 的定语从句,即"from where…absorb". **译文:**"把四甲基硅烷(TMS)加入此溶液或液体化合物中,四甲基硅烷的所有质子在相同的磁场强度下吸收而产生尖锐的单峰,此单峰位于高磁场,而有机化合物的多数质子从高场开始吸收。"

Exercises

1. Fill in the blanks with proper words or phrases.

    Nuclear magnetic resonance was ____ described and ____ in molecular beams by Isidor Rabi in 1938, by extending the Stern-Gerlach experiment, and in 1944, Rabi was ____ the Nobel Prize in Physics for this work. In 1946, Felix Bloch and Edward Mills Purcell ____ the technique for use on liquids and ____, for which they ____ the Nobel Prize in Physics in 1952. Yevgeny Zavoisky ____ observed nuclear magnetic resonance in 1941, well before Felix Bloch and Edward Mills Purcell, but dismissed the results as not ____.

    A. awarded   B. likely   C. expanded   D. shared

    E. measured   F. first   G. reproducible   H. solids

2. Match the following explanation with the proper word.

    ____ axis, ____ usable, ____ match, ____ fundamental, ____ impose

    A. relating to the most basic and important parts of something

    B. the imaginary line around which a large round object

    C. a situation in which something is suitable for something else, so that the two things work together successfully

    D. compel to behave in a certain way

    E. capable of being put to use

Lesson 29
NUCLEAR MAGNETIC RESONANCE(II)

Chemical Shift

As indicated previously, protons which have the same environment in a given molecule absorb at the same applied field strength, and protons which have different environments absorb at different applied field strengths. For the most part, magnetically equivalent protons are also chemically equivalent protons, and we can estimate how many sets of proton signals the nmr spectrum of a compound will exhibit from a knowledge of its structure. Thus, we would expect methyl alcohol to exhibit two sets of signals, ethyl alcohol three sets, and n-propyl alcohol four sets.

$$CH_3\text{—}O\text{—}H \qquad CH_3\text{—}CH_2\text{—}O\text{—}H \qquad CH_3\text{—}CH_2\text{—}CH_2\text{—}O\text{—}H$$
$$b\phantom{\text{—}O\text{—}}a \qquad \phantom{CH_3\text{—}}c\phantom{CH_2\text{—}}b\phantom{\text{—}O\text{—}}a \qquad \phantom{CH_3\text{—}}d\phantom{CH_2\text{—}}c\phantom{CH_2\text{—}}b\phantom{\text{—}O\text{—}}a$$

It is also important to keep in mind that differences in stereochemistry can cause differences in magnetic (and chemical) properties of protons. Thus, we might expect the NMR spectrum of vinyl bromide to exhibit three sets of proton signals and that of propylene dichloride four sets.

Sometimes we observe fewer sets of signals in the NMR spectrum of a compound than we anticipate on the basis of the numbers of non-equivalent protons present. The reason for this is that the magnetic environments of two or more non-equivalent protons might not be different enough to cause a noticeable separation of their nmr signals[1]. We have actually observed such a situation in the NMR spectrum of toluene (Fig. 3). Although the three hydrogen atoms of the methyl group are equivalent to one another, the five hydrogen atoms bonded to the aromatic ring are not all equivalent; there are two equivalent ortho hydrogen atoms, two equivalent meta hydrogen atoms and one para hydrogen atom. For this molecule it so happens that the chemical shifts for the ortho, meta and para hydrogen atoms are so nearly the same that only one aromatic proton signal is observed. This is not the case for all aromatic compounds, or even for a majority of them. In theory, we might have expected four sets of proton signals for toluene. However, the resolving power of an NMR spectrometer, good as it is in most applications of such spectroscopy, is not sufficiently good to distinguish among the ortho, meta and para hydrogens of toluene, so that we see but two signals in the spectrum of this compound[2].

There are several generalizations which can be made with regard to the position at which various protons absorb in the nmr spectra of organic compounds:

1. An increase in the electronegativity of a substituent causes a downfield shift of the proton signal. For example, the protons of a methyl group which is bonded to a saturated carbon atom absorb at $\delta=0.67 \sim 1.83$ ppm, those[3] of a methyl group bonded to nitrogen absorb at $\delta=2.16 \sim 3.33$ ppm and those of a methyl group bonded to oxygen absorb at $\delta=3.33 \sim 4.00$ ppm.

2. When a molecule is placed in a magnetic field, an induced circulation is set up among its electrons about a given atom. This circulation generates an induced field which is directed opposite to the magnetic field at the atom concerned, and since the field felt by the proton is thereby diminished, the proton is said to be shielded[4]. However, the circulation of π electrons about nearby functional groups generates a field that can either oppose or reinforce the applied field at the proton, depending on its location. If the induced field opposes the applied field, the proton is shielded, but if the induced field reinforces the applied field, the proton is deshielded. A shielded proton requires a higher applied field strength and a deshielded proton requires a lower applied field strength to provide the effective field strength at which absorption occurs. Thus, in benzene and other aromatic compounds where the protons are deshielded, the pmr signal appears at $\delta=6.0 \sim 8.5$ ppm. On the other hand, in alkynes, where the acetylenic protons are shielded, the pmr signal appears at $\delta=2 \sim 3$ ppm.

3. The proton of a methine group ($-\overset{|}{\underset{|}{C}}-H$) generally absorbs downfield in relationship to the protons of a methylene group ($-CH_2-$), which, in turn, absorb downfield relative to methyl protons.

4. The occurrence of hydrogen bonding usually causes a downfield shift of the pmr signal.

Spin-Spin Splitting

Let us consider the NMR spectrum of ethyl iodide. There are two sets of signals, one for the methyl protons and the other for the methylene protons. However, the signal for the methyl protons is split into three equally spaced absorption peaks because of a magnetic interaction with the methylene protons. This is known as spin-spin splitting, and the effect is a "first-order" one[5]. In like manner, the signal for the methylene protons is split into four equally spaced peaks owing to the spin-spin splitting effect of the methyl protons. Furthermore, several of the peaks in both sets of signals are further split as the result of a "second-order" spin-spin splitting.

Coupling Constants

The distance between peaks in a doublet, triplet or multiplet is known as the coupling constant, J, and is measured in cps. The coupling constant is a measure of the effectiveness of spin-spin coupling, and its value does not change with a change in the radio frequency used. When necessary, absorption attributable to chemical shift can be distinguished from that attributable to first order spin-spin splitting by taking the spectrum at a different radio frequency. Peak separations due to first-order spin-spin splitting remain constant, but those due to chemical shifts change.

The numerical values of J vary widely, and, the larger the value of J, the stronger the coupling is said to be. Protons on the same atom and on adjacent atoms usually couple with one another, but protons on nonadjacent atoms usually do not couple strongly unless there are intervening multiple bonds. Spatial features of the system under consideration are also of importance. In short, the magnitude of J depends markedly on the structural relationships between the coupled protons.

New words and Expressions

signal[ˌsignəl]n. 信号
stereochemistry[ˌsteriəˈkemistri]n. 立体化学
vinyl bromide 溴代乙烯
propylene dichloride 1,2-二氯丙烷,二氯化丙烯
chemical shift 化学位移
applied field [əˈplaidˈfi:ld] 外加(磁)场
magnetically[mægˈnetikəli]ad. 磁性地,磁性上
anticipate[ænˈtisipeit]vt. 预期,预想
noticeable[ˈnəutisəbl]a. 显明的,显著的
situation[ˌsitjuˈeiʃən]n. 位置
resolving power[riˈzɔlviŋˈpauə] 分辨率,分辨能力
induce [inˈdju:s]vt. 感应,引诱
circulation[ˌsə:kjuˈleiʃən]n. 运行,循环
direct[diˈrekt]vt. 指向,指引,指示方向
concerned[kənˈsə:nd]a. 有关的,涉及到的
diminish[diˈminiʃ]v. 减少,缩小
shield[ʃi:ld]v. 屏蔽,防御
reinforce[ˌri:inˈfɔ:s]vt. 增强,加强
deshielded [di:ˈʃi:ldid]a. 去屏蔽的
acetylenic[əˌsetiˈlinik]a. 乙炔的
methine[ˈmeθain]n. 甲川
split[split]v. 裂分,裂开
spin-spin splitting 自旋-自旋裂分
space[speis]vt. 隔开,分隔

first-order spin-spin splitting
　一级自旋-自旋裂分
second-order spin-spin splitting
　二级自旋-自旋裂分
equivalent proton[iˈkwivələnt ˈprəutɔn]
　等性质子
coupling[ˈkʌpliŋ]n. 偶合
　~ constant 偶合常数
doublet[ˈdʌblit]n. 二重峰,双峰;电子对
triplet[ˈtriplit]n. 三重峰
multiplet[ˈmʌltiplit]n. 多重峰
cps=cycles per second 周/秒,赫
effectiveness[iˈfektivnis]n. 效力
numerical[nju:ˈmerikəl]a. 数字的
intervene [ˌintəˈvi:n]vi. 插入,介于其间
magnitude[ˈmægnitju:d]n. 数量,长度

前　缀

stereo-[ˈsteriə-] 立体,空间
　stereochemistry, stereoisomer
　stereospecific(立体专一性的),
　stereometer(体积仪)
magni-[ˈmægni-, mægˈni-] 大,巨大
　magnitude, magnify,
　magnification

词　组

for the most part　多半,大部分(来说)
it so happens that…　碰巧,正巧
this is not the case　情况不是这样,
　　情况并非如此
good to (+inf.)　可以
opposite to…　与……相反
said to be…　据说是,可以说是,称做是
relative to…　相对于,以……为基准
in(a)like manner　同样
distinguished from…　与……不同
in short　简言之

Notes

1. The reason for this is that the magnetic environments of two or more nonequivalent protons might not be different enough to cause a noticeable separation of their nmr signals. 该句是复合句,其中带有表语从句"that the…… nmr signals."从句中的 enough to…signals 是修饰 different 的结果状语。**译文:**"其理由是两个或更多非等性质子的磁环境,不会不同到足以导致它们的核磁信号明显地分开。"

2. However, the resolving power of an nmr spectrometer, good as it is in most applications of such spectroscopy, is not sufficiently good to distinguish among the ortho, meta and para hydrogens of toluene, so that we see but two signals in the spectrum of this compound. 该句为复合句,主句是"the resolving…spectrometer +is not … of toluene"."so that…compound"是结果状语从句,此处 but 即 only,"good as……spectroscopy"是让步状语从句,as 引导的让步从句特点是常把表语提前,即表语+as+句子其它成分。**译文:**"然而,尽管核磁共振波谱仪在大部分使用中很好,可是它的分辨率不能更有效地区分甲苯中的邻位,间位和对位氢,结果我们在该化合物的谱图中仅看到两个信号。"

3. those: the protons

4. "This circulation generates an induced field which is directed opposite to the magnetic field at the atom concerned, and since the field felt by the proton is thereby diminished, the proton is said to be shielded." 系并列复合句,第一个分句为"This circulation…atom concerned".其中带有修饰 field 的定语从句"which is … atom concerned"。第二个分句是"and since the…shielded",其中带有一个原因状语从句,即"since the…diminished"。**译文:**此感应环流产生了指向与有关原子磁场相反方向的感应磁场,并且由于质子所感应的场因此而减弱,此质子可认为是被屏蔽了。"

5. one: spin-spin splitting

6. 注意:卤化物的译名易发生错误
　　chloroethane=ethyl chloride 氯乙烷
　　dichloroethane=ethylene (di)chloride 二氯乙烷
　　chloroethylene=vinyl chloride 氯乙烯
　　dichloropropane=propylene (di)chloride 二氯丙烷

7. 核磁谱及其它波谱中的各类峰一般由词尾为"le"的形容词+词尾"t"构成名词,如:

singlet 单峰, doublet 二重峰, triplet 三重峰,
quadruplet 四重峰, quintuplet 五重峰, multiplet 多重峰.

Exercises

1. Fill in the blanks with proper words or phrases.

   The most ___ used nuclei are ¹H and ¹³C, ___ isotopes of many other elements can be studied by high-field NMR spectroscopy ___. A key feature of NMR is that the resonance ___ of a particular simple substance is usually ___ proportional to the strength of the applied magnetic ___. It is this feature that is exploited in imaging techniques; if a sample is ___ in a non-uniform magnetic field then the resonance frequencies of the sample's nuclei ___ on where in the field they are located.

   A. frequency   B. commonly   C. as well   D. field
   E. depend   F. although   G. placed   H. directly

2. Match the following explanation with the proper word.

   ___ noticeable, ___ induce, ___ reinforce, ___ coupling, ___ numerical

   A. to give support to an opinion, idea, or feeling, and make it stronger

   B. expressed or considered in numbers

   C. capable or worthy of being perceived

   D. something that connects two things together

   E. to persuade someone to do something

Lesson 30
A MAP OF PHYSICAL CHEMISTRY

Three Theories of Physical Chemistry

Physical chemistry rests on three main conceptual theories: quantum mechanics, statistical mechanics, and thermodynamics.

Quantum mechanics describes the motions and energetics of microscopic particles. These are particles too small to be seen with a light microscope, such as electrons, nuclei, atoms, and molecules. Given an expression for the potential energy of interaction of any group of these particles, quantum mechanics allows us, at least in principle, to compute properties of the particles such as the probabilities that they have certain positions and momenta, and the quantized energy levels accessible to them.

The modifying phrase "in principle" is crucial to the accuracy of the preceding sentence, because computations on all but the simplest systems bog down within a few steps in fearful mathematical complexity. Some idea of the very limited power of quantum mechanics to make predictions emerges from the list of real systems for which it gives the exact descriptions: the hydrogen atom and the H_2^+ ion.[1] The list ends there. To be fair, excellent approximate descriptions have been given for many small atoms and molecules. Calculations of the structure and properties of a single water molecule can now be made in good agreement with experiment, but the computation for a group of ten water molecules is a frontier problem, and calculations of the properties of sodium and chloride ions surrounded by water molecules are at the limit of—and perhaps beyond—the capacity of current methods and computers.

In contrast to quantum mechanics, thermodynamics deals with the interdependence of macroscopic properties, such as the temperature, pressure, and concentrations in an aqueous solution of hemoglobin and oxygen. This general theory has two facets useful in applications to the life sciences. First, thermodynamics places strict limitations on the interconversion of different forms of energy. Since the interconversion of heat, chemical energy, electrical energy, mechanical energy, and light is basic to life, it is clear that thermodynamics must be at the center of any physical description of living matter. The second important facet of thermodynamic theory is that it allows us to predict the position of any chemical equilibrium, knowing only the properties of the substances involved in the reaction. This is of great importance in understanding metabolism. In addition to these two facets, thermodynamics is such a familiar theory to physical chemists that its concepts are borrowed for use in other less fundamental theories, such as that of chemical kinetics. This is why we take up thermodynamics at the start of the book, and treat it thoroughly enough so that before long you may regard it as an old friend.

Statistical mechanics bridges the microscopic realm of quantum mechanics and the macroscopic realm of thermodynamics. This theory tells us how to use the energies of the individual particles described by quantum mechanics to arrive at the properties of the macroscopic system. An example of an application of statistical mechanics is the computation of the pressure(a macroscopic property) of liquid argon as a function of its temperature and volume. Values of the pressure, in good agreement with experiment, can be calculated if one starts with an accurate potential energy function for the interaction of two argon atoms. Such a potential energy function can be obtained from quantum mechanics. Today, similar calculations on liquid water are a research problem in statistical mechanics. Although it is possible to use statistical mechanics to relate some characteristics of polymer chains to observable properties, to treat a solution of hemoglobin by rigorous methods, and to obtain reasonable answers, is not possible at the present time.

It should be clear from the few examples given for quantum mechanics and statistical mechanics, that biochemists or life scientists must resort to experiments if they are to learn much about the systems that attract their interest. But this does not mean that quantum and statistical mechanics are of little use in their work. On the contrary, they constitute the framework within which nearly every experimental result must be interpreted. It is the interpretive power of these theories, in addition to their predictive power, that makes them so useful to scientists. Our discussions of the principles of quantum mechanics and of statistical mechanics form the foundation for the discussions of experimental methods in many of the other chapters.

Why Mathematics?

The newcomer to the study of physical chemistry finds one feature common to all its divisions: the language of mathematics. Some newcomers, particularly from the life sciences, wonder whether this is necessary, or whether it is employed as a means to keep down the number of newcomers. We believe that there are three reasons why mathematics forms an essential part of physical chemistry. The first is precise definition. When a quantity is defined by an equation, one may dispute the wisdom of the definition, but rarely is there any dispute about its meaning[2]. This is much less true with verbal definitions, as the thriving activity of our law courts and diplomatic missions attests[2]. The second benefit of mathematics is the possibility of mathematical deduction. Once scientific concepts have been defined in mathematical terms, manipulation of symbols can produce results that are not obvious in the original concepts. The third reason for using mathematics is more subtle, and though vital in science, it sometimes seems unimportant to students. From mathematical expressions, scientists can estimate the probable error in a quantity they are calculating or measuring. This estimate can be crucial, because few scientific experiments lead to a clear "yes" or "no" answer; more often the answers are of the form "maybe yes" or "perhaps no." To be sure of the conclusion, one must be able to judge the size of the error in the measurements and to know just how big is the "perhaps."

The authors realize that many newcomers to physical chemistry experience some initial difficulty with the mathematical language, and have endeavored to be careful guides and interpreters. Section 2~10 is designed to help readers whose calculus is rusty or so newly acquired that it needs breaking in.

Model and Reality in Physical Chemistry

Physical chemistry may be defined as "the science that describes the world in terms of atoms, molecules, and energy." As we noted previously, the physical chemist casts his or her descriptions of matter and energy in terms of the three main conceptual theories. Each of these theories is a model of reality. We use these models in thinking about the world because they are consistent with many observations, but no serious scientist would take these models as reality itself; each model is at best a simplified representation of reality.

The relationship of models to reality has concerned thinkers since Plato. One school of thought, the Idealists, of whom the philosophers Plato and Hegel (1770~1831) are the most notable examples, holds that concepts such as space, order, and causation are attributes of the human mind which organize our perceptions of the world. This school believes that models preexist in our minds. A second school is that of the Empiricists, as exemplified by the philosopher David Hume (1711~1776). They believe that our sense organs record and define the order that exists in the world about us. Space, time, forces, causation, and so forth are attributes of this world that exist independently of us.

Most modern scientists, if forced to state their views on the question, might take a position between these extremes. Experience and observation are essential for understanding the world, but the complexity of external reality is so great that to comprehend it we must describe it in terms of simplified models. Great scientific minds produce these models, and the models are extended and refined by our fitting them to observations. This interplay of model and reality, of our minds with the external world, is part of the excitement of physical chemistry, and of science in general.

New words and Expressions

quantum mechanics['kwɔntəm mi'kæniks] n. 量子力学
statistical[stə'tisitikəl]mechanics n. 统计力学
thermodynamics[ˌθəːmədai'næmiks]n. 热力学
energetics[enə'dʒetiks] n.pl. 动能学
microscopic[ˌmaikrə'skɔpik] a. 微观的
microscope['maikrəuskəup]n. 显微镜
probability[prɔbə'biliti]n. 或然率, 概率
momentum[məu'mentəm]n. 动量
accessible[ək'sesəbl]a. 能接近的, 可以进入的
approximate[ə'prɔksimeit]a. 近似, 大概
frontier['frʌntjə]n. 前沿, 前线
macrosopic[ˌmækrə'skɔpik]a. 宏观的, 肉眼可见的
interdependence[ˌintədi'pendəns]n. 相互依赖

hemoglobin[hiːmou'gləubin]n. 血红蛋白
facet['fæsit]n. 方面
limitation[ˌlimi'teiʃən]n. 限制
interconversion[ˌintə(ː)kən'vəːʃən]n. 互变, 相互转化
metabolism[me'tæbəlizəm]n. 新陈代谢
realm[relm]n. 领域
observable[əb'zəːvəbl]a. 观察得出的
rigorous['rigərəs]a. 严格的, 精确的
resort[ri'zɔt]vi. 求助
framework['freimwəːk]n. 构架, 结构
division[di'viʒən]n. 划分, 区分
dispute [dis'pjuːt]vt. 争论
thrive[θraiv]vi. 兴旺, 繁荣
attest[ə'test]vt. 证明, 证实, n. 作证

deduction[di'dʌkʃən]n. 推断,推论
subtle['sʌtl]a. 巧妙的,微妙的,微细的
vital['vaitl]a. 极重要的,不可缺的
estimate['estimeit]vt. 评价,判断
conclusion[kən'klu:ʒən]n. 结论
endeavor[in'devə]n. 努力,尽力
reality[ri(:)'æliti]n. 真实
school of thought n. 学派
causation[kɔ:'zeiʃən]n. 起因
perception[pə'sepʃən]n. 感性认识,观念
observation [ˌɔbzə(:)'veiʃən]n. 观测
comprehend[ˌkɔmpri'hənd]vt. 包括,领悟

词组

be crucial to… 对…是决定性的
bog down 陷入困境
emerge from 从…中显出
in good agreement with… 与…很好地吻合
on the contrary 相反
be consistent with… 与…一致
independently of 独立于…
be essential for… 对…是基本的
take up 处理,从事

Notes

1. "Some idea of …… and the H + 2 ion." 为复合句,主句为"Some idea of…real systems","for which…H_2^+ ion"为修饰"real systems"的定语从句,从句中 it 指 quantum mechanics.**译文**:"从量子力学给出表列真实体系以准确的描述(如像对氢原子和 H_2^+ 离子的描述)这一事实出发,显示出量子力学在进行预言时,其能力十分有限。"

2. "When a quantity……about its meaning." + "This is much……missions attests."**译文**:"只用一个(数学)方程去定义一个(物理)量时,人们可能会议论该定义是否合理,而很少对它的含义提出质疑。用文字定义该物理量是很难准确的,就像我们法院和外交使团作证时所富有的灵活性。"

Exercises

1. Fill in the blacks with proper words or phrases.
    Physical chemistry is the ___ of macroscopic, atomic, subatomic, and particulate ___ in chemical systems in ___ of the principles, practices, and concepts of physics such ___ motion, ___, force, time, thermodynamics, quantum chemistry, statistical mechanics, analytical dynamics and ___ equilibrium.
    Physical chemistry, in ___ to chemical physics, is predominantly a macroscopic or supra-molecular science, as the ___ of the principles on which it was founded relate to the bulk rather than the molecular/atomic structure alone.
    　　A. terms　　　B. study　　　C. majority　　　D. chemical
    　　E. contrast　　F. phenomena　G. as　　　　　H. energy

2. Match the following explanation with the proper word.
    ___ approximate, ___ rigorous, ___ vital, ___ estimate, ___ reality
    A. extremely important and necessary for something to succeed or exist
    B. what actually happens or is true, not what is imagined or thought
    C. not quite exact or correct
    D. careful, thorough, and exact
    E. an approximate calculation of quantity or degree or worth

Lesson 31
THE CHEMICAL THERMODYNAMICS

The branch of science which includes the study of energy transformations is called thermodynamics. Basic to thermodynamics are two "laws" derived from experience, which can be stated as follows:

1. Energy can neither be created nor destroyed—the energy of the universe is constant.
2. The entropy of the universe is always increasing.

These generalizations are statements of the first and second laws of thermodynamics. The laws and the meaning of entropy will be discussed and expanded upon in this lesson. It will be shown that energy transformations on a macroscopic scale—that is, between large aggregates of atoms and / or molecules can be understood in terms of a set of logical principles. Thus thermodynamics provides a model of the behavior of matter in bulk. The power of such a model is that it does not depend on atomic or molecular structure. Furthermore, conclusions about a given process, based on this model, do not require details of how the process is carried out. Applied to chemistry, thermodynamics provides criteria for predicting whether a given reaction can occur. If a reaction is feasible, the extent to which it will proceed under a given set of conditions can be predicted. One great value of thermodynamics is that it is possible to use data from experiments which can be conveniently carried out to arrive at conclusions about experiments which are difficult or even impossible to perform.

Systems. Initial States. Final States

At the outset, it is worthwhile to define some terms which are customarily used in discussing the interactions of energy with matter. That portion of matter which is being investigated is called the system. All other objects in the universe which may interact with the system are called the surroundings. For example, 1 liter of a 1 M aqueous solution of sodium chloride may be under investigation. The solution's container would be considered part of the surroundings. A system is described by identifying its constituents and their quantities, the temperature, the pressure, and perhaps some other relevant conditions, such as the physical states of the substances involved. A complete description of the system defines its state. The initial state of a system is its state before it undergoes a change, and the final state describes the system after a change has occurred. In going from the initial state to the final state, a system may exchange energy with its surrounding, and/or its components may change composition; but there can be to change in the total mass of the system. No matter can be lost to or gained from the surroundings.

The properties of a system which uniquely define the state of the system are called thermodynamic properties or state functions. For example, consider a system consisting of one mole of an ideal gas. The state of the system is specified by giving any two of the properties pressure, volume, and temperature. As was described in the previous chapter, the pressure, P, volume, V, and temperature, T, of n moles of an ideal gas are related by the equation

$$PV=nRT$$

This equation, expressing the relationship of its state functions, is called the equation of state for an ideal gas. The equations of state for real gases, liquids, solids and solutions are more complicated than that for an ideal gas. When a system undergoes a change in state, the change in value of any state function depends only on the initial and final states of the system and not on how the change is accomplished. Indeed, as will be shown, the importance of state functions lies in the fact that for a given system, the change in their values can be obtained by considering only the initial and final states of the system.

When a system undergoes a reaction described by a chemical equation, the description of the reactants defines the initial state of the system, and the corresponding description of the products defines the final state.

The properties of a system which do not depend on the quantity of matter present are called intensive properties. For example, density, pressure, and temperature are intensive properties. Properties which are proportional to the quantity of matter in the system are called extensive properties. The mass of a sample is an extensive property.

Heat Capacity

The quantity of heat required to raise the temperature of 1 gram of any substance 1 degree Celsius is called the specific heat capacity of that substance, or more simply its specific heat. The quantity of heat necessary to raise the temperature of 1 mole of a substance 1 degree Celsius is called its molar heat capacity.

The First Law of Thermodynamics

Any system in a given state will possess a given quantity of energy, called its internal energy, E. Internal energy is an extensive property. By either releasing energy or by absorbing energy, a system may change from an initial state where its internal energy is E_1 to a different (final) state where its internal energy is E_2. The change in internal energy is

$$\triangle E=E_2-E_1$$

It is seldom necessary (and of little practical use) to know the individual values of energies E_1 and E_2. The difference in energy between two states is of prime importance and is usually conveniently determined. The change (increase or decrease) of energy in a given system was determined from its mass, its heat capacity, and the change in temperature. At no time was the total energy of the system considered, only the gain or loss of heat was found.[1]

Energy may be transferred into or out of a system in forms other than heat. For example a chemical system may transfer mechanical energy through expansion of a gaseous product. With proper experimental arrangement, electrical energy may be obtained from a chemical system. It is customary to denote all forms of transferred energy other than heat as work, w. Thus when a system changes from one state to another, the change in its internal energy is given by.

$$\triangle E = q + w$$

where q is the heat absorbed by the system and w is the work done on the system. The relationship $\triangle E = q + w$ is a mathematical statement of the first law of thermodynamics—energy can neither be created nor destroyed.

Enthalpy

In the laboratory, many chemical reactions are carried out in open containers. When a reaction takes place in contact with the atmosphere, the volume of the system will change in such a way that the final pressure of the system equals the atmospheric pressure. Since the atmospheric pressure does not usually change significantly over a period of hours, a reaction occurring in an open vessel may be considered to be a constant pressure process. Any change in the volume of the reaction system would result in work being done against the constant pressure of the atmosphere.

Free Energy and Entropy—Criteria for Spontaneous Change

A major objective of chemists is to understand and control chemical reactions—to know whether or not under a given set of conditions two substances will react when mixed, to predict the extent to which a given reaction will proceed before equilibrium is established, and to determine whether or not a given reaction will be endothermic or exothermic.

The enthalpy change in a chemical reaction is a measure of the difference in energy content of the products and reactants. It is tempting to assume that exothermic reactions will proceed spontaneously upon mixing the reactants and that endothermic reactions do not occur spontaneously. However, there are endothermic reactions which do proceed spontaneously.

The transfer of heat energy from an object at a higher temperature to one at a lower temperature is a familiar example of a spontaneous process. It must be recalled that heat is a unique form of energy in that at constant temperature, heat can not be completely converted to any other form of energy.[2] The heat content, or enthalpy, of any system must be considered in two parts:

1. That which is free to be converted to other forms of energy.

2. That which is necessary to maintain the system at the specified temperature and thus is unavailable for conversion.

It is customary to solve the equation representing the second law of thermodynamics explicitly for $\triangle G$ as follows:

$$\triangle G = \triangle H - T\triangle S \ (\text{constant } T)$$

and to apply the following criteria for spontaneous change and for the equilibrium state:

1. If ΔG is negative, the given process may occur spontaneously.

2. If ΔG is positive, the indicated process cannot occur spontaneously; instead the reverse of the indicated process may occur.

3. If ΔG is zero, neither the indicated process nor the reverse process can occur spontaneously. The system is in a state of equilibrium. The indicated process is said to be a reversible one because a very small change in conditions can make $\triangle G$ either positive or negative.

New words and Expressions

create [kri(ː)'eit] vt. 产生, 创造
destroy [dis'trɔi] vt. 消灭, 破坏
universe ['juːnivəːs] n. 宇宙, 整体
logical ['lɔdʒikəl] a. 逻辑的, 合理的, 能推理的
provide [prə'vaid] vt. 供给, 提供
feasible ['fiːzəbl] a. 易实现的, 可实现的
initial state 始态, 起始状态
final state 终态, 终止状态
worthwhile ['wəː'hwail] a. 值得的
customary ['kʌstəməri] a. 通常的, 习惯的
customarily ['kʌstəmərili] ad. 通常, 习惯上
surroundings [sə'raundiŋs] n. 环境
relevant ['relivənt] a. 有关的, 中肯的
physical state 物态
uniquely [juː'niːkli] ad. 唯一地, 独特地
state function 状态函数
accomplish [ə'kɔmpliʃ] vt. 达到, 完成, 实现
intensive property [in'tensiv'prɔpəti]
   强度性质
extensive property [iks'tensiv 'prɔpəti] 广延性质, 量度性质
heat capacity [hiːt kə'pæsiti] 热容
Celsius ['selsjəs] n. 摄氏分度法, 摄氏度;

a. 摄氏的
specific heat capacity 比热容
specific heat [spi'sifik hiːt] 比热
molar heat capacity   摩尔热容
internal energy   内能
significantly [sig'nifikəntli] ad. 值得注目的
vessel ['vesl] n. 器皿, 容器
constant pressure process 恒压过程
chang in enthalpy=enthalpy change 焓变
endothermic ['endəu'θəːmik] a. 吸热的
energy content   内能
tempt [tempt] vt. 诱, 引诱, 引起
recall [ri'kɔːl] vt. 想起, 忆起, 恢复
heat content   热函, 热含量
explicit [iks'plisit] a. 明确的, 清楚的;
   −ly, ad.

前　缀

endo − [ˌendəu−] 吸, 桥, 内
   endothermal, endocyclic,
   endo-configuration, endocrine
exo− ['eksəu−] 放, 外, 向外
   exothermic, exoelectric,
   exoconfiguration, oxohormone

uni-['juːni-] 单,一
 unicolor, uniform, universal

词 组

expand (up)on 阐述,详谈
in bulk 整个地,成块地
under a given set of conditions
 在给定的条件下
arrive at conclusions
arrive at a conclusion }得出结论

at the outset 起先,当初
of…importance 具有重要意义,
     有……重要性
at no time 从不,决不
in contact with… 和……接触
over a period of… 在……时期内
free to (+ *inf.*) 能够自由……
identical with 和……等同,和……一样
as follows 如下所述

Notes

1. At no time was the total energy of the system considered, only the gain or loss of heat was found. 该句是并列句,第一个分句"At no…considered"中由于 at no time 否定词组的存在,所以部分倒装,该分句主语是"the total energy of the system."谓语是 was considered.
2. "It must be recalled that heat is a unique form of energy in that at constant temperature, heat cannot be completely convented to any other form of energy." 为主从复合句,主句是"It must… unique form of energy","in that at… of energy"为 in that(=because)引导的原因状语从句。

Exercises

1. Fill in the blacks with proper words or phrases.

 Thermodynamics is the ___ of physics that has to do with heat and ___ and their relation to energy and work. The ___ of these quantities is governed by the four ___ of thermodynamics, irrespective of the composition or ___ properties of the material or system in question. The laws of thermodynamics are explained in ___ of microscopic constituents by statistical mechanics. Thermodynamics applies to a wide variety of topics in science and ___, especially physical chemistry, chemical engineering and ___ engineering.

  A. specific   B. branch   C. laws   D. behavior,
  E. terms   F. mechanical   G. engineering   H. temperature

2. Choose a proper prefix to make the follow words match the meanings.
 ___ valent 等价的, ___close 围绕, ___ cyclic 内环的, ___ electric 放电的,
 ___ isomer 立体异构体, ___ magnetic 电磁的, ___ catalysis 生物催化,
 ___ analysis 常量分析

  A. stereo-   B. equi-   C. macro-   D. en-
  E. exo-   F. bio-   G. electro-   H. endo-

Lesson 32
CHEMICAL EQUILIBRIUM AND KINETICS

A major objective of chemists is to understand chemical reactions-to know whether under a given set of conditions two substances will react when mixed, to determine whether a given reaction will be exothermic or endothermic, and to predict the extent to which a given reaction will proceed before equilibrium is established. An equilibrium state, produced as a consequence of two opposing reactions occurring simultaneously, is a state in which there is no net change as long as there is no change in conditions. In this lesson it will be shown how one can predict the equilibrium states of chemical systems from thermodynamic data, and conversely how the experimental measurements on equilibrium states provide useful thermodynamic data. Thermodynamics alone cannot explain the rate at which equilibrium is established, nor does it provide details its of the mechanism by which equilibrium is established. Such explanations can be developed from considerations of the quantum theory of molecular structure and from statistical mechanics.

To appreciate fully the nature of the chemical equilibrium state, it is necessary first to have some acquaintance with the factors which influence reaction rates. The factors which influence the rates of a chemical reaction are temperature, concentrations of reactants (or partial pressures of gaseous reactants), and presence of a catalyst. In general, for a given reaction, the higher the temperature, the faster the reaction will occur. The concentrations of reactants or partial pressure of gaseous reactants will affect the rate of reaction; an increase in concentration or partial pressure increases the rate of most reactions. Substances which accelerate a chemical reaction but which themselves are not used up in the reaction are called catalysts.

Dynamic Equilibrium

In many cases, direct reactions between two substances appear to cease before all of either starting material is exhausted. Moreover, the products of chemical reactions themselves often react to produce the starting materials. For example, nitrogen and hydrogen combine at 500° C in the presence of a catalyst to produce ammonia:

$$N_2 + 3H_2 \rightarrow 2NH_3$$

At the same temperature and in the presence of the same catalyst, pure ammonia decomposes into nitrogen and hydrogen:

$$2NH_3 \rightarrow 3H_2 + N_2$$

For convenience, these two opposing reactions are denoted in one equation by use of a double arrow:

$$N_2 + 3H_2 \rightleftharpoons 2NH_3$$

The reaction proceeding toward the right is called the forward reaction; the other is called the reverse reaction.

If either ammonia or a mixture of nitrogen and hydrogen is subjected to the above conditions, a mixture of all three gases will result. The rate of reaction between the materials which were introduced into the reaction vessel will decrease after the reaction starts; because their concentrations are decreasing. Conversely, after the start of the reaction the material being produced will react faster, since there will be more of it. Thus the faster forward reaction becomes slower, and the slower reverse reaction speeds up. Ultimately the time comes when the rates of the forward and reverse reactions become equal, and there will be no further net change. This situation is called equilibrium. Equilibrium is a dynamic state because both reactions are still proceeding; but since the two opposing reactions are proceeding at equal rates, no net change is observed.

All chemical reactions ultimately proceed toward equilibrium. In a practical sense, however, some reactions go so far in one direction that the reverse reaction cannot be detected, and they are said to go to completion. The principles of chemical equilibrium apply even to these, and it will be seen that for many of them, the extent of reaction can be expressed quantitatively[1].

Equilibrium Constants

Equilibrium is a state of dynamic balance between two opposing processes. For a general reaction at a given temperature,

$$A + B \rightleftharpoons C + D$$

at the point of equilibrium, the following ratio must be a constant:

$$K = \frac{[C][D]}{[A][B]}$$

The constant, K, is called the equilibrium constant of the reaction. It has a specific value at a given temperature. If the concentration of any of the components in the system at equilibrium is changed, the concentrations of the other components will change in such a manner that the defined ratio remains equal to K as long as the temperature does not change[2]. The equilibrium constant expression quantitatively defines the equilibrium state.

More generally, for the reversible reaction

$$aA + bB \rightleftharpoons cC + dD$$

the equilibrium constant expression is written as follows:

$$K = \frac{[C]^c [D]^d}{[A]^a [B]^b}$$

By convention, the concentration terms of the reaction products are always placed in the numerator of the equilibrium constant expression. It should be noted that the exponents of the concentration terms in the equilibrium constant expression are the coefficients of the respective species in the balanced chemical equation.

Chemical Kinetics

When a system is in the equilibrium state, the rate of the forward reaction is identical to the rate of the reverse reaction. It is important to know just how fast a reactant is being used up in a process, or the speed with which a product is being formed. It is also important to have detailed information about rates of reactants in order to test theories and mechanisms for various kinds of chemical processes.

Experiments show that a number of reaction variables affect reaction rates:

Temperature. The rates of chemical reations are temperature-dependent. Therefore, it is common practice when studying the rate in the laboratory to carry out reactions at constant temperature(isothermally), thus eliminating one variable[3].

Pressure and Volume. Volume and pressure are important in a kinetic consideration of gas phase reactions. Usually, volume is fixed by running the reaction in a container of fixed dimensions. For solid and liquid state reactions, pressure is usually atmospheric, and the volume of the reacting system is relatively unimportant because there is little change in volume.

Concentration. At any particular temperature, the rates of most chemical reactions are functions of the concentrations of one or more of the components of the system. In practice, it is usually the concentration of the reactants that are used in determining the over-all rates of reaction[4].

Catalyst. Any substance that affects the rate of a chemical reaction but cannot be identified as a product or reactant is said to be a catalyst. A catalyst may accelerate or decelerate the rate, but we usually refer to decelerating catalysts as inhibitors.

The order of a chemical reaction is given by the number of atomic or molecular species whose concentrations directly determine the reaction rate.

The rate of hydrolysis of ethyl acetate in water is directly proportional to the ethyl acetate concentration. Therefore, the reaction is said to be first order.

New words and Expressions

kinetics[kaiˈnetiks] n. 动力学的
equilibrium state 平衡(状)态
opposing[əˈpəuziŋ] n. 相反,相对,反抗
datum[ˈdeitəm](复 data[ˈdeitə])n. 资料, 论据
molecular structure 分子结构

quantum[ˈkwɔntəm] n. 量子
quantum theory 量子论
acquaintance[əˈkweintəns] n. 熟悉,认识 (with)
partial pressure[ˈpɑːʃəl ˈpreʃə] 分压
affect [əˈfekt] vt. 影响

accelerate [æk'seləreit] vt. 加速
dynamic [dai'næmik] a. 动力学的
cease ['si:s] v. 停止,终止
forward reaction 正反应
subject [səb'dʒekt] vt. 遭受,蒙受(to)
specific value 比值
numerator ['nju:məreitə] n. (分数中的)分子
exponent [eks'pəunənt] n. 指数
test [test] vt. 检验,验证,试验
isothermal [ˌaisəu'θə:məl] a. 等温线的
decelerate [di:'seləreit] v. 减速,减慢
order ['ɔ:də] n. 级数
rate law 速度定律

前 缀

kin(e)- [kain-, kin-] 运动
kinetics, kinemometer (流速计),
　kinetin (激动素)

equi- [i'kwi-, 'i:kwi-] 相等,平等
　equilibrium, equimolar, equivalent

词 组

as a consequence of… 由于…(结果)
as long as… 只要
not…,…nor 不……,也不
have acquaintance with… 熟悉,认识
use up 消耗掉,用完
in a practical sense 实际上
in such a manner that… 以这样的方式
　以致……
by convention 按照惯例
it is common practice (+inf.) 通常的做
　法是……
over-all 总的,全部的
an educated guess 有根据的推测

Notes

1. The principles of chemical equilibrium apply even to these, and it will be seen that for many of them, the extent of reaction can be expressed quantitatively. 该句为 and 连接的并列复合句,第一个分句中的 these 指上句中提到的反应,即"some reactions go so… to go to completion"。**译文**:"化学平衡的原理甚至也适用于被认为是进行完全的反应,显然许多这类反应其反应程度可以定量地表达。"

2. "If the concentration of any of the components in the system at equilibrium is changed, the concentrations of the other components will change in such a manner that the defined ratio remains equal to K as long as the temperature does not change." 系主从复合句,if 引导的是条件状语从句,主句是"the concentrations of the other…句末",主句中 in such a manner 为方式状语,"that the defined…not change"为修饰 manner 的定语从句,此定语从句中又带有一个条件状语从句,即"as long as…not change"。**译文**:"如果平衡时体系中任一组分的浓度改变,那么只要温度不变其它组分的浓度将以保持相等 K 的固定比例的方式变化。"

3. "Therefore, it is common practice when studying the rate in the laboratory to carry out reactions at constant temperature (isothermally), thus eliminating one variable." 系主从复合句,主句是"it is common practice to carry out…句末",其中"thus eliminating one variable"是结果状语,一般 thus+现在分词的短语在句中常表示结果。"when studying the rate in the laboratory"为时间状语。

4. In practice, it is usually the concentration of the reactants that are used in determining the overall rates of reaction. 此句是强调句型。句中强调了主语"the concentration of the reactions"。译文:"实际上,一般正是反应物的浓度,用以测定反应的总速率。"

Exercises

1. Fill in the blanks with proper words or phrases.

   The equilibrium constant of a chemical reaction is the ___ of its reaction quotient at ___ equilibrium, a state approached by a dynamic chemical system ___ sufficient time and has no measurable tendency towards ___ change. For a given set of reaction ___, the equilibrium constant is independent of the ___ analytical concentrations of the reactant and ___ species in the mixture. Thus, given the initial composition of a system, known equilibrium constant values can be used to ___ the composition of the system at equilibrium.

   A. chemical   B. further   C. initial   D. determine
   E. product   F. conditions   G. value   H. after

2. Match the following explanation with the proper word.

   ___ opposing, ___ acquaintance, ___ affect, ___ cease, ___ decelerate

   A. to stop doing something or stop happening

   B. somebody or something familiar with

   C. completely different from others

   D. reduce the speed of something

   E. to do something that produces an effect or change in something or in someone's situation

Lesson 33
THE RATES OF CHEMICAL REACTIONS

Introduction

In this chapter we look into how chemical reactions occur. The principal aspect we examine is the rate of a reaction, and we shall see how it depends on the temperature and the concentrations of the species that are present.

There are two main reasons for studying the rates of reactions. The first is the practical importance of being able to predict how quickly a reaction mixture will move its equilibrium state: the rate might depend on a number of factors under our control, such as the temperature, the pressure, and the presence of a catalyst, and, depending on our aims, we may be able to make the reaction proceed at an optimum rate. For instance, in an industrial process it might be economical for the reactions to proceed very rapidly; but not so rapidly as to produce an explosion. By contrast, in a biological process it may be appropriate for a reaction to proceed only slowly, and to be switched on and off at the demand of some activity.

The second reason for studying reaction rates (which, as we shall see, is closely bound up with the first) is that the study of rates can reveal the mechanisms of reactions. The term 'mechanism' has two connotations in this context. The first is the analysis of a chemical reaction into a sequence of elementary steps. For example, we might discover that the reaction of hydrogen and bromine proceeds by a sequence of steps involving the fission of Br_2 into two bromine atoms, the attack of one of these atoms on H_2, and so on. The statement of all the elementary steps constitutes the statement of the mechanism of the reaction. The other meaning of mechanism relates to the individual steps themselves, and concerns their detailed nature. In this sense 'mechanism' concerns what happens as a bromine atom approaches and attacks a rotating, vibrating, hydrogen molecule.

The first type of analysis of mechanism is the central feature of classical chemical kinetics, and we concentrate on it in this chapter. The second type of analysis, called chemical dynamics, had to await the technological advances that made available molecular beams for the study of individual molecular collisions, and is discussed in the next chapter. The dividing line between chemical kinetics and chemical dynamics is not clear cut: crude models of individual reaction steps were built on the basis of kinetic analyses, and we see something of this in the present chapter.

Empirical chemical kinetics

The basic data of chemical kinetics are the concentrations of the reactants and products as functions of time. The method selected for monitoring the concentrations depends on the nature of the species involved in the reaction, and on its rapidity.

Many reactions go to completion (that is, attain thermodynamic equilibrium) over a period of minutes of hours, and may be monitored by classical techniques. One of the following methods is often chosen.

(1) *Pressure changes.* A reaction in the gas phase might result in a change of pressure, and so its progress may be monitored by recording the pressure as a function of time. An instance of this is the decomposition of nitrogen(V)oxide, N_2O_5, according to

$$2N_2O_5(g) \rightarrow 4NO_2(g) + O_2(g)$$

For every mole of N_2O_5 destroyed, 5/2 moles of gaseous products are formed, and so the pressure of the system increases during the course of the reaction. This method is inappropriate for reactions that leave the overall pressure unchanged, and for reactions in solution.

(2) *Spectroscopy.* A technique that is available even when no pressure change occurs is the spectroscopic analysis of the mixture. For instance, the reaction

$$H_2(g) + Br_2(g) \rightarrow 2HBr(g)$$

can be followed by monitoring the intensity of absorption of visible light by the bromine.

(3) *Polarimetry.* When the optical activity of a mixture changes in the course of reaction, it can be monitored by measuring the angle of optical rotation. This is a historically important method because its application to the hydrolysis of sucrose was the first significant study of the rate of a reaction (by Wilhelmy in 1850).

(4) *Electrochemical methods.* When a reaction changes the number or nature of ions present in a solution, its course may be followed by monitoring the conductivity of the solution. One very important class of reactions consists of those occurring at electrodes, and we examine them in Chapter 30.

(5) *Miscellaneous methods.* Other methods of determining composition include mass spectrometry and chromatography. In order to employ these techniques, a small amount of the reaction mixture is bled from the reacting system at a series of times after the initiation of the reaction, and then analysed.

There are three main ways of applying these analytical techniques.

(1) *Real time analysis.* In this method the composition of the system is analysed while the reaction is in progress.

(2) *Quenching.* In this method the reaction is frozen after it has been allowed to proceed for a certain time, and then the composition is analysed by any suitable technique. The quenching can normally be achieved by lowering the temperature suddenly, but this is suitable only for reactions that are slow enough for there to be little reaction during the time it takes to cool the mixture.

(3) *Flow method.* In this method, solutions of the reagents are mixed as they flow together into a chamber, Fig 4. The reaction continues as the thoroughly mixed solutions flow through the outlet tube, and observation of the composition at different positions along the tube (for example,

by spectroscopy) is equivalent to observing the reaction mixture at different times after mixing. Reactions that are complete within a few milliseconds can be observed with this technique, but its principal disadvantage is that large volumes of solutions are necessary. The method has been improved, and a modification, the stopped-flow method, is in wide use.

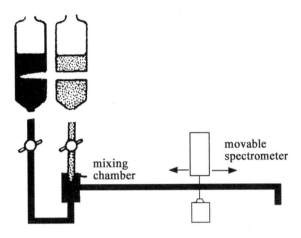

Figure 4. Apparatus used in the flow technique

The rates of chemical reactions have been found to depend very strongly on the temperature, and many follow the Arrhenius rate law, that the rate is proportional to $\exp(-E_a/RT)$, where E_a is the activation energy of the reaction. The implication of this observation for experimental studies is that the temperature of the reaction mixture must be held as constant as possible throughout the course of the reaction, otherwise the observed reaction rate will be a meaningless average of rates at different temperatures. This requirement puts severe demands on the design of the experiment. Gas phase reactions, for instance, are often carried out in a vessel held in contact with a substantial block of metal, and liquid phase reactions, including flow reactions, must be carried out in an efficient thermostat.

The general result of these experiments is that the rates of chemical reactions depend on the composition of the reaction mixture, and most depend exponentially on the temperature. The next few sections look at these observations in more detail.

The rates of reactions

Suppose the reaction of interest is of the form

$$A+B \rightarrow \text{products}, P$$

and that the concentrations of the species $A, B,$ and P are [A], [B], and [P] respectively. The rate of the reaction can be expressed as the rate of change of the concentration of any of the species. Thus the rate of formation of product is $d[P]/dt$ and the rate of destruction of A is $d[A]/dt$. In the present case

$$d[P]/dt = -d[A]/dt = -d[B]/dt$$

because a B molecule must be destroyed for every A molecule destroyed, and in the process a P molecule is formed. Any of these derivatives can serve as the definition of the rate of reaction, the only care needed is to keep the signs correct.

New words and Expressions

explosion[iks'pləuʒən] n. 爆炸
connotation[ˌkɔnəu'teiʃən] n. 涵义
dynamics[dai'næmiks] n. 动力学
beam[bi:m] n. 束
decomposition[ˌdi:kɔmpə'ziʃən] n. 分解
polarimetry[ˌpəulə'rimitri] n. 旋光测定法
sucrose['sju:krəus] n. 蔗糖
conductivity[ˌkɔndʌk'tiviti] n. 导电率
electrode [i'lektrəud] n. 电极
miscellaneous[misi'leinjəs] a. 各种各样的
bleed[bli:d] vt. 放出(液、浆等)
tube[tju:b] n. 管
millisecond['miliˌsekənd] n. 毫秒
thermostat['θə:məstæt] n. 恒温器
exponential[ˌekspəu'nenʃəl] a. 指数的

词 组

look into… 窥视,调查
by contrast(with) （和…）相比之下
be appropriate for… 适于
be bound up with … 与…有密切联系
be suitable for … 适于作…
be proportional to… 与…成正比
raise to a (third) power 自乘到(三次)幂
analysis of … into… 将…分解成
be followed by … 用…方法跟踪
be equivalent to (prep.)… 与…等价的
put demands on … 对…提出要求
at (three)times the rate of … 以(三)倍于
  …的速度

Exercises

1. Fill in the blacks with proper words or phrases.
   Chemical kinetics, also known as reaction ____, is the study of ____ of chemical processes. Chemical kinetics ____ investigations of how different experimental conditions can ____ the speed of a chemical reaction and yield ____ about the reaction's mechanism and ____ states, as well as the construction of mathematical ____ that can ____ the characteristics of a chemical reaction.
   A. information   B. kinetics   C. describe   D. transition
   E. includes   F. models   G. rates   H. influence

2. Match the following explanation with the proper word.
   ____ dynamics, ____ beam, ____ decomposition, ____ electrode, ____ bleed
   A. a small piece of metal or a wire that is used to send electricity through a system or through a person's body
   B. the process of decay that takes place when a living thing changes chemically
   C. drain of liquid or steam
   D. a line of light shining from the sun, a lamp etc.
   E. the branch of mechanics concerned with the forces that change or produce the motions of bodies

Lesson 34
NATURE OF THE COLLOIDAL STATE

The earliest electric light bulbs, such as those introduced by Thomas Edison, used as a light source a hot carbon filament inside an evacuated glass bulb. These were subsequently superseded by the metal tungsten as filament material and the bulbs were filled with a gas (at first nitrogen but later mostly argon with a little nitrogen), to reduce the amount of metal evaporation at the high temperatures involved. To produce an efficient bulb it was necessary to understand the details of what was happening at the metal surface in contact with the gas. The most prolific and effective researcher in this area, in the first decades of this century, was Irving Langmuir, who made an extensive study of the adsorption of gases onto metal surfaces and how the amount adsorbed was affected by the nature of the metal and the gas pressure, composition, and temperature.

From that early work a huge area of study has developed, spurred on by the scientific and technological problems involved in the electronics industry, first through the development of the electronic valve or vacuum tube and subsequently the transistor. There is now a wide variety of tools available with which to examine the details of the surface structure and the way it is changed by reaction with an adsorbing molecule[1]. To do justice to that area would, however, take us too far afield. Some discussion of gas adsorption on solids is given in the standard physical chemistry texts (Atkins 1982); we will concentrate here on a few aspects of relevance to colloid chemistry.

One important application of the study of adsorption is to the determination of the total surface area of a finely divided solid. A catalyst, for example, is often produced in the form of a fine powder with the active material on its surface. Its effectiveness is then determined largely by how much area there is to adsorb the reactants and generate the products. The area of a solid can be measured if we can determine the number of gas molecules of known size required to form a single layer on the surface.

Before it became possible to study the microscopic details of the adsorption process and the molecular interactions which were going on at a metal surface, the main insights were provided by examining the macroscopic behaviour, in particular how fast molecules of gas became adsorbed and how much was adsorbed as a function of gas pressure and temperature[2]. Note that we use the term adsorption to indicate that the molecule is stuck to the surface, as distinct from absorption where molecules would go into the interior of a substance.

It is also useful to distinguish two types of adsorption: physical and chemical, referred to as physisorption and chemisorption respectively. In the first case, the gas molecule is bound to the solid surface by physical forces, chiefly the van der Waals force. In the second case a chemical bond is formed between the gas (called the adsorbate) and the underlying solid (called the adsorbent or substrate). A physisorbed gas molecule may have its bond structure slightly distorted but it retains its identity. A chemisorbed molecule normally breaks down to some extent and the fragments then interact with the surface atoms of the solid. The energy involved in the latter case (usually of the order of several hundred kilojoules per mole) is significantly higher than for physisorption (where the energy is more like 30-60 kJ mol^{-1}). Physical adsorption increases as the temperature is lowered, whereas chemical adsorption normally decreases at low temperature because the activation energy required to break chemical bonds is no longer available.

Gases which are strongly bonded and relatively inert, like nitrogen, are more likely to be physisorbed on a metal surface. Oxygen, on the other hand, is more usually chemisorbed to form an oxide layer on a metal surface. Sometimes a gas has a special relation with a particular metal, as is for instance the case with hydrogen gas on palladium. There the interaction is so strong that the gas dissociates to individual atoms which then penetrate into the metal crystal.

The mechanics of adsorption

When a clean metal surface is brought into contact with a gas at a reasonable pressure, say of the order of 0.01 atm, the gas molecules will bombard the surface in large numbers. The number of collisions per unit area, at normal temperatures, can be calculated from the kinetic theory (Atkins 1982) and is about $3 \times 10^{23} \mathrm{cm}^{-2}\mathrm{s}^{-1}$ at this pressure. Since each square centimetre of surface contains about 10^{15} atoms, each of those atoms will be struck by a gas molecule about 10^8 times per second. Some of those molecules simply bounce back off the surface, but at any particular pressure a certain fraction will remain. The extent of adsorption is measured by the coverage θ, which is given by

$$\theta = \frac{\text{Number of surface sites occupied}}{\text{Total number of surface sites}}$$

The rate at which the first adsorbed layer is built up decreases as the layer nears completion. There is also a tendency for the most highly active sites on the surface to be filled up first, and this too makes the rate fall off as the surface fills up. Nevertheless, we would expect a solid to come to equilibrium with the surrounding gas in less than a second, unless the gas is able to penetrate into the interior of the solid.

The experimental data on gas adsorption is collected by placing a solid surface (often in the form of a powder) in an evacuated chamber and heating it for some time, whilst continuing to pump out any evolved gases. When the surface is considered to be 'clean', the sample is taken to the (usually lower) temperature of the experiment. Known amounts of the adsorbing gas are then allowed into the sample chamber and the residual gas pressure is measured after equilibrium has been established. This allows one to estimate the amount adsorbed as a function of the equilibrium pressure.

We will be concerned not only with gas adsorption but also with adsorption from solution, where the situation is very different although the result is often much the same. When considering the adsorption of a solute from solution onto a solid surface, we must recognize that the surface is already covered with the solvent molecules which may interact more or less strongly with it. For systems involving water that interaction is usually very strong. An adsorbate molecule therefore has to compete with a water molecule for a place at the surface and will only be adsorbed if the free energy of the system is lowered by that adsorption. The rate at which adsorption equilibrium is attained can be very much slower in this case, partly because it depends on diffusion of the solute to the surface and partly because a competitive reaction or exchange process is usually involved.

New words and Expressions

colloidal[kəˈlɔidl] *a.* 胶质的,胶态的
filament[ˈfiləmənt] *n.* 灯丝
evacuate[iˈvækjueit] *vt.* 抽真空
bulb[bʌlb] *n.* 球状物,灯泡,烧瓶
supersede[ˌsju:pəˈsi:d] *vt.* 代替
prolific[prəˈlifik] *a.* 富有成果的,丰富的
spur[spə:] *vi.* 驱赶(on)
transistor[trænˈzistə] *n.* 晶体管
afield[əˈfi:ld] *ad.* 向野外;离题,离谱,离开
layer[ˈleiə] *n.* 层
insight[ˈinsait] *n.* 见识
behaviour[biˈheivjə] *n.* 行为,特性
physisorption[fiziˈsɔ:pʃən] *n.* 物理吸附
chemisorption[kemiˈsɔ:pʃən] *n.* 化学吸附
the van der Waals force 范德华力
adsorbate[ædˈsɔ:beit] *n.* 被吸附物,吸附质
distorte[disˈtɔ:t] *vt.* 使扭曲
penetrate[ˈpenitreit] *vt.* 渗入

bombard[bɔmˈbɑ:d] *vt.* 轰击
strike[straik] *vt.* 碰,撞,攻击
bounce[bauns] *vi.* 反弹(off)
evolve[iˈvɔlv] *vt.* 离析,放出
attain[əˈtein] *vt.* 获得,达到
site[sait] *n.* 点,位置
valve[vælv] *n.* 真空管;阀

词组

be distinct from … 不同于…
be referred to as … 被称之为…
do justice to … 公平对待…
bring into contact with … 使其与…接触
bounce back off … 从…上反弹回来
fall off 降低,衰减
penetrate into … 渗入…之中
compete with … for … 与…为…而竞争

Notes

1. "There is now …… an adsorbing molecule."为一复合句,主句是"There is …and the way",主句中"with which… the way"为一不定式短语,作主句主语tools的定语,而此短语中有两个宾

语,即the detals 和the way,而"it is …molecule"是修饰the way的定语从句。**译文:**"目前有许多工具可用来考察(固体)表面结构的细节和表面结构通过与吸附分子的反应而变化的方式。"

2."Before it became…… pressure and temperature."系复合句,主句为"the main…句末",主句中"by examining…句末"系动名词短语,其中动名词examining带有一个直接宾语和两个宾语从句,它们均是主句的状语。"Before it … at a metal surface"为时间状语从句,此从句又是一个复合句,其中"which were…surface"是修饰adsorption process 和 interactions的定语从句。

译文:"在研究在金属表面上进行的吸附过程和分子的相互作用的微观细节成为可能之前,这些主要的见解就已通过考察宏观行为,特别是气体分子如何迅速地被吸附以及与气体压力和温度相关的被吸附气体分子的多少而获得了。"

Exercises

1. Fill in the blanks with proper words or phrases.

    In ____, a colloid is a ____ in which one substance of microscopically dispersed ____ particles is suspended throughout another substance. Sometimes the ____ substance alone is called the colloid; the term colloidal suspension refers to the ____ mixture. To qualify ____ a colloid, the mixture must be ____ that does not settle or would take a very long time to ____ appreciably.

      A. dispersed   B. chemistry   C. overall   D. mixture,

      E. insoluble   F. settle   G. one   H. as

2. Match the following explanation with the proper word.

      ____ evacuate, ____ supersede, ____ layer, ____ absorbate, ____ attain

      A. take the place or move into the position

      B. to succeed in achieving something after trying for a long time

      C. create a vacuum in a bulb, flask, reaction vessel

      D. a material that has been or is capable of being absorbed

      E. an amount or piece of a material or substance that covers a surface or that is between two other things

Lesson 35
ELECTROCHEMICAL CELLS

Early studies of the chemical behavior of electricity led to the discovery and isolation of a number of new elements: sodium, potassium, barium, strontium, and magnesium. It was observed that a relationship existed between the amount of chemical change taking place and the amount of electricity used in the process. In 1833, Michael Faraday published two fundamental laws of electrolysis based on electrochemical experiments:

(1) The chemical action of a current of electricity is in direct proportion to the absolute quantity of electricity which passes.

(2) The atoms of substances which are equivalent to each other in their ordinary chemical actions have equal quantities of electricity naturally associated with them.

Faraday studied the electrolysis of the blue solution that results from dissolving copper sulfate ($CuSO_4$) in water. Copper metal was deposited on the negative electrode (cathode) while oxygen gas bubbled off at the positive electrode (anode). The mass of copper (the increased weight of the cathode) turned out to be dependent on: (1) the magnitude of the electric current (measured in amperes); (2) the length of time the current was allowed to run (measured in seconds). The mass of copper deposited was found to be directly proportional to current and time; doubling current or time doubled mass; doubling both current and time quadrupled mass:

$$mass \propto (current) \cdot (time)$$

But current strength (amperes) is defined as the quantity of charge (usually measured in coulombs) transferred per unit of time (seconds):

$$current = charge/time$$

Thus, the product of current and time gives the total charge that has traveled through during the entire electrolysis:

$$mass \propto (charge/time) \cdot time = mass \propto charge$$

and we have arrived at the same place Faraday was led to, his first law of electrolysis[1].

The mass of an element freed from its chemical combination at an electrode (copper from copper sulfate) during electrolysis is proportional to the quantity of charge that traveled through the electrode.

Faraday then proceeded to measure the masses of different elements liberated from chemical combination by identical amounts of electric charge. What he found was that the amount deposited depended on the element's atomic weight and its combining capacity or valence. We have now arrived at Faraday's second law of electrolysis:

If you divide the atomic weight by the combining capacity, you will have calculated the amount of an element freed of its chemical combination by 96,500 coulombs of electric charge. Examination of the data in Table 17 provides illustrations of Faraday's second law. In each example, the mass liberated is equal to the atomic mass divided by the element's combining capacity.

A Galvanic cell is used to produce electrical energy from a chemical reaction. The chemical change that takes place is a change of oxidation state or a change of concentration. The driving force or potential of the chemical reaction is measured in electrical terms, the volt being the unit of measurement. Two metals immersed into a solution of acid, base, or salt will produce a voltage between them. The greater the difference of oxidation potential between the two metals, the greater the magnitude of the measured voltage.

1. Electrolysis of Water

Set up the electrolysis apparatus. To make it work, add several drops of concentrated sulfuric acid (H_2SO_4) to the water in the apparatus.

Allow a current to pass through the apparatus until there is a noticeable amount of gas in each burette. Be sure to measure the time that current is flowing, as well as the amount of current.

Measure and record the volumes of hydrogen and oxygen produced in the apparatus.

2. Electroplating

The quantity of electricity measured in coulombs that is equivalent to the total charge on one mole of electrons is called the faraday[2].

$$Q = I \times t \tag{1}$$

Table 17. Amounts of elements liberated from their chemical combinations during electrolysis by passage of 96500 coulombs of charge.

Element	Atomic weight	Combining capacity	Amount of element liberated(g)
Hydrogen	1.008	1	1.008
Oxygen	16.00	2	8.00
Aluminum	26.98	3	8.99
Copper	63.54	2	31.77
Sodium	22.99	1	22.99
Potassium	39.10	1	39.10
Barium	137.34	2	68.67
Strontium	87.62	2	43.81
Magnesium	24.30	2	12.15

where Q is coulombs, I is current in amperes, and t is time in seconds; also,

$$Q = m/Z \tag{2}$$

where m is the mass (in grams) of material deposited at an electrode and Z is the number of electrons transferred in the process. Therefore,

$$m = Z \times I \times t \tag{3}$$

In order to determine the number of electrons that are transferred in neutralizing an ion, we must unite chemical half-reactions.

Dry two copper plates thoroughly and weigh them to within 0.001g on an analytical balance. Place them in a 1.0 M solution of copper sulfate.

Determine the electrochemical equivalent for copper. This is done by measuring the amount of current that flowed and the time required to deposit a quantity of copper onto one plate from the copper ions in solution (or to dissolve a quantity of copper from one plate into solution).

Record the current every 30 sec for at least 30 min.

3. Galvanic Cells

Just as in electrolysis or electroplating, reactions occur at both electrodes. Usually, two half-reactions are involved:

at the anode:
$$Zn^0 \rightarrow Zn^{2+} + 2e^- (oxidation)$$

at the cathode:
$$Cu^{2+} + 2e^- \rightarrow Cu^0 (reduction)$$

A voltaic or Galvanic cell do not connect the voltmeter immediately. A salt bridge, usually of KC1 solution, is added to the system to allow ions to migrate from one area to another as solutions are depleted of ions during the process[3].

Clean each metal strip thoroughly with acid or sandpaper and place a zinc strip in one beaker and a copper strip in the other beaker.

Add about 50 ml of zinc solution to the beaker containing the zinc strip and an equal amount of copper solution to the other beaker.

Connect the voltmeter long enough[4] to take a reading of the voltage and to determine the direction of electron flow. If the voltmeter needle registers below zero, the electrodes need to be reversed.

You should have enough data to determine the electromotive force (EMF) of your cell, which was a zinc-copper cell. The symbol for this type of cell, in which zinc was oxidized and copper was reduced, is $Zn/Zn^{2+}//Cu^{2+}/Cu$.

By changing one electrode and solution, keeping the Cu^{2+}/Cu couple, you can prepare an EMF series for metals.

New words and Expressions

electrochemical [iˈlektrəuˈkemikəl] a. 电化学的
electrochemical cell 化学电池
electricity [ilekˈtrisiti] n. 电,电流,电学

electrolysis [ilekˈtrɔlisis] n. 电解
current [ˈkʌrənt] n. (水,气)流,电流
ampere [ˈæmpɛə] n. 安培
double [ˈdʌbl] n.v.a.ad. 加倍,二倍,双重

quadruple['kwɔdrupl] v.n.a.ad. 四倍,四重
current strength   电流强度
coulomb['ku:lɔm] n. 库伦
entire[in'taiə] a. 全部的,整个的
galvanic cell[gæl'vænik sel] 原电池,自发电池
potential[pə'tenʃəl] n. 电位,电势
volt[vəult,vɔlt] n. 伏特
immerse[i'mə:s] vt. 浸入
voltage['vəultidʒ] n. 电压
oxidation potential 氧化电位
faraday['færədi] n. 法拉第
variable power['vɛəriəbl'pauə] 可变动力,可变电源
ammeter['æmitə] n. 安培计,电表
corrode[kə'rəud] v. 腐蚀
salt bridge   盐桥
voltaic[vɔl'teiik] a. 伏特的,伏打的
deplete[di'pli:t] vt. 用尽,枯竭
connect[kə'nekt] vt. 接通,连接

voltameter[vɔl'tæmitə] n. 伏特计,电压表
strip[strip] n. 长条
sandpaper['sænd,peipə] n. 砂纸
register['redʒistə] vt. 指示
electromotive force=EMF   电动势
EMF series   电动序

前缀和后缀

volt(a)-['vɔlt(i,ə)-,vɔl't(æ,ei)-] 伏特
  voltage,voltaic,voltameter
-ode[-əud] 表示"路","通路","极","管"
  electrode,anode,cathode,diode triode

词 组

in direct proportion to… 与……成正比
equivalent to…   相当于,等于
associated with … 伴随……,与……有关
turn out to be (+a.) 结果弄清楚是,原来是
travel through 通过,贯穿
sure to (+inf.) 切记,务必,一定

Notes

1. "Thus, the product of current and time gives the total charge that has traveled through during the entire electrolysis:mass∝(charge/time)·time=mass∝charge
   and we have arrived at the same place Faraday was led to, his first law of electrolysis." 为并列复合句,第一个分句是"Thus,the product…entire electrolysis",其中"that has …electrolysis"为修饰 the total charge 的定语从句。第二个分句是"we have … law of electrolysis",其中"Faraday was led to"为 the same place 的定语从句,"this first law of electrolysis"为 the same place 的同位语。**译文**:"于是,电流和时间的乘积给出完全电解时所通过的总电量:质量∝(电量/时间)·时间 = 质量∝电量,因此我们达到法拉第导出的相同情况,即法拉第电解第一定律。"

2. "The quantity of electricity measured in coulombs that is equivalent to the total charge on one mole of electrons is called the faraday." 为主从复合句"that is …of electrons"为修饰 coulombs 的定语从句,主句主语是 The quantity of electricity,谓语是 is called。**译文**:"以相当于每摩尔电子总电荷 – 库伦所测得的电量称做法拉第。"

3. "A salt bridge, usually of KCl solution   is added to the system to allow ions to migrate from one area to another as solutions are depleted of ions during the process." 系主从复合句,主句是"A

salt …one area to another", "as solutions…the process"为as引导的时间状语从句。**译文**:"通常由氯化钾溶液组成的盐桥加入系统中,让离子从一个区域向另一个区域迁移,直到(作用时)耗尽溶液中的离子。"

4.long enough:enough在修饰形容词、副词或动词时,一般放在这些词的后面,作"足够"解。

Exercises

1. Fill in the blanks with proper words or phrases.

   Electrochemistry is the ____ of physical chemistry that studies the relationship ____ electricity, as a measurable and quantitative ____, and identifiable chemical change, with ____ electricity considered an outcome of a particular chemical ____ or vice versa. These reactions involve electric charges moving between ____ and an electrolyte. Thus, electrochemistry deals ____ the interaction between electrical ____ and chemical change.

   A. change B. between C. branch D. electrodes,

   E. either F. with G. phenomenon H. energy

2. Match the following explanation with the proper word.

   ____ electrolysis, ____ entire, ____ immerse, ____ corrode, ____ deplete

   A. to gradually make something weaker or destroy it completely

   B. to put someone or something deep into a liquid so that they are completely covered

   C. to reduce the amount of something that is present or available

   D. a chemical decomposition reaction produced by passing an electric current through a solution containing ions

   E. constituting the full quantity or extent; complete

Lesson 36
BOILING POINTS AND DISTILLATION

Boiling Points of Pure Liquids

Any given liquid, when admitted into a closed evacuated space, evaporates until the vapor attains a certain definite pressure, which depends only upon the temperature. This pressure, which is the pressure exerted by the vapor in equilibrium with the liquid, is the vapor pressure of the liquid at that temperature. As the temperature increases, the vapor pressure of a typical liquid x increases regularly as shown by the generalized vapor pressure-temperature curve BC, in Fig.5.

At the temperature, Tp, where the vapor pressure reaches 101.3kPa. X begins to boil and Tp is called the NORMAL BOILING POINT of X. Every liquid which does not decompose before its vapor pressure reaches 101.3kPa, has its own characteristic boiling point. In general, the boiling point of a substance depends upon the mass of its molecules and the strength of the attractive forces between them. For a given homologous series, the boiling points of the member compounds rise fairly regularly with increasing molecular weight.

Polar liquids tend to boil higher than nonpolar liquids of the same molecular weight, and associated polar liquids usually boil considerably higher than nonassociated polar compounds.

The boiling point is a characteristic constant that is widely used in the identification of liquids. Because of its marked dependence upon pressure and its rather erratic response to impurities, however, it is generally less reliable and useful in characterization and as a criterion of purity than is the melting point for solids[1].

Boiling Points of Solutions

The normal boiling point of any solution is the temperature at which the total vapor pressure of the solution is equal to 101.3kPa. The effect of any solute, Y, on the boiling point of X will depend, then, upon the nature of Y. If Y is less volatile than X, then the total vapor pressure of the solution is lower, at any given temperature, than the vapor pressure of pure X.

Such a case is represented by curve B′C′, in which the experimentally determined values for the vapor pressures of a solution are plotted against temperature. The vapor pressure of the solution does not reach 101.3kPa until a temperature Tp′ is attained. In other words, the presence of the less volatile solute raises the boiling point of X from Tp to Tp′. A solution of sugar or salt in water is a familiar example of this type of slution.

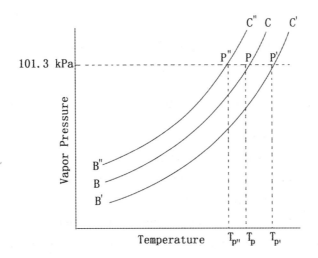

Figure 5.Generalized vapor pressure diagrams for a pure liquid (BC),
for a solution in which the solute is less volatile than the solvent (B'C'), and for a solution
in which the solute is more volatile than the solvent(B"C").

On the other hand, if Y is more volatile than X, then the total vapor pressure of the solution is higher than that of pure X, as shown by curve B"C". The vapor pressure of such a soulution reaches 101.3kPa at temperature Tp"; hence the effect of the more volatile solute is to lower the boiling point of X from Tp to Tp". A solution of acetone in water is an example of this type.

In any solution of two liquids X and Y, the molecules of X are diluted by molecules of Y, and, conversely, the molecules of Y are diluted by molecules of X. You would therefore expect the vapor pressure due to X to be less than that of pure X; in fact, you might predict that the PARTIAL PRESSURE due to X would be proportional to the molecular concentration of X.

Similarly, the partial pressure of Y might be expected to be proportional to the molecular concentration of Y. This is, in fact, the relationship which holds for so-called ideal solutions. It is expressed in Raoult's Law: the partial pressure of a component in a solution at a given temperature is equal to the vapor pressure of the pure substance multiplied by its mole fraction in solution. In symbols, for a solution of components X and Y,

$$P_x = P_x^0 N_x$$

where P_x=the partial pressure of X in solution,

P_x^0= the vapor pressure of pure X at that temperature,

N_x=the mole fraction of X in the solution.

Similarly,

$$P_y = P_y^0 N_y$$

where P_y=the partial pressure of Y in solution,

P_y^0=the vapor pressure of pure Y at that temperature,

N_y=the mole fraction of y in solution.

The total pressure, P_T, of the solution would be the sum of the partial pressures of X and Y.

$$P_T = P_x + P_y$$

Temperature-Composition Diagram for Solutions which Follow Raoult's Law.

These facts are represented graphically in Fig. 6 which shows a typical temperature-composition diagram. This diagram is a temperature-composition plot of the experimentally determined values for the system benzene-toluene, but it is representative of the plot for all solutions which are described by Raoult's Law.

Boiling points (ordinates) are plotted against composition expressed as mole fractions (abscissae). Pure benzene (100 percent X) boils at 80.1° (point A) and pure toluene (100 percent Y) at 110.6° (point B). All mixtures of the two boil at intermediate temperatures, as shown by the liquid (lower) curve. This curve shows the temperature at which a mixture of benzene and toluene of any given composition begins to boil. The vapor (upper) curve represents the composition of the vapor in equilibrium with the liquid at any given temperature.

For example, consider the changes which occur when a 20 mole per cent benzene—80 mole per cent toluene solution (represented by point P) is heated. At 101.6°, corresponding to point L_1, the liquid begins to boil. The first trace of vapor which is formed is, of course, in equilibrium with the liquid at 101.6°. It has the composition 38 mole per cent benzene—62 mole per cent toluene, as represented by point V_1, and is therefore considerably richer than the liquid in benzene.

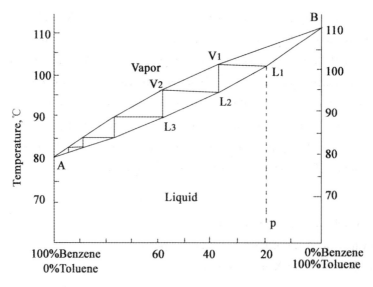

Figure 6. Temperature-Composition diagram of the system benzene-toluene.

As the distillation proceeds, the concentration of toluene in the liquid phase and the boiling point increase continuously, following the values represented by L_1B. Finally, at the end of the distillation, the liquid phase is pure toluene boiling at 110.6℃. Similarly, the vapor becomes progressively richer in toluene also, following V_1B. Always, however, it is richer in benzene than is the liquid with which it is in equilibrium, as shown by the points of intersection of any horizontal line with the vapor and liquid curves[2].

Obviously, a single simple distillation could never separate a 20:80 molar mixture of benzene and toluene into the pure components. But now consider what would be accomplished if the first trace of vapor formed by distillation of the mixture were cooled and condensed. It would, of course, form liquid corresponding to point L_2 of composition 38 mole per cent benzene—62 mole per cent toluene at 94.5℃.

Now, if liquid L_2 were distilled, the first trace of vapor formed would have the composition 59 mole per cent benzene—41 mole per cent toluene (point V_2) and when cooled would condense to liquid at 88℃ (point L_3). This, in turn, may be further enriched in benzene as indicated on the graph.

New words and Expressions

pure[pjuə] *a.* 纯的
admit[əd'mit] *vt.* 通入
vapor pressure['veipə 'preʃə] 蒸气压
nonpolar[nɔn'pəulə] *a.* 非极性的
erratic[i'rætik] *a.* 不规律的,不稳定的
response[ris'pɔns, rəs'pɔns] *n.* 感应,反应,回答
reliable[ri'laiəbl] *a.* 可靠的,可信赖的
hold[həuld] *vi.* 有效,适用
ideal solution[ai'diəl sə'lju:ʃən] 理想溶液
Raoult's Law 拉乌尔定律
mole fraction 摩尔分数
total pressure[təutl'preʃə] 总压
fractional distillation 精密分馏
diagram['daiəgræm] *n.* 图,图表
proceed[prə'si:d] *vi.* 进行,继续进行,开始

intersection[ˌintə'sekʃən] *n.* 交叉,相交
enrich[in'ritʃ] *vi.* 富集(in)

前缀

dia-[ˌdaiə-, dai'æ-] 表示"通过,横切中间,完全"之意
　　diagram, diameter, diagonal(对角线)
　　diamagnetic(抗磁的)
pro-[prə-] 在前,居先,向前,代替
　　proceed, promotor, propagation,
　　program(程序), prochiral(前手性)

词组

tend to (inf.)　倾向于……,有助于……
multiplied by…　被……乘,乘以……
rich in …　富有……(的),富含……

Notes

1. "Because of its marked dependence upon pressure and its rather erratic response to impurities, however, it is generally less reliable and useful in characterization and as a criterion of purity than is the melting point for solids." 为主从复合句,主句是"Because of … of purity",其中 "Because of … to impurities"为表示原因的状语, however为插入语,主句中主语 it 指前面的

the boiling point."than is … for solids"为比较状语从句。**译文**:"然而,由于对压力的显著依赖性和对杂质的相当无规律的反应,沸点用在鉴别和作为纯度的判据方面,较之固体的熔点一般不太可信和较少使用。"

2. "Always, however, it is richer in benzene than is the liquid with which it is in equilbrium, as shown by the points of intersection of any horizontal line with the vapor and liquid curves."为主从复合句,主句是"it is richer in benzene","than is the liquid"是比较状语从句,"with which it is in equilibrium"是修饰the liquid的定语从句,此从句中和主句中的it均指the vapor."as shown…句末"可看作是非限制性定语从句,从句中省去了is,as代表前面整个主句的意思。**译文**:"然而,蒸气较之其处于平衡的液体总是富含苯的,正如任一水平线与蒸气曲线和液体曲线的交点所示。"

Exercises

1. Fill in the blacks with proper words or phrases.

   Distillation is the process of separating the components or ____ from a liquid mixture by using ____ boiling and condensation. Distillation may ____ in complete separation, or it may be a separation that increases the concentration of selected components in the ____. In either case, the process uses differences in the ____ of the components. In industrial chemistry, distillation is a unit operation of practically ____ importance, but it is a ____ separation process, not a chemical reaction.

     A. partial   B. physical   C. substances   D. universal
     E. result   F. selective   G. volatility   H. mixture

2. Match the following explanation with the proper word.

     ____ admit, ____ erratic, ____ reliable, ____ proceed, ____ enrich

     A. worthy of being depended on

     B. to allow someone or something to enter

     C. to improve the quality or number of something

     D. likely to perform unpredictably

     E. to continue to do something that has already been planned or started

Lesson 37
EXTRACTIVE AND AZEOTROPIC DISTILLATION

Extractive and azeotropic distillation have the common feature that a substance not normally present in the mixture to be separated is deliberately introduced into the system in order to increase the difference in volatility of the most hard‑to‑separate components. Extractive distillation can be defined as distillation in the presence of a substance which is relatively non‑volatile compared to the components to be separated, and which, therefore, is charged continuously near the top of the fractionating tower, so that an appreciable concentration is maintained on all plates in the tower below its entry. Azeotropic distillation can be defined as distillation in which the added substance forms an azeotrope with one or more of the components in the feed, and by virtue of this is present on most of the plates in the tower above its entry at an appreciable level of concentration[1].

These separation methods find their principal applications in the separation of mixtures whose components boil too close together for the economical use of simple fractionating equipment. These separation methods are particularly applicable when the components to be separated differ in chemical type. The theoretical principles involved are well documented, and will not be further considered here. The processes differ in the means used to maintain the desired solvent concentration on the plates of the tower. In extractive distillation the high concentration of solvent is maintained by virtue of its nonvolatility, and by the fact that it is charged at a high point in the tower. The solvent is, necessarily, removed from the base of the principal tower. In azeotropic distillation, most of the solvent is taken off overhead, with relatively small amounts (ideally, none) drawn off with the bottoms[2].

Extractive distillation is generally more flexible than azeotropic distillation, a greater variety of solvents and a wider range of operating conditions are available; and the concentration of solvent may be controlled by heat and material balances rather than by the accident of azeotrope composition. Furthermore, since vaporisation of the solvent is not required, heat loads are usually considerably less. It has been mainly used for the separation of toluene, not benzene, but is mentioned here for the sake of completeness.

The use of azeotropic distillation as a means of separation of BTX components from other non‑aromatic hydrocarbons has been known and employed for some thirty years[3]. Acetone is used as an entrainer to purify benzene from similar‑boiling non‑aromatic hydrocarbons. Toluene can be separated by the use of either methanol, or of methyl ethyl ketone. Ethyl benzene may be separated from styrene either by isobutanol, or by 1-nitropropane.

In a 1966 review paper, further information was made available. Fifty-eight possible entraining agents for separating ethyl benzene (bpt 136.2℃) and para-xylene (bpt 138.4℃) have been examined. It would appear that 2-methyl butanol is the most successful agent, requiring a column with only 48 per cent of the number of theoretical plates required if no entrainer was used.

The separation of para- and meta-xylenes (bpts 138.2 and 139.2℃ respectively) is much more difficult. Of thirty-five entrainers examined the best appears to be 2-methylbutanol, but the change of relative volatility is only from 1.020 to 1.029, and hence it can be safely concluded that azeotropic distillation for the separation of the para- and meta-xylenes is not an economic proposition.

Last, and easiest of the C_8 aromatics is the separation of meta-from ortho-xylene[4] (bpts 139.2 and 144.5℃ respectively). Twenty-eight entrainers were examined, the best being formic acid, requiring a column containing only 70 per cent of the theoretical plates required if no entrainer is used.

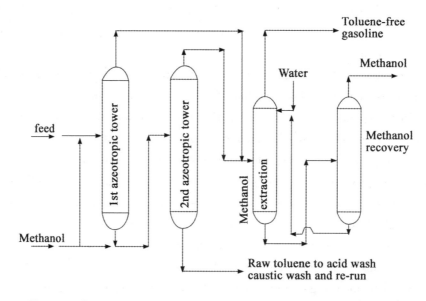

Figure 7. Concentration of toluene azeotropic distillation with methanol.

Two commercial processes have been developed for the separation of toluene using azeotropic distillation; one using an aqueous solution of methyl ethyl ketone and water and the other using methanol. Both processes operate on a narrow boiling range concentrate. The equilibrium relationships are well established. One commercial plant has been built using the methanol azeotrope which was developed by Socony-Vacuum Oil Co. Fig. 7 gives the toluene concentration step. The azeotropic system consists of two towers each containing 36 trays operating at 30 p.s.i.g. The toluene concentrate is mixed with 75 vol per cent of anhydrous methanol for a feed containing 25 vol per cent toluene, and charged to the first tower. Cold reflux of 0.6 to 0.7 vols/vol of feed is employed.Practically complete recovery of toluene is obtained in the form of a 70 vol per cent concentrate from the bottom methanol.Cold reflux of the order 4.0 to 4.5 vols/vol of feed is employed, and toluene recovery from this tower averages about 95 per cent, while overall yields of about 80 percent of the toluene present in the initial feed are obtained.Losses are distributed mainly in the azeotrope overhead and in the acid treating and re-running stages.

The MEK process, developed by Union Oil Company, has had one commercial plant operating, and handles a wide variety of straight-run and cracked stocks. A preconcentrate feedstock with a toluene content in the range 35~85 vol per cent is charged, together with the solvent (90 vol per cent MEK plus 10 vol per cent H_2O), to the 22nd tray of a 50-tray tower.The volumetric ratio of solvent+water/hydrocarbon charge varies between 2.0 to 3.0 to 1 depending upon the boiling range and type of non-aromatics present.Cold reflux in the ratio of 0 to 1 reflux/feed is employed.The azeotropic distillate is composed of about 75 vol per cent solvent.A small amount of water-rich phase separates from it in the reflux drain(80 vol per cent water, 20 vol per cent MEK) and is recycled back to the feed[5].95~97 per cent of the toluene present in the feed is recovered as bottoms from the azeotropic tower as 99 per cent purity toluene on a solvent-free basis.The removal of MEK from the azeotropic distillate is accomplished by four stages of water-extraction thus reducing the ketone content to approximately 1 vol per cent.The washed material is then fractionated in an auxiliary tower to remove the remaining ketone.Recovery of MEK from the water extracts is done by distillation of the MEK-water azeotrope from the excess water in a tower with only a few trays and a low reflux ratio.The azeotrope thus obtained is then fed directly to the primary azeotrope tower together with the toluene pre-concentrate feed.The crude toluene so produced, after removal of MEK is treated with sulphuric acid, and followed by redistillation to produce nitration-grade toluene.Overall recovery of toluene is around 90 per cent.

New words and Expressions

extractive[iks'træktiv] a.抽提的,萃取的
extractive distillation 萃取蒸馏
azeotropic[ˌeiziː'trɔpik] a.共沸的
azeotropic distillation 共沸蒸馏
feed[fiːd] v.n.进料,供料,原料
hard-to-separate 难分离的
charge[tʃɑːdʒ] vt.投料,装料,进料
fractionating tower[ˈfrækʃəˈneitiŋˈtauə]
 精馏塔
bottoms[ˈbɔtəms] n.釜渣,釜(残)液
plate[pleit] n.板,塔板

azeotrope[ˈeiziːətrəup] n.共沸物,共沸混合物
equipment[iˈkwipmənt] n.设备,装备
document[ˈdɔkjumənt] vt.证明
overhead[ˌəuvəˈhed] n.塔顶馏出物
flexible[ˈfleksəbl] a.易操作的,易适应的
material balance 物料平衡
vaporisation=vaporization
 [ˌveipəraiˈzeiʃən] n.蒸发
load[ləud] n.v.负荷,负载,装填
bpt=b.p.=boiling point
BTX=benzene-toluene-xylene

苯－甲苯－二甲苯
entrainer [inˈtreinə] n. 携带剂,夹带剂
entraining agent   携带剂
theoretical plate   理论塔板
proposition [ˌprɔpəˈzeiʃən] n. 意见;见解,事情
p.s.i.g=pounds per square inch guage
　[ˈpaunds pə:skwɛə intʃ geidʒ]
　磅／平方英寸（表压）
cold reflux   冷回流
overall yield   总产率
recover [riˈkʌvə] v. 回收
recovery [riˈkʌvəri] n. 回收,回收率
acid wash   酸洗
caustic [ˈkɔ:stik] a. 苛性的
              n. 苛性碱
caustic wash   碱洗
MEK=methyl ehtyl ketone   甲乙酮
straight-run   直馏馏分,直馏
stock [stɔk] n. 原料,材料
preconcentrate [ˌpriˈkɔnsentreit] n. 预(先)浓缩物
tray [trei] tower   盘式塔
solvent-free   无溶剂
auxiliary [ɔ:gˈziljəri] a. 辅助的

reflux ratio   回流比

前　缀

extra- [eksˈtræ(ei,ə)-] 在外,外面,离开,
　extractive, extractor, extractant
　extraneous
vapori- [ˌveipəˈri-,ˈveipəri-] 水蒸气,雾气
　vaporization, vaporizer, vaporimeter,
　vaporish
en- [in-, en-] 置于…之内,进入…之中
　entrainer, entrance, enclose
　encapsulate(胶囊,包封), enhydrous

词　组

by virtue of …   根据,借助于,由于,因为
take off   流出,馏出
draw off…   拿出,排出,移去
for the sake of …   为……起见,由于……缘故
feed…into…   把……送入
the ratio of…to…   ……与……之比;……比……
on…basis   以…根据,依据…
charged with…   充有,装满

Notes

1. "Azeotropic distillation can be defined as distillation in which the added substance forms an azeotrope with one or more of the components in the feed, and by virtue of this is present on most of the plates in the tower above its entry at an appreciable level of concentration." 系主从复合句,主句是"Azeotropic…as distillation","in which the…the feed"为修饰 distillation 的定语从句,"and by virtue of …句末"为原因状语从句。**译文**："共沸蒸馏可定义为加入的物质与原料中的一个或多个组分形成共沸混合物的蒸馏,据此加入的物质以显著的浓度存在于它的入口上部塔中的大多数塔盘上。"

2. 该句中 with relatively…bottoms 是独立主格结构,作状语,说明伴随状况。

3. for some thirty years：这里 some+数字作"约""左右"解。

4. 该句中 easiest 作表语,所以前面不加 the。

5. A small amount of water-rich phase…is recycled back to the feed. 该句系简单句,句中 it 指上句的 the azeotropic distillate. **译文**:"在回流排放口处,少量的富含水相由共沸馏出物中分离出来并循环回原料中。

Exercises

1. Fill in the blanks with proper words or phrases.

   In chemical engineering, azeotropic distillation usually ____ to the specific ____ of adding another component to generate a new, lower-boiling azeotrope that is ____. The addition of a material separation ____ , such as benzene to an ethanol/water mixture, ____ the molecular and eliminates the azeotrope. Added in the liquid phase, the new component can alter the activity ____ of various compounds in different ways ____ altering a mixture's relative volatility.

   A. technique   B. interactions   C. thus   D. heterogeneous,
   E. changes   F. coefficient   G. refers   H. agent,

2. Match the following explanation with the proper word.

   ____ equipment, ____ load, ____ recover, ____ caustic, ____ auxiliary

   A. a large quantity of something that is carried by a vehicle, person etc.
   B. to get back something
   C. the tools, machines etc. that you need to do a particular job or activity
   D. someone or something that acts as assistant
   E. any chemical substance that burns or destroys living tissue

Lesson 38
CRYSTALLIZATION

General Methods of Crystallization. Since the days of the earliest alchemists, solids have been purified by crystallization from suitable solvents. Today this technique still stands as the most useful method for the purification of solid substances.

As commonly practiced, purification by crystallization depends upon the fact that most solids are more soluble in hot than in cold solvents. The solid to be purified is dissolved in the hot solvent at its boiling point, the hot mixture is filtered to remove all insoluble impurities, and then crystallization is allowed to proceed as the solution cools. In the ideal case, all of the desired substance separates in nicely crystalline form and all the soluble impurities remain dissolved in the mother liquor. Finally, the crystals are collected on a filter and dried. If a single crystallization operation does not yield a pure substance, the process may be repeated with the same or another solvent.

The great beauty of crystallization as a purification technique lies in the fact that the orientation of molecules in a crystal lattice is an extremely delicate and selective process. Only infrequently do different substances crystallize in the same lattice. At times, the desired solid can be crystallized selectively from a solution also saturated with other solid imupricies simply by the careful introduction of a tiny seed crystal. In such cases, the molecules of the desired compound leave the solution to take positions in the crystal lattice, while the mother liquor remains saturated, or even becomes supersaturated, with respect to the foreign materials.

A solid solute may, of course, be crystallized by spontaneous evaporation of solvent from a saturated solution. Occasionally, this is used as a method of purification. Evaporation should proceed very slowly to avoid formation of a crust of impure solid at the evaporating surface. In general, this method is less effective than the classical crystallization technique.

Selection of Solvent. *Similia similibus solvunter* (like dissolves like) was a watchword among the alchemists and medieval iatrochemists. It is still perhaps the best three-word summary of solvent behavior; a detailed study of the relationship between structure and solvent action becomes highly complex. In the final analysis, the best way to find a suitable recrystallization solvent for a given substance is by experimental trial. A few helpful and reasonably valid generalizations may, however, speed the process.

No nonionic compound dissolves appreciably in water unless its molecules are ionized in water solution or can co-associate with water molecules through hydrogen bonding. Thus, hydrocarbons and their halogen derivatives are virtually insoluble in water. Compounds whose molecules contain functional groups [such as alcohol, aldehyde, ketone, carboxylic acid, and amide groups], which can form hydrogen bonds with water, are soluble in water unless the ratio of the total number of carbon atoms to such functional groups in the molecule exceeds 4 or 5. Then the solubility falls off rapidly. Thus, acetamide is souble in water, but caproamide is not. In fact, it is a very general rule that as any homologous series is ascended, the solubilities and all other physical properties of the members tend to approach those of the parent hydrocarbon.

The hydrogen bonds or hydrogen bridges with which we are concerned in dealing with the water solubilities of organic compounds are almost exclusively those in which hydrogen links oxygen atoms to oxygen or nitrogen atoms[1].

For the purposes of discussing solubility behavior, we may conveniently divide solutes into three classes; (1) those that are associated in the liquid state; (2) those that are not associated but can co-associate with water; (3) those that are neither associated themselves nor capable of co-associating with water.

Aldehydes, ketones, esters and tertiary amines (R_3N), and similar oxygen- and nitrogen-containing compounds, can co-associate with water, even though they, themselves, are nonassociated becauese of the lack of a hydrogen-bonding hydrogen. In the co-association of such nonassociated compounds with water, the hydrogen-bonding hydrogen must be supplied by the water molecules. With aldehydes, for example, the bonding would be represented as

$$\begin{array}{c} H O \\ | /\backslash \\ R-C=O\cdots H H \end{array}$$

On the other hand, compounds such as alcohols, carboxylic acids, amides, and primary and secondary amines (RNH_2 and R_2NH), whose molecules possess hydrogen atoms capable of forming hydrogen bonds are themselves associated in the liquid state, and can also co-associate with water. Theoretically, at least, such co-association can proceed through the hydrogen of water or through the hydrogen attached to oxygen or nitrogen in the organic molecule.

Finally, hydrocarbons and their simple halogen derivatives neither are associated nor can they co-associate with water.

Most organic compounds which lack a hydrogen-bonding hydrogen atom dissolve readily in ether, benzene, ligroin, and other typical nonassociated solvents simply by a process of molecular mixing. Organic compounds, which themselves are associated in the liquid state, are likewise reasonably soluble in such solvents, unless (1) they have two or more hydrogen-bonding functional groups per molecule, approaching a ratio of one such group for each carbon atom, or (2) unless they are solids with high melting points. Thus, n-propyl alcohol and caproic acid are soluble in ether, but glycerol, with one hydroxyl group for each of the three carbon atoms, adipic acid, with a melting point of 153℃, and glucose [$HOCH_2(CHOH)_4CHO$], with five hydroxyl groups per six carbon atoms, are not.

Ether and benzene are quite similar in solvent action. Ether is, in general, a better solvent than benzene for associated compounds, and both are better than petroleum ether and ligroin. Petroleum ether is similar to ligroin, but has somewhat weaker solvent action.

New words and Expressions

alchemist[ˈælkimist] n. 炼金术士,炼丹家
purification[ˌpjuərifiˈkeiʃən] n. 提纯
liquor[ˈlikə] n. 液体
mother liquor   母液
delicate[ˈdelikit] a. 精密的,严谨的,周密的
infrequently[inˈfri:kwəntli] ad. 偶尔
tiny[ˈtaini] a. 微小的,极小的
seed[si:d] n. 晶种,种子
~ crystal 晶种
supersaturate[ˌsju:pəˈsætʃəreit] vt. 过饱和
foreign material   异质物料,外来杂质
similia similibus solvunt(拉丁语)=like dissolves like   相似溶解相似
watchword[ˈwɔtʃwə:d] n. 口号,格言
medieval[ˌmediˈi:vəl] a. 中世纪的
iatrochemist[aiˌætrəˈkemist] n. 医疗化学家
recrystallization[ˈri:kristəlaiˈzeiʃən] n. 重结晶
trial[ˈtraiəl] n. 试验,考验,努力
nonionic[ˌnɔnaiˈɔnik] a. 非离子的
co-associate[kəu əˈsəuʃieit] a. 共缔合的
hydrogen bonding   氢键
virtually[ˈvə:tjuəli] ad. 几乎,差不多
acetamide[ˌæsiˈtæmaid] n. 乙酰胺

caproamide[ˌkæprəˈæmaid] n. 己酰胺
caproic acid[kəˈprəuik æsid] 己酸
deal[di:l] vi. 研究,论及,涉及(with)

前 缀

acet(o)-[ˈæsit(əu)-] 乙酰,乙川
   acetamide, acetanilide acetonitrile, acetophenone
capr(i,o)-[ˈkæpr(i,ə)-, kəˈpr(i,ou)-]
（只用于己、辛、酸、及其衍生物）
   caproic acid, caprylic acid, capric acid, caprolactam, capryl, caprinitrile(癸腈)
gluco-[ˈglu:kou(ə)-] (=glyco-)甜的
   glucose, gluconic acid, glucoside, gluconolactone(葡糖酸内酯)

词 组

with respect to…   关于,相对于,就……而论
in the final(last)analysis   总之,归根结底
together with…   以及,和……一起
concerned in   涉及到,与……有关

Notes

1. "The hydrogen bonds or hydrogen bridges with which we are concerned in dealing with the water solubilities of organic compounds are almost exclusively those in which hydrogen links oxygen atoms to oxygen or nitrogen atoms."系主从复合句, "with which we are …organic compouds"是修饰主句主语"The hydrogen bonds or hydrogen bridges"的定语从句, "in which hydrogen…nitrogen atoms"又是修饰表语those的定语从句。译文: "我们涉及到的研究水溶性有机化合物的氢键和氢桥,几乎完全是氢把氧原子和氧,或氢把氧原子和氮原子加以连接。"

Exercises

1. Fill in the blanks with proper words or phrases.

    Crystallization is the (natural or ___) process by which a solid forms, where the atoms or molecules are highly ___ into a structure known as a crystal. Some of the ___ by which crystals form are ___ from a solution, freezing, or more rarely ___ directly from a gas. Attributes of the resulting crystal depend largely on factors such as ___, air pressure, and in the ___ liquid crystals, time of fluid ___.

A. temperature	B. organized	C. evaporation	D. case
E. precipitating	F. deposition	G. artificial	H. ways

2. Match the following explanation with the proper word.

    ___ purification, ___ delicate, ___ supersaturate, ___ trial, ___ virtually

    A. needing to be dealt with carefully or sensitively in order to avoid problems or failure

    B. the act of cleaning by getting rid of impurities

    C. a process of testing to find out whether something works effectively and is safe

    D. almost

    E. to add to a solution beyond saturation

Lesson 39
MATERIAL ACCOUNTING—— THE LAW OF CONSERVATION OF MASS REALLY WORKS

In attempting to carry out a reaction in the laboratory, every chemistry student has learned that a vital step in the procedure deals with the calculation of the amount of starting materials required and the amount of product that is expected to be produced. Such calculations are also required when chemical reactions are carried out on an industrial scale. However, the format of the calculations differs from those that the chemist normally utilizes in the laboratory. The calculations differ because chemists in the laboratory normally work with closed systems, whereas most industrial processes are invariably open systms. Open systems are those in which there is a flow of materials into and out of the process equipment. In such systems heat or work energy may be added to certain process units and may be removed from others not only to satisfy operational requirements, but also to operate most economically[1]. The complete accounting of all mass and energy in such chemical processes is referred to as a material and energy balance.

The material balance is frequently referred to as a mass balance or weight balance in industrial chemistry practice. Therefore, these terms are used interchangeably. A material balance can be effected without an energy balance, but an energy balance requires a knowledge of the mass and composition of all streams. This combination, the material and energy balance, is one of the most powerful tools used in the sequence of steps necessary to bring a chemical reaction from the idea stage to a viable large-scale commercial process[2]. It is also an essential tool in the effective evaluation of the day-to-day operation of an existing chemical process. The concept of the material balance is so simple, however, that the student is prone to assume erroneously that he can apply these balances skillfully without much training in their application. Thus, it is the purpose of this lesson to discuss the basic principles of material balances and how they can be applied to industrial problems. Energy balance concepts are considered in next lesson.

In general, there are two types of chemical processes with which the industrial chemist deals, the "batch process" and the "continuous process". In the batch process the chemicals are added to the processing vessel in one operation and then the process is carried out. In some cases the products are removed during this period, but in others they are removed after the processing is completed. In the continuous process the chemical charge and the products enter and leave continuously. The material balance can be applied to either a batch or a continuous process. The biggest difference in applying the material balance to these two processes involves the element of time. A material balance applied to a batch process usually does not include a time variable. The balance in such a case is generally made over a complete cycle, which involves merely the processing of a single charge. In the case of a continuous process, however, the time variable must enter the material balance. The balance must be made over a specified period of time.

The material balance is based upon the concept of the law of conservation of matter, which in effect says that, except for situations involving nuclear reactions, atoms are neither created nor destroyed. Atoms that enter a system must either accumulate in the system or must leave. This observation leads to the balance expressed in Eq.1, which is valid for all atomic species in the system.

$$\begin{Bmatrix} \text{Accumulation of} \\ \text{atomic species j} \\ \text{within the system} \end{Bmatrix} = \begin{Bmatrix} \text{total atomic} \\ \text{species j entering} \\ \text{system} \end{Bmatrix} - \begin{Bmatrix} \text{total atomic} \\ \text{species j} \\ \text{leaving system} \end{Bmatrix} \qquad (1)$$

By summing over all the atomic species entering and leaving a system, the total material balance is obtained:

$$\begin{Bmatrix} \text{Total accumulation} \\ \text{within the system} \end{Bmatrix} = \begin{Bmatrix} \text{total mass} \\ \text{entering system} \end{Bmatrix} - \begin{Bmatrix} \text{total mass} \\ \text{leaving system} \end{Bmatrix} \qquad (2)$$

when there is no accumulation within the system, Eq.2 reduces to the following:

$$\begin{Bmatrix} \text{Total mass} \\ \text{entering system} \end{Bmatrix} = \begin{Bmatrix} \text{total mass} \\ \text{leaving system} \end{Bmatrix} \qquad (3)$$

Inherent in the formulation of each of the above balances is the concept of a system for which the balance is made[3]. By system we mean any arbitrary portion or whole of a process as set out specifically by the chemist for analysis.

In most continuous processes that the chemist encounters, the mass accumulation term will be zero; that is, the process is in a steady-state.

One of the many useful applications of the material balance is during the pilot plant stage when the performance of a given process is being investigated. If each item of the input and output streams is measured, the material balance will serve to check the accuracy of the experimental measurements. If a balance is not obtained within the desired accuracy, it is apparent that one or more of the measurements is in error. In many of the processes with which the chemist will deal, it will be impossible or uneconomical to measure directly all the items of input and output. However, if sufficient other data are available, by making a material balance on the process, it is possible to get the needed information about the quantities and compositions at the inaccessible locations.

Another use of the material balance is to make calculations for prospective processes. For example, if we choose to prepare ethylamine by the ethyl alcohol-ammonia reaction, we must devise a plan for routing the various materials throughout the emerging process. Information about the quantities of the various materials (reactants, products, unconverted reactants, waste, etc.) routed through the process can be used to determine process equipment sizes, and the specific paths through which the materials are routed will determine the types of separation problems that must be solved[4]. Therefore, the key to the successful design and development of a chemical process begins with an understanding of the material flow in the process.

New words and Expressions

account[ə'kaunt] v.计算,衡算
material accounting 物料衡算
conservation[ˌkɔnsə(ː)'veiʃən] n.守恒
　~ of mass 质量守恒
format['fɔːmæt] n.格式,形式
closed system　封闭体系
open system　敞开体系
invariably[in'vɛəriəbli] ad.不变地,总是
work energ　静压能
process unit　工艺设备
energy balance　能量平衡(衡算)
mass balance=weight balance 物料衡算,物料平衡
interchangeably[ˌintə'tʃeindʒəbli] ad.可互相变换地,可交替地
stream [striːm] n.流体,流,流动
idea[ai'diə] n.概念,计划,幻想
viable['vaiəbl] a.实际的,可实现的
prone[prəun] a.有……之倾向,易于……的,习惯于
skillfully['skilfəli] ad.熟练的,巧妙地
batch process[bætʃ 'prəuses] 分批法,间歇法(过程)
continuous process　连续化过程,连续法

enclose[in'kləuz] vt.封入,包围
steady['stedi] state 稳态,守恒状态
input['in-put] n.进料(量),输入
output['autput] n.产量,出料
equate[i'kweit] vt.使相等
pilot plant['pailət plɑːnt]中间工厂,试验工厂
item['aitəm] n.条,项目
check[tʃek] v.核对,检查
inaccessible[ˌinæk'sesəbl] a.不能进入的,不能到达的
prospective[prə'spektiv] a.预期的
devise[di'vaiz] vt.设想,计划
route[ruːt] vt.定路线,输送(through)

词　组

day to day　日常,每日
make over　进行完,经过
the law of conservation of matter
　物质守恒定律,物质不灭定律
in effect　实际上
by summing over　总计,概括起来
set out…　提出,陈述
in error　弄错了的

Notes

1. "In such systems heat or work energy may be added to certain process units and may be removed from others not only to satisfy operational requirements, but also to operate most economically." 为简单句,该句主语是"heat or work energy",有两个以and连接的谓语,"not only…but also"连接的两个不定式短语作目的状语。**译文:** "在这样的体系中,不仅为了满足操作的需要,而且为了更经济地操作,可以给一些过程单元施以热或静压能,给另一些单元移去热或静压能。"

2. This combination, the material and energy balance, is one of the most powerful tools used in the

sequence of steps necessary to bring a chemical reaction from the idea stage to a viable large-scale commercial process. 该句中 This combination 是主语,"the material…balence"是其同位语。"used in …process"是过去分词短语作定语,修饰 tools。**译文**:"物料平衡和能量平衡的结合是把化学反应从理想状态变成实际的大规模工业生产过程的必要步骤和最强有力的工具之一"。

3. Inherent in the formulation of each of the above balances is the concept of a system for which the balance is made. 该句系倒装句,强调表语,所以把表语放在句首。**译文**:"对于已建立平衡的体系的概念是每个上述平衡公式的内在联系。"

4. Information about the quantities of the various materials (reactants, products, unconverted reactants, waste, etc.) routed through the process can be used to determine process equipment sizes, and the specific paths through which the materials are routed will determine the types of separation problems that must be solved. 该句是并列复合句,"Information…equipment sizes"是第一个分句,其中"routed through the process"是修饰 materials 的过去分词短语,"and the specfic paths…be solved"是第二个分句,"through which…routed"是修饰 paths 的定语从句,which 指 the specific paths, 此从句又带有一个修饰 separation problems 的定语从句,即"that must be solved"。**译文**:"关于(输送)通过该过程的各种物料(反应物,产物,未转化的反应物,废料等)的数量信息,可用来确定工艺设备的大小,并且物料通过的特定路径将决定必须加以解决的分离操作的类型。"

Exercises

1. Fill in the blanks with proper words or phrases.

   The ___ of conservation of mass states that for any system closed to all transfers of ___ and energy, the mass of the system must remain constant ___ time. For example, in ___ reactions, the mass of the chemical ___ before the reaction is equal to the mass of the components after the reaction. Thus, ___ any chemical reaction and low-energy thermodynamic ___ in an isolated system, the total mass of the reactants must be ___ to the mass of the products.

   A. equal   B. chemical   C. law   D. processes
   E. over   F. during   G. components   H. matter

2. Match the following explanation with the proper word.

   ___ conservation, ___ interchangeable, ___ prone, ___ enclose, ___ route

   A. preventing something from being lost or wasted

   B. likely to do something or suffer from something, especially something bad or harmful

   C. send documents or materials to appropriate destinations

   D. to surround something, especially with a fence or wall, in order to make it separate

   E. capable of replacing or changing places with something else

Lesson 40
THE LITERATURE MATRIX OF CHEMISTRY

The literature of chemistry and chemical technology is a rich and vast knowledge resource through which we can interact with those who have shaped our past and who are shaping our present.

On accepting the invitation to write this book, I hoped to achieve the following objectives:

To give the reader an appreciation of the value of the literature matrix and the vital role it has played in the progress of chemistry and technology.

To delineate the scope and content of the literature matrix so that the reader can interact with and gain access to it effectively.

To orient the book to students majoring in chemistry and chemical engineering and to scientists and engineers employed by the chemical industry in research and development and in plant operations.

Whereas a minority of chemists and chemical engineers affect the literature as authors, all are affected by the literature as readers and users. Reading and using the literature are not only a tradition; they are a necessity if we are to maintain scientific growth (self-education), relate facts (idea seeking), and establish background information for new research programs (insurance against repeating what has already been done).

Too many graduates leave the educational environment with the belief that learning goes on in academic buildings and nowhere else. To limit thinking within the bounds of formal education and training makes us artisans and our science an art, and courts technical obsolescence within a decade[1].

The chemical literature offers professional chemists and chemical engineers an opportunity for continuous, lifelong self-education. Ideally, every course in science and engineering curriculums should train students to utilize the literature for self-education. The student should be taught not only segmented disciplines, but also how to learn science and technology that is changing rapidly in directions that cannot be anticipated easily.

The amount of information to be taught has increased so much that most professors find little time to teach the literature. Furthermore, chemistry and chemical technology are increasingly segmented into new disciplines and subdisciplines, such as polymer chemistry, material science, and environmental science. The need to teach electronics and computer science in addition to the new discipilines have forced the elimination of courses in literature, history, and philosophy of chemistry from the majority of curriculums. Of the approximately 2000 colleges and universities that grant degrees in chemistry, only a few offer courses in the literature of chemistry and still fewer in the history and philosophy of chemistry. Paradoxically, when the chemical literature was relatively small, the literature and history of chemistry were considered to be important components of the curriculum, and a high percentage of colleges and universities had courses in those subjects. Many textbooks written for students of the late nineteenth and the first three decades of the twentieth century emphasized the literature and history of chemistry. Unfortunately, this is no longer the case.

The twentieth century has been a period of rapid growth in the chemical industry and in governmental laboratories, in research and development funding by both the chemical industry and the federal government, and in the numbers of chemists and chemical engineers. The result has been a correspondingly rapid growth of the literature in a multitude of fragmented disciplines and subdisciplines. The size, growth, and complexity of the literature became such in the twentieth century, and particularly since 1940, that a multitude of information services were created and a number of guide books were written to aid the user of the literature. One of the best-known of these was *A Guide to the Literature of Chemistry* by E.J.Crane and A.M.Patterson, published in 1927 by John Wiley & Sons. This book enjoyed wide use as a text for courses in the literature of chemistry, as did the second edition (1957) by Crane, Patterson, and E.B.Marr. Two other highly regarded and much used texts were *Chemical Publications- Their Nature and Use* by M.G.Mellon (1928, 1940, and 1948; McGraw-Hill) and *Library Guide for the Chemist* by B.A.Soule (1938, McGraw-Hill).

Another response to the size, growth and complexity of the literature was the appearance of a new specialist, the chemical information scientist, and a new subdiscipline of chemistry, chemical information science—now a well-established career for thousands of chemists. Although the activities engaged in by these chemists are taught in colleges and universities, the courses are not a part of the chemistry curriculum, nor do they constitute a curriculum for chemical information science[2]. Chemical information scientists edit and write technical material, translate, index, abstract, search the literature, design information systems, and relate the literature to the needs of an environment. As computers became increasingly important in processing information, chemical information scientists played an important role in employing this new tool for computerized information systems and services.

In the nineteenth century the literature of chemistry consisted of personal contacts, lectures, correspondence, books, and a few journals. As late as World War I it was not very difficult for a chemist to read practically everything of importance published in chemistry. Thereafter it became increasingly difficult, and by the 1930s it was impossible to read everything of importance.

Today, the chemical literature consists of books, encyclopedias, treatises, data compilations, handbooks, patents, journals, abstract journals, trade literature, government publications, market research reports, and a variety of computer-based information services. Although a part of this literature matrix is discussed in books by Crane, Patterson, and Marr; Mellon; Soule; and others, the character of the literature has changed radically since these books were written. The present book includes the earlier literature, which is still of importance for retrospective searching, and the significant traditional literature and information services, which are essential for maintaining current awareness and for retrospective searching.

Books, encyclopedias, treatises, data compilations, etc., are the vade mecum of students in all subject areas, especially in science and engineering. These are the subjects covered in Chapters 1, 2, and 3.

Books are the major resource utilized in the educational process, and one who has not learned how to use school and public libraries can hardly claim to be educated. The books one acquires during the academic years are but a drop in the ocean of literature, and this drop evaporates rather quickly into obsolescence. Throughout one's professional career it is important to gain familiarity with a large number of books, encyclopedias, treatises, etc., and with sources that give information about new books. The survey of books in Chapter 1 is neither definitive nor all-inclusive; it is, however, highly selective, based on my own use of many of the books listed or on the evaluations of others. Year of publication is not given for every book because many undergo periodic revision; the reader should seek the latest edition.

Familiarity with treatises and encyclopedias, such as those listed and discussed in Chapter 2, is a sine qua non for all practicing chemists and engineers. Organic chemists cut their teeth on Beilstein and Houben-Weyl, and inorganic chemists on Gmelin. The most important general reference book today is the Kirk—Othmer *Encyclopedia of Chemical Technology.* Considerable searching and learning time is saved by knowing and consulting these encyclopedias and treatises. Like most other tools of chemistry, expertise comes with frequent use.

Every chemist and chemical engineer should have a personal copy of a single-volume handbook, such as Lange's, Perry's, or CRC's, and should be aware of and frequently consult comprehensive works, such as Landolt−Börnstein, *International Critical Tables*, and the special data compliers of critical data discussed in Chapter 3.

Patents constitute an integral resource of the chemical literature. They have a unique literary form, written to satisfy legal requirements, and very unlike that used for reports or journal literature. Most important, they are an essential and useful source of chemical technology, and they play a critical role in the conduct of research and development in the chemical industry. The number of abstracts of basic patents published in *Chemical Abstracts*, which very recently has been in excess of 60 000 per year, gives an indication of the size of this literature. Chapter 4 discusses patents as a resource and how to use them.

Journal literature has been the fastest growing segment of the chemical literature. Whereas books, encyclopedias, and treatises discuss the past events of chemistry, journals record the current happenings. Journals came into existence in the seventeenth century as a better and faster communication medium than letter, pamphlet, or book, and slowly evolved into the dominant medium for reporting and communicating activities in the laboratory. As chemistry was increasingly subdivided into specialties and the products of chemical industries increasingly used by other industries, the journal literature grew to cover these areas of interest. In addition to reporting current research, journals were introduced to review the literature and to serve the needs of trade groups. As the number of journals in chemistry, chemical technology, and allied fields increased to the present 20 000 ~ 40 000, the journal literature also became the dominant resource for retrospective searching. Chapter 5 puts this vast resource into perspective, identifying the important journals by period of publication, country of publication, and—for journals currently published—subject area covered. Because there are so many journals being published today, only the major journals in various subdisciplines are listed.

In the nineteenth century, the size and complexity of the journal literature already prompted the introduction of *Pharmaceutisches Centralblatt* (1830) [name changed to *Chemisches Zentralblatt* (1856)]. the first journal in chemistry consisting wholly of abstracts. It was followed by Science Abstracts in 1989 and *Chemical Abstracts* in 1907. There are now over 1500 indexing / abstracting publications, covering the journal, report, patent, and book literature. Chapter 6 discusses these services as they evolved over the centuries. Chapter 7 is a fairly detailed discussion of *Chemical Abstracts* and Chapter 8 of other indexing/abstracting services.

The introduction of computer-based information services, the subject of Chapter 9, added a new dimension to the chemical literature matrix. Since the mid-1960s, online data bases have become a major tool for searching the chemical literature. They are a tool, however, that requires knowledge of the contents and limitations of data bases and how to access the data bases through the systems imposed by data base brokers. Despite the increasing number of data bases available online from various data base brokers, most chemists use the new tool mostly through intermediaries. Terminals, a maze of operating systems, programming languages, and intermediaries are still barriers between the scientist and the computer. But there is hope that through proper management computers and telecommunications can bring the literature of chemistry to us according to our specific needs and in our terms.

Of the various components of the literature matrix, computer-based information services are undergoing the most rapid changes. For about two months in 1981, I was given an opportunity to update the present book before it went to the publisher. I added new information to every chapter, but the greatest number of changes and additions were in the chapter on computer-based information services. Data base producers especially are undergoing changes toward improving their products. A salient example of the changing nature of computer-searchable information systems is Chemical Abstracts Service, whose products CASIA, CA Condensates, and Patent Concordance, no longer produced, have been replaced by far superior products, CA BIBLIOFILE and CA SEARCH.

Most scientists and engineers prefer the dreams of the future above the history of the past[3]. But our constantly evolving present is a consequence of our expanding past, and the best of today's chemistry will be a part of the past. The study of history is more than a luxury. The past has played a tremendous role in shaping our literature and continues to be a part of our total knowledge. The chemical literature, although young and vigorous, is steeped in history. Those who have made this history and literature are discussed in Chapter 10. Chapter 11 traces the development of American chemistry and its literature from colonial times to the twentieth century. Only deceased chemists are considered in this chapter; otherwise chapter 11 would have been the longest in the book.

Communications has been the most pervasive force in the creation and maintenance of chemistry and chemical technology. The printing press was the magic wand that made this possible. But like the "sorcerer's apprentice," we do not know how to control the outpouring; we seem to have more documents than we want or need; or else, the ones we want and need are diluted with too many which are not relevant. Information storage and retrieval are still based on the printing press and paper, the filing cabinet, and library shelves. For the reader, microfiche and microfilm, introduced to save space and costs, are inferior to paper. Despite the progress that has been and is being made in computerized information systems and in telecommunications, most of us will continue to read and file printed documents for some time to come. But the paperless society is on the horizon and will be part of our future.

New words and Expressions

literature ['litəritʃə] n. 文献
matrix ['meitriks] n. 摇篮,来源,源泉
delineate [di'linieit] vt. 叙述,描绘
bound [baund] n.(pl.) 界线,范围
artisan [ˌɑti'zæn] n. 工匠
obsolesence [ˌɔbsə'lesns] n. 萎缩,过时
curriculum [kə'rikjuləm] n. 全部课程,必修课程
segment ['segmənt] vt. 分割
discipline ['disiplin] n. 学科
  subdiscipline n. 分支,分科
paradoxical [ˌpærə'dɔksikəl] a. 反常的,反论的
funding [ˌfʌdiŋ] n.(提供)基金
multitude ['mʌltitjuːd] n. 许多,大量
encyclopedia [enˌsaiklou'piːdjə] n. 百科全书
treatise ['triːtiz] n. 论文
patent ['peitənt] n. 专利
retrospective [ˌretrou'spektiv] a. 追溯的
awareness [ə'wɛənis] n. 意识
vade mecum ['veidi 'miːkəm] n. 手册,袖珍指南
sine qua non [ˌsaini kwei 'nɔn] n. 必要条件
consult [kən'sʌlt] vt. 请教
legal ['liːgəl] a. 法律上的
conduct ['kɔndəkt] n. 进行,实施
abstract ['æbstrækt] n. 萃取物,摘要
pamphlet ['pæmflit] n. 小册子
pharmaceutisch   药物(法文)
Centralblatt = Zentralblatt(德文) 摘要,文摘
evolve [i'vɔlv] vt.vi. 发展,进化

online ['ɔnlain] n. 联机,在线(计算机术语)
data base   n. 数据库
broker ['broukə] n.(买卖)中介人经纪人
intermediary [ˌintə(ː)'miːdjəri] n. 媒介
maze [meiz] n. 迷宫,迷津
update [ʌp'deit] vt.n. 现代化
salient ['seiljənt] a. 显著的,突出的
concordance [kən'kɔːdəns] n. 词汇索引
luxury ['lʌkʃəri] n. 奢华,享受
tremendous [tri'mendəs] a. 非常重要的
pervasive [pəː'veisiv] a. 扩大的,渗透的
maintenance ['meintinəns] n. 保持
wand [wɔnd] n. 短杖,棒,棍
sorcerer ['sɔːsərə] n. 魔术师
apprentice [ə'prentis] n. 学徒
outpour [aut'pɔː] vt.vi. 泻出,流出
retrieval [ri'triːvəl] n.(可)修补
file [fail] n. 文件
cabinet ['kæbinit] n. 橱,柜,小室
microfiche ['maikrəufiːʃ] n. 缩微胶片

词 组

affect (vt.) ……as……   以……身份影响(作用)于……
insurance against……   确保免于……
engage in……   从事于……
offer……an opportunity for……   为……提供一个干……的机会
gain familiarity with……   熟悉……
a multitude of……   很多
be segmented into……   被分割为……

a sine qua non for…… 对……是一个绝对必要的条件
cut one's teeth on… 从……处开始懂事
be aware of …… 意识到
in excess of…… 超过
put……into perspective 给予透彻的说明
be inferior to…… 不如
on the horizon 刚冒出地平线
evolve……into 演化……成……
be steeped in…… 埋头于……
gain access to……接近……

Notes

1. "To limit thinking……within a decade"系简单句,to limit 引导的不定式短语作主语,句中有两个并列的谓语,即 makes 和 courts。**译文**:"局限于正规教育和正规培训范围的想法将使人们成为工匠,使我们的科学成为一种技艺,要不了十年,便会导致技术的过时"。
2. "Although the activities……information science"为复合句,"The courses……句末"为主句,主句是由两个并列句构成的复合句,"Although……universities"为让步状语从句,其中 the activities 为主语,谓语是 are taught。**译文**:"尽管在高等院校讲授着这些化学家所从事的工作,但这些课程不算是化学学科的必修课,也不构成化学信息科学的必修课"。
3. "Most scientists……of the past"该句为简单句,句中 above,意思是"超出,超出……之外"。**译文**:"大多数科学家和工程师更喜欢对未来的憧景超出过去历史之外。"

Exercises

1. Fill in the blanks with proper words or phrases.

   In academics and scholarship, a reference or bibliographical reference is a ___ of information provided in a ___ or bibliography of a ___ work such as a book, article, ___, report, oration or any other ___ type, specifying the written work of ___ person used in the creation of that text. The ___ purpose of references is to allow readers to ___ the sources of a text, either for validity or to learn more about the subject.

    A. text   B. written   C. piece   D. primary
    E. essay   F. examine   G. another   H. footnote

2. Match the following explanation with the proper word.

   ___ literature, ___ funding, ___ patent, ___ conduct, ___ tremendous

    A. a set of exclusive rights granted by a state to an inventor for a limited period of time in exchange for the public disclosure of an invention

    B. all the books, articles etc. on a particular subject

    C. to carry out a particular activity or process, especially in order to get information or prove facts

    D. money that is provided by an organization for a particular purpose

    E. very big, fast, powerful etc.

延伸阅读

APPENDIX I

COMMON LABORATORY EQUIPMENT
化学实验室常用仪器

Any suitable type of apparatus may be used for the experiments described in this manual. The authors recommend, however, that standard taper equipment be employed because it is so easily adaptable to the needs of various laboratory procedures that a few pieces serve many purposes. Furthermore, in addition to its advantage in convenience and in saving much of the student's time, its use introduces the student to modern research techniques.

The recommended list of apparatus is shown below. Unless otherwise indicated, all ground joints are of the 24/40 size.

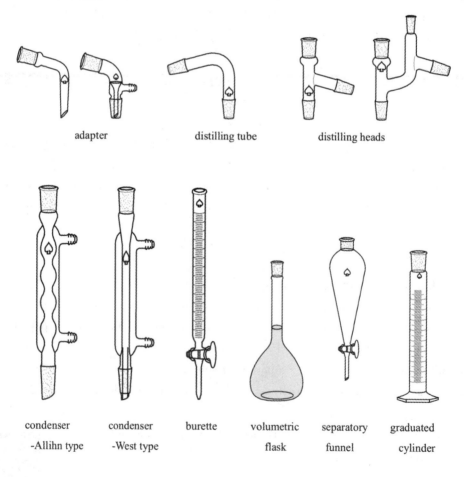

adapter　　　distilling tube　　　distilling heads

condenser　　condenser　　burette　　volumetric　　separatory　　graduated
-Allihn type　-West type　　　　　　flask　　　　funnel　　　cylinder

CHEMISTRY ENGLISH

boiling flask

boiling flask
-2-neck

boiling flask
-3-neck

filter flask
(suction flask)

Erlenmeyer
flask

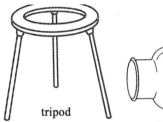
tripod

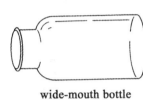
wide-mouth bottle

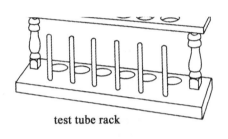
test tube rack

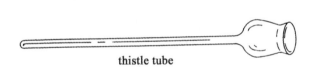
thistle tube

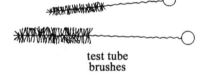
test tube brushes

pipette   clamp

wire-gauze

file

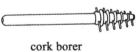
cork borer

forceps

porcelain crucible

watch glass

evaporation dish

porcelain spoon and spatula

clay triangle

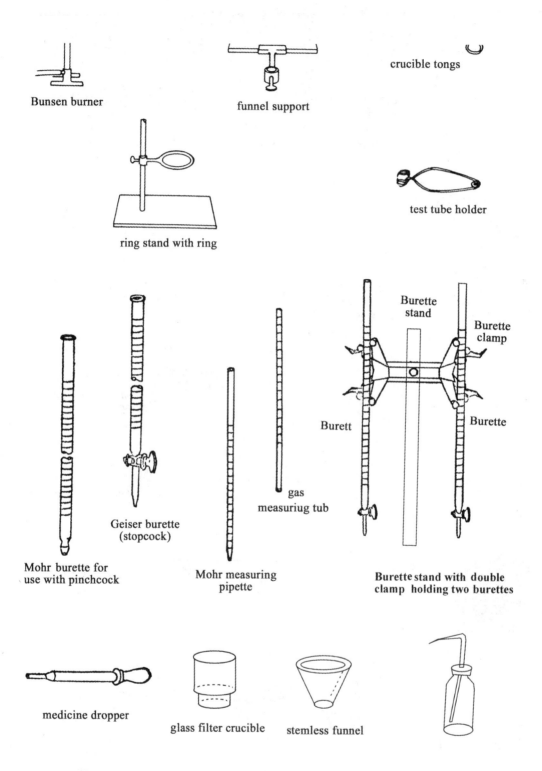

New words and Expressions

adapter [əˈdæptə] n. 接液管
air condenser [kənˈdensə] 空气冷凝管
beaker [ˈbiːkə] n. 烧杯

boiling flask [ˈbɔiliŋ flɑːsk] 烧瓶
boiling flask-3-neck 三颈烧瓶
Büchner funnel [ˈbuːknəˈfʌnl] 布氏漏斗

burette(=buret)[bjuəˈret] n. 滴定管
　～ clamp[ˈklæmp]滴定管夹
　～ stand[stænd]滴定管架(台)
　～ stand with double clamp holding two burettes　持两个滴定管的双夹滴定管架(台)
burner[ˈbə:nə] n. 炉子,燃烧器,灯头
Bunsen burner[ˈbunsn ˈbə:nə] 本生灯;煤气灯
Claisen distilling head[kˈlæsən disˈtiliŋ hed] 克莱森蒸馏头,减压蒸馏头
clamp[ˈklæmp] n. 夹子,夹板
clamp holder [ˈhəuldə] 持夹器
clay triangle [klei ˈtraiæŋgl] 泥三角
cork borer [kɔ:k ˈbɔ:rə] 软木钻孔器
condenser[kənˈdensə] n. 冷凝器
condenser-Allihn type[ˈælin taip] 球型冷凝管
condenser-west type [west taip]直型冷凝管
crucible[ˈkru:sibl] n. 坩埚
　～ tongs[tɔŋs] 坩埚钳
　～ with cover 带盖的坩埚
distilling head[disˈtiliŋ hed] 蒸馏头
distilling tube[tju:b] 蒸馏管
Erlenmeyer flask[ˈerlənmaiə flɑ:sk]n.三角烧瓶,锥型瓶,艾伦迈耶烧瓶
evaporating dish (porcelain)[iˈvæpəreitiŋ diʃ]蒸发皿(瓷的)
extension[iksˈtenʃən] clamp 铁夹子,延伸夹
file [fail] n.锉刀
filter flask(=suction flask)[ˈfiltə flɑ:sk] 吸滤瓶,抽滤瓶
flame spreader [feim spredə] 火焰扩张器,鱼尾灯头
Florence[ˈflɔrəns] flask 平底烧瓶,佛罗伦萨烧瓶
forceps[ˈfɔ:seps] n. 镊子
fractionating column[ˈfrækʃəˈneitiŋ ˈkɔləm] 分馏柱,精馏柱

funnel[ˈfʌnl] n. 漏斗
　～ support 漏斗架
gas measuring tube[ˈmeʒəriŋ tju:b] 气体量管
Geiser[ˈgaisə] burette (stopcock)[ˈstɔpkɔk] 酸滴定管,活塞式滴定管
glass filter crucible 玻璃过滤坩埚
graduated cylinder[ˈgrædjueitid ˈsilində] 量筒
Griffin beaker[ˈgrifin ˈbi:kə] 烧杯,格里芬烧杯
ground joint[graund dʒɔint]磨口接头
Hirsch funnel [həʃ ˈfʌnl]赫尔什漏斗,赫氏漏斗
long-stem funnel[lɔŋstem ˈfʌnl] 长颈(柄)漏斗
medicine dropper[ˈdrɔpə] 滴管,医用滴管
Mohr[mɔə] burette 莫尔滴定管,碱式滴定管
Mohr measuring pipette[ˈmeʒəriŋ piˈpet] 莫尔吸量管,吸量管
mortar[ˈmɔ:tə]n.研钵
pestle[ˈpesl,ˈpestl] n.研杵,研槌;v.研碎
pinch[pintʃ] clamp 弹簧夹
pinchcock[pintʃˈkɔk] n. 弹簧节流夹
pipette(=pipet)[piˈpet] n. 吸液管,吸量管
plastic squeeze bottle[skwi:z ˈbɔtl]塑料挤压瓶,塑料洗瓶
porcelain[ˈpɔ:slin] a .n. 瓷的,瓷制的,瓷器
　～ crucible 瓷坩埚
　～ spoon and spatula 瓷勺和刮铲
reducing bushing[riˈdju:siŋ buʃiŋ]减压套管,减压衬圈
ring stand[riŋ stænd] 铁架台,铁环架
ring clamp　铁环夹
rubber pipette bulb[ˈrʌbə piˈpet bʌlb] 橡胶吸球,吸耳球
screw[skru:] clamp 螺旋夹
separatory funnel[ˈsepərətəri ˈfʌnl]分液漏

斗
standard taper['stændəd 'teipə] equipment 标准锥度仪器,标准接口仪器
stemless[s'temlis] funnel 无颈(柄)漏斗
stirring rod['stə:riŋ rɔd] 搅拌棒
stopcock['stɔpkɔk] n. 活塞,活栓,龙头
test tube 试管
test tube brush 试管刷
test tube holder 试管夹
test tube rack 试管架

Thiele['θi:l] melting point tube 提勒熔点管
Thistle['θisl] tube 蓟头漏斗
transfer pipette 移液管
tripod['traipɔd] n. 三角架
volumetric['vɔlju'metrik] flask 容量瓶
watch glass 表面皿,表玻璃
wide-mouth bottle 广口瓶
wing top[wiŋ tɔp] 鱼尾灯头
wire-gauze(=wire mesh) (石棉)铁丝网

APPENDIX II

SPECIALIST HETEROCYCLIC NOMENCLATURE
杂环化合物的命名

Rule B-1 Extension of Hantzsch-Widman System

1.1—Monocyclic compounds containing one or more hetero atoms in a three- to ten-membered ring are named by combining the appropriate prefix or prefixes from Table I (eliding "a" where necessary) with a stem from Table II. The state of hydrogenation is indicated either in the stem, as shown in Table II, or by the prefixes "dihydro-", "tetrahydro-", etc., according to Rule B-1.2.

TABLE I. In decreasing order of priority

Element	Valence	Prefix	Element	Valence	Prefix
Oxygen	II	Oxa	Antimony	III	Stiba*
Sulfur	II	Thia	Bismuth	III	Bisma
Selenium	II	Selena	Silicon	IV	Sila
Tellurium	II	Tellura	Germanium	IV	Germa
Nitrogen	III	Aza	Tin	IV	Stanna
Phosphorus	III	Phospha*	Lead	IV	Plumba
Arsenic	III	Arsa*	Boron	III	Bora
Mercury	II	Mercura			

* When immediately followed by "-in" or "-ine", "phospha-" should be replaced by "phosphor-", "arsa-" should be replaced by "arsen-" and "stiba-" should be replaced by "antimon-". In addition, the saturated six-membered rings corresponding to phosphorin and arsenin are named phosphorinane and arsenane. Further exceptions: borin is replaced by borinane.

TABLE II

No. of members in the ring	Rings containing nitrogen Unsaturation (a)	Saturation	Rings containing no nitrogen Unsaturation (a)	Saturation
3	-irine	-iridine	-irene	-irane(e)
4	-ete	-etidine	-ete	-etane
5	-ole	-olidine	-ole	-olane
6	-ine(b)	(c)	-in(b)	-ane(d)
7	-epine	(c)	-epin	-epane
8	-ocine	(c)	-ocin	-ocane
9	-onine	(c)	-onin	-onane
10(f)	-ecine	(c)	-ecin	-ecane

(a) Corresponding to the maximum number of non-cumulative double bonds, the hetero elements having the normal valences shown in Table I.

(b) For phosphorus, arsenic, antimony and boron, see the special provisions of Table I.

(c) Expressed by prefixing "perhydro" to the name of the corresponding unsaturated compound.

(d) Not applicable to silicon, germanium, tin and lead. In this case, "perhydro-" is prefixed to the name of the corresponding unsaturated compound.

(e) The syllables denoting the size of rings containing 3, 4 or 7~10 members are derived as follows: "ir" from tri, "et" from tetra, "ep" from hepta, "oc" from octa, "on" from nona, and "ec" from deca.

(f) Rings with more than ten members are named by replacement nomenclature (cf. Rule B-4).

* It is necessary to elide the final "a" of the prefix when followed immediately by a vowel, e.g. ox(a)azole.

Examples:

Oxirane   Aziridine   2H-Azepine

1.2—Heterocyclic systems whose unsaturation is less than the one corresponding to the maximum number of non-cumulative double bonds are named by using the prefixes "dihydro-", "tetrahydro-", etc.

In the case of 4- and 5-membered rings, a special termination is used for the structures containing one double bond, when there can be more than one non-cumulative double bond.

No. of members of the partly saturated rings	Rings containing nitrogen	Rings containing no nitrogen
4	-etine	-etene
5	-oline	-olene

Examples:

Δ^3-1,2-Azarsetine   3-Silolene

1.3—Multiplicity of the same hetero atom is indicated by a prefix "di-" "tri-", etc., placed before the appropriate "a" term (Table I).

Example:

1,3,5-Triazine

1.4—If two or more kinds of "a" terms occur in the same name, their order of citation is in order of their appearance in Table I of Rule B-1.1.

Examples:

*As exceptions, Greek capital delta (Δ), followed by superscript, locant(s), is used to denote a double bond in a compound named according to Rule B-1.2 if its name is preceded by locants for hetero atoms; and also to denote a double bond uniting components in an assembly of rings (cf. Examples to Rules A-52.1 and C-71.1) or in conjunctive names.

1.51-The position of a single hetero atom determines the numbering in a monocyclic compound.

Example:

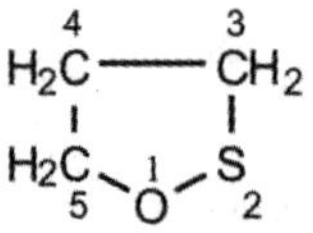

1,2-Oxathiolane 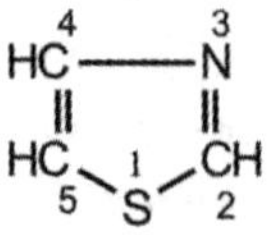   1,3-Thiazole

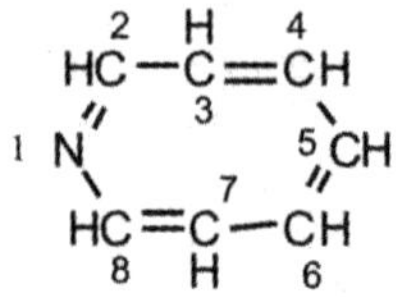

Azocine

1.52-When the same hetero atom occurs more than once in a ring, the numbering is chosen to give the lowest locants to the hetero atoms.

Example:

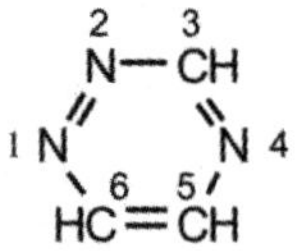

1,2,4-Triazine

1.53-when hetero atoms of different kinds are present, the locant 1 is given to a hetero atom which is as high as possible in Table I. The numbering is then chosen to give the lowest locants to the hetero atoms.

Example:

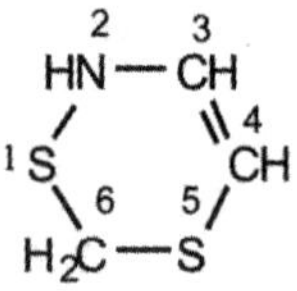

2H,6H-1,5,2-Dithiazine
(not: 1,3,4- Dithiazine)
(not: 1,3,6- Dithiazine)
(not: 1,5,4- Dithiazine)

The numbering has to begin with a sulfur atom. The choice of this atom is determined by the set of locants which can be attributed to the remaining hetero atoms of any kind.

As the set 1,2,5 is lower than 1,3,4 or 1,3,6 or 1,5,4 in the usual sense, the name is 1,5,2-dithiazine.

Rule B-2. Trivial and Semi-trivial Names

2.11-The following trivial and semi-trivial names constitute a partial list of such names which are retained for the compound and as a basis of fusion names.

Parent Compound   Radical Name

 Thiophene   Thioenyl (2-shown)

 Furan   Furyl (3-shown)

 Pyran (2H-shown)   Pyranyl (2H-pyran-3-yl shown)

 Isobenzofuran   Isobenzofuranyl (1-shown)

 Chromene (2H-shown) Chromenyl (2H-Chromen-3-yl shown)

 2H-Pyrrole   2H-Pyrrolyl (2H-Pyrrol-3-yl shown)

 Pyrrole   Pyrrolyl (3-shown)

 Imidazole   Imidazolyl (2-shown)

 Pyrazole   Pyrazolyl (1-shown)

 Isothiazole   Isothiazolyl (3-shown)

 Isoxazole   Isoxazolyl (3-shown)

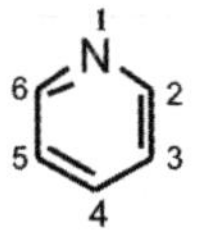

Pyridine

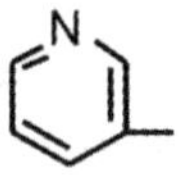

Pyridyl (3-shown)

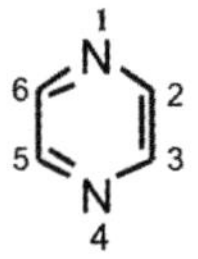

Pyrazine

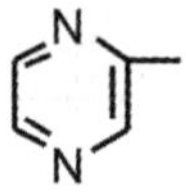

Pyrazinyl

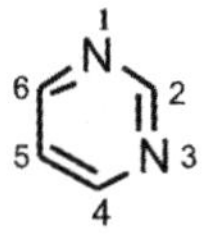

Pyrimidine

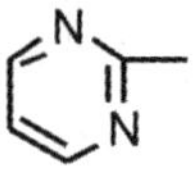

Pyrimidinyl (2-shown)

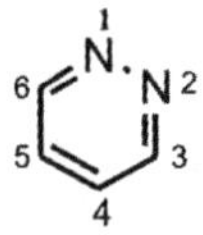

Pyridazine

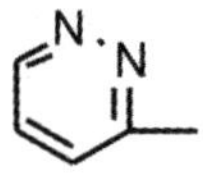

Pyridazinyl (3-shown)

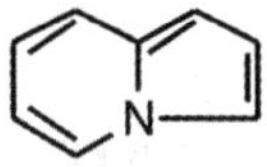

Indolizine

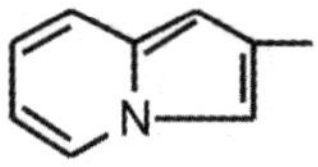

Indolizinyl (2-shown)

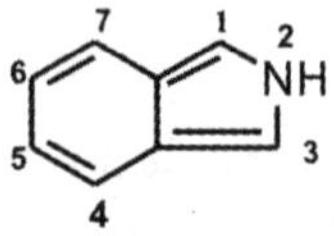

Isoindole

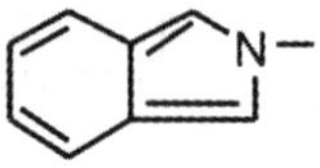

Isoindolyl (2-shown)

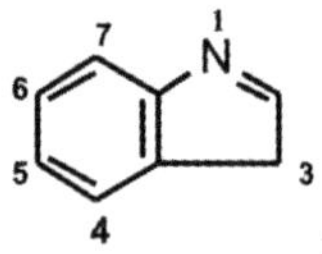

3H-Indole

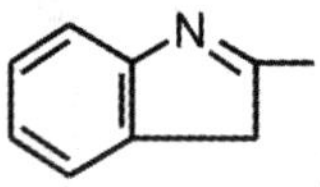

3H-Indolyl (3H-Indol-2-yl shown)

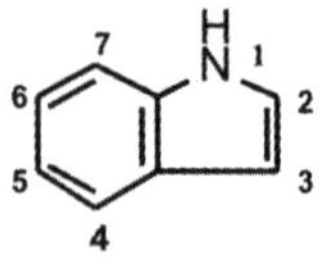

Indole

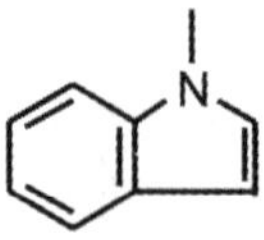

Indolyl (1-shown)

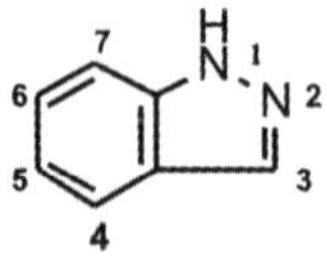

1H-Indazole

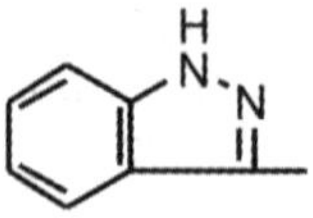

Indazolyl (1H-Indazol-3-yl shown)

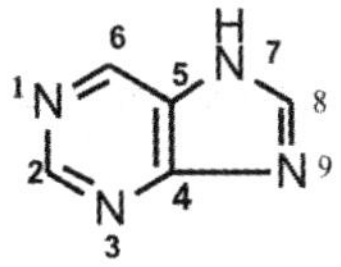

Purine

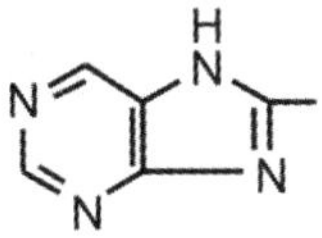

Purinyl (8-shown)

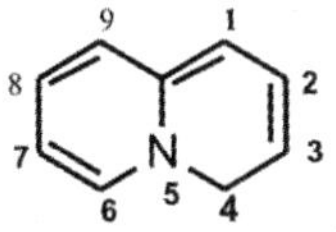

4H-Quinolizine

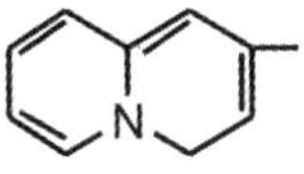

4H-Quinolizinyl (4H-Quinolizin-2-yl shown)

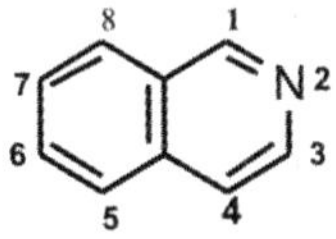

Isoquinoline

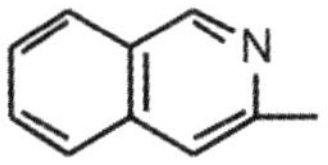

Isoquinolyl (3-shown)

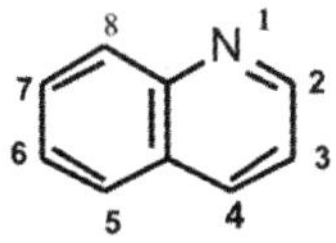

Quinoline

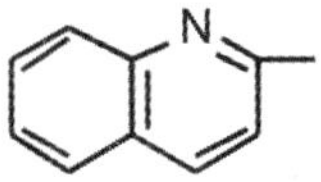

Quinolinyl (2-shown)

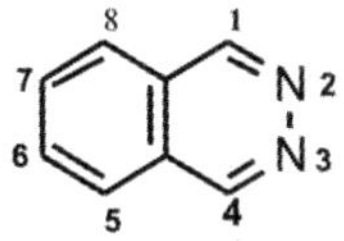

Phthalazine

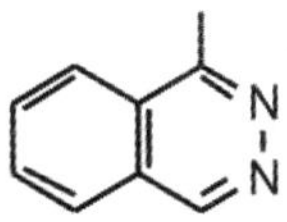

Phthalazinyl (1-shown)

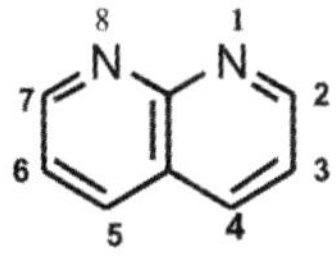

Naphthyridine (1,8-shown)

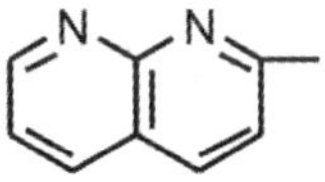

Naphthyridinyl
(1,8-Naphthyridin-2-yl shown)

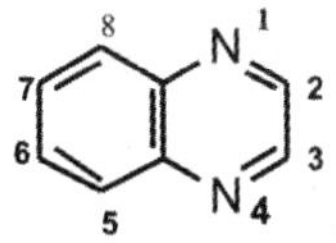

Quinoxaline

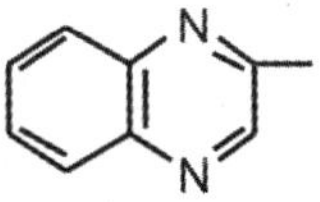

Quinoxalinyl (2-shown)

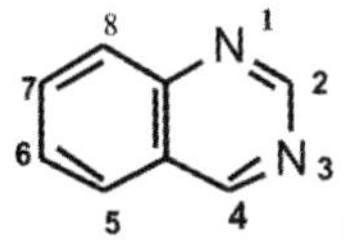

Quinazoline

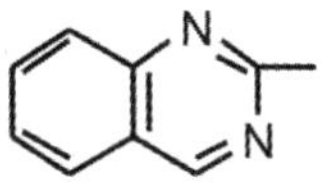

Quinazolinyl (2-shown)

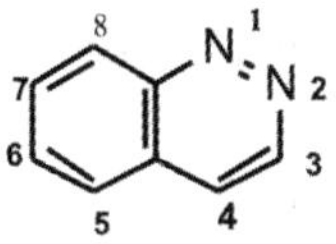

Cinnoline

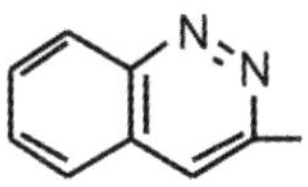

Cinnolinyl (3-shown)

New words and Expressions

heterocyclic [ˌhetərə'saiklik] a. 杂环的
extension [iks'tenʃən] n. 伸展，扩大
stem [stem] n. 词干
phosphorin ['fɔsfɔrin] n. 磷杂环己烯
arsenin ['ɑ:zənin] n. 砷杂环己烯
phosphorinane ['fɔsfɔrinein] n. 磷杂环己烷
arsenane [ɑ:sənein] n. 砷杂环己烷
borin ['bɔrin] n. 硼杂环己烷
borinane ['bɔrinein] n. 硼杂己环
provision [prə'viʒən] n. 规定，供应
syllable ['siləbl] n. 音节
oxazole ['ɔksəzəul] n. 噁唑
oxirane ['ɔksirein] n. 环氧乙烷，氧杂环丙烷
aziridine ['æziridi:n] n. 氮丙啶，氮杂环丙烷
2H-azepine ['æzəpi:n] n. 2H-氮杂环庚烯
Δ3-1,2-azarsetine [æz'ɑ:zəti:n] n. Δ3-1,2-氮杂砷杂环丁烯
3-silolene ['siləuli:n] n. 硅杂啉，硅杂环戊-3-烯
1,3,5-triazine [trai'æzi:n] n. 1,3,5-三嗪，1,3,5-三氮杂苯
oxathiolane [ɔksə'θaiəlein] n. 噁噻烷
triazole [trai'æzəul] n. 三唑
azocine ['æzəusi:n] n. 氮杂环辛四烯
2H,6H-1,5,2-dithiazine [ˌdaiθai'æzi:n] n. 2H,6H-1,5,2-二硫杂氮杂环己烯
fusion ['fju:ʒən] n. 熔化，熔合，合成
parent compound 母体化合物
thiophene ['θaiəfi:n] n. 噻吩
thienyl ['θaiənil] n. 噻吩基
furan ['fjuərən] n. 呋喃
furyl ['fjuəril] n. 呋喃基
pyran ['pairæn] n. 吡喃
pyranyl ['pairənil] n. 吡喃基
isobenzofuran [aiθəˌbenzəu'fjuərən] n. 异苯并呋喃
isobenzofuranyl [aiθəˌbenzəu'fjuərənil] n. 异苯并呋喃基
chromene ['krəumi:n] n. 苯并吡喃
chromenyl ['krəuminil] n. 苯并吡喃基
pyrrole [pi'rəul] n. 吡咯
pyrrolyl [pi'rəulil] n. 吡咯基
imidazole [ˌimi'dæzəul] n. 咪唑
imidazolyl [ˌimi'dæzəulil] n. 咪唑基
pyrazole ['pairəzəul] n. 吡唑
pyrazolyl ['pairəzəulil] n. 吡唑基
isothiazole [aisə'θaiəzəul] n. 异噻唑
isothiazolyl [aisə'θaiəzəulil] n. 异噻唑基
pyridine ['piridi:n] n. 吡啶
pyridinyl ['piridinil] n. 吡啶基
pyrazine ['pirəzi:n] n. 吡嗪
pyrazinyl ['pirəzinil] n. 吡嗪基
pyrimidine [pai'rimidi:n] n. 嘧啶
pyrimidinyl [pai'rimidinil] n. 嘧啶基
pyridazine [ˌpairi'dæzi:n] n. 哒嗪
pyridazinyl [ˌpairi'dæzinil] n. 哒嗪基
indolizine [in'dɔlizi:n] n. 中氮茚
indolizinyl [in'dɔlizinil] n. 中氮茚基，4—N—吲哚基
isoindole [aisə'indəul] n. 异吲哚
isoindolyl [aisə'indəlil] n. 异吲哚基
indole ['indəu(ɔ)l] n. 吲哚，氮杂茚
indolyl ['indɔlil] n. 吲哚基
indazole ['indəzəul] n. 吲唑
indazolyl ['indəzɔlil] n. 吲唑基
purine ['pjuəri:n] n. 嘌呤
purinyl ['pjuərinil] n. 嘌呤基
quinolizine [kwi'nɔlizi:n] n. 喹嗪
quinolizinyl [kwi'nɔlizinil] n. 喹嗪基
isoquinoline [aisə'kwinəli:n] n. 异喹啉
isoquinolyl [aisə'kwinəlil] n. 异喹啉基
quinoline ['kwinəli:n] n. 喹啉
quinolyl ['kwinəlil] n. 喹啉基
phthalazine ['fθæləzi:n] n. 2,3—二氮杂萘

phthalazinyl[ˈfθæləzinil]n.2,3—二氮杂萘基
naphthyridine[ˌnæfθəˈridiːn]n.1,5—二氮杂萘
naphthyridinyl[ˌnæfθəˈridinil]n.1,5—二氮杂萘基
quinoxaline[kwiˈnɔksəliːn]n.1,4—二氮杂萘
quinoxalinyl[kwiˈnɔksəlinil]n.1,4—二氮杂萘基
quinazoline[kwinəˈzəuliːn]n.1,3—二氮杂萘
quinazolinyl[kwnəˈzəulinil]n.1,3—二氮杂萘基
cinnoline[ˈsinəliːn]n.1,2—二氮杂萘
cinnolinyl[ˈsinəlinil]n.1,2—二氮杂萘基

前缀

Oxa-[ɔksə-]氧杂,噁
Thia-[θaiə-]硫杂,噻
Selena-[ˈselinə-]硒杂
Tellura-[ˈteljuərə-]碲杂
Aza-[ˈæzə-]氮杂,吖
phospha-[ˈfɔsfə-]磷杂
Arsa-[ˈɑːsə-]砷杂,砒
Stiba-[ˈstibə-]锑杂,
Bisma-[ˈbizmə-]铋杂
Sila-[ˈsilə-]硅杂
Germa-[ˈdʒəmə-]锗杂
Stanna-[ˈstænə-]锡杂
Plumba-[ˈplʌmbə-]铅杂
Bora-[ˈbɔrə-]硼杂
Mercura-[ˈməkjurə-]汞杂

后　缀

-irine[-iriːn]含氮不饱和三员杂环化合物词尾
-ete[-it]不饱和四员杂环化合物词尾
-ole[-əul]不饱和五员杂环化合物词尾
-ine[-iːn]含氮不饱和六员杂环化合物词尾
-epine[-epiːn]含氮不饱和七员杂环化合物词尾
-ocine[-əusiːn]含氮不饱和八员杂环化合物词尾
-onine[-əuniːn]含氮不饱和九员杂环化合物词尾
-ecine[-esiːn]含氮不饱和十员杂环化合物词尾
-iridine[-iridiːn]含氮三员饱和杂环化合物词尾
-etidine[-etədiːn]含氮四员饱和杂环化合物词尾
-olidine[-əulədiːn]含氮五员饱和杂环化合物词尾
–irene[–iriːn,–əriːn]不含氮不饱和三员杂环化合物词尾
–irane[–irein,–ərein]不含氮饱和三员杂环化合物词尾
–etane[–etein,–ətein]不含氮饱和四员杂环化合物词尾
–olane[–ɔlein]不含氮饱和五员杂环化合物词尾
–in[–in]不含氮不饱和六员杂环化合物词尾
–ane[–ein,–ən]不含氮饱和六员杂环化合物词尾
–epin[–epin,–əpin]不含氮不饱和七员杂环化合物词尾
–epane[–epein,–əpein]不含氮饱和七员杂环化合物词尾
–ocin[–ɔsin]不含氮不饱和八员杂环化合物词尾
–ocane[–ɔkein]不含氮饱和八员杂环化合物词尾
–onin[–ɔnin]不含氮不饱和九员杂环化合物词尾
–onane[–ɔnein]不含氮饱和九员杂环化合物词尾
–ecin[–esin,–əsin]不含氮不饱和十员杂环化合物词尾
–ecane[–ekein,–əkein]不含氮饱和十员杂环化合物词尾

APPENDIX III

ABBREVIATIONS FOR COMMON ORGANIC COMPOUNDS
常用有机化合物英文缩写

英文缩写	英文名称	中文译名
AA	acetylacetone	乙酰丙酮
ABR	acrylnitrile-butadiene-rubber	丁腈橡胶
ABS	acrylonitrile-butadiene-styrene	聚丙烯腈-丁二烯-苯乙烯树脂
Ac	acetyl	乙酰基
ACS	acrylonitrile-chlorizate ethylene-styrene	丙烯腈-氯化乙烯-苯乙烯共聚物
acac	acetylacetonate	乙酰丙酮化合物
AER	anion exchange resin	阴离子交换树脂
AIBN	azobisisobutyronitrile	偶氮二异丁腈
aq.	aqueous	水的,含水的
AR	acrylic rubber	丙烯酸酯橡胶
9-BBN	9-borabicyclo[3.3.1]nonane	9-硼杂双环[3.3.1]壬烷
BD/AN	butadiene-acrynitrile rubber	丁腈橡胶
Bipy, Bpy	2,2′-bipyridyl	2,2′-联吡啶
BMA	polybutylmethacrylate	聚甲基丙烯酸丁酯
Bn	benzyl	苄基
BP, BPO	benzoyl peroxide	过氧化苯甲酰
Bp	biphenyl	联苯
BOC, Boc	t-butoxycarbonyl	叔丁氧羰基
BR	butadiene rubber	聚丁二烯橡胶
B/S	butadiene-styrene	丁苯橡胶
BTX	benzene-toluene-xylene	苯—甲苯—二甲苯
Bu	n-butyl	正丁基
Bz	benzoyl	苯甲酰基
Cat.	catalyst, catalytic, catalyzed	催化剂,催化的
Cbz	carbobenzyloxy	苄氧羰基
CD	2-chlorobutadiene	2-氯丁二烯
CER	cation exchange resin	阳离子交换树脂
cis-BR, PB, CPBR	cis-polybutadiene rubber	顺式聚丁二烯橡胶,顺丁胶
CMC	carboxymethyl cellulose	羧甲基纤维素
COD	cylcooctadiene	环辛二烯
COT	cyclooctatetraene	环辛四烯
Cp	cyclopentadienyl	环戊二烯基

CP	chloroparaffin	氯化石蜡
CPVC	chlorinated polyvinyl chloride	氯化聚氯乙烯
CR	chloroprene rubber	氯丁橡胶
CSA	10-camphorsulfonic acid	10-樟脑磺酸
CTAB	cetyltrimethylammonium bromide	十六烷基三甲基溴化铵
CTAC	cetyltrimethylammonium chloride	十六烷基三甲基氯化铵
2,4-D	2,4-dichlorophenoxyacetic acid	2,4-D;2,4-二氯苯氧乙酸
DABCO	1,4-diazabicyclo[2.2.2]octane	1,4-二氮杂双环[2.2.2]辛烷
DAP	dialkyl phthalate	邻苯二甲酸二烷基酯
DAS	dialkyl sebacate	癸二酸二烷基酯
dba	dibenzylidene acetone	二亚苄基丙酮,1,5-二苯基-1,4-戊二烯-3-酮
DBN	1,5-diazabicyclo[4.3.0]non-5-ene	1,5-二氮杂双环[4.3.0]壬-5-烯
DBP	dibutyl phthalate	邻苯二甲酸二丁酯
DBS	dibutyl sebacate; dodecyl benzene sulfonate	癸二酸二丁酯;十二烷基苯磺酸盐
DBU	1,8-diazabicyclo[5.4.0]undec-7-ene	1,8-二氮杂双环[5.4.0]十一-7-烯
DCC	1,3-dicyclohexylcarbodiimide	二环己基碳二亚胺
DCE	1,2-dichloroethane	1,2-二氯乙烷
DCPD	dicyclopentadiene	双环戊二烯,二聚环戊二烯
DDB	dodecyl benzene	十二烷基苯
DDBS	dodecyl benzene sulfonate	十二烷基苯磺酸盐
DDM	dodecyl mercaptan	十二烷基硫醇
DDP	didecyl phthalate	邻苯二甲酸二癸酯
DDQ	2,3-dichloro-5,6-dicyano-1,4-benzoquinone	2,3-二氯-5,6-二氰基-1,4-苯醌
DDVP	O,O-dimethyl-2,2-dichlorovinylphosphate	敌敌畏
DEA	diethylamine	二乙胺
DEAD	diethyl azodicarboxylate	偶氮二甲酸二乙酯
DEG	diethylene glycol	二甘醇,二乙二醇
Dibal-H	diisobutylaluminum hydride	二异丁基铝氢化物
DIBP	diisobutyl phthalate;	邻苯二甲酸二异丁酯
DIOP	diisooctyl phthalate	邻苯二甲酸二异辛酯
DIPB	diisopropylbenzene	二异丙苯
Diphos(dppe)	1,2-bis(diphenylphosphino)ethane, ethylenebis(diphenylphosphino)	1,2-二(二苯膦基)乙烷
diphos-4(dppb)	1,4-bis(diphenylphosphino)butane	1,4-二(二苯膦基)乙烷
DMAP	4-dimethylaminopyridine	4-二甲氨基吡啶
DME	dimethoxyethane	1,2-二氧基乙烷,乙二醇二甲醚
DMF	N,N'-dimethylformamide	N,N'-二甲基甲酰胺
DMSO	dimethyl sulfoxide	二甲基亚砜
DMT	dimethyl terephthalate	对苯二甲酸二甲酯
DOP	dioctyl phthalate	邻苯二甲酸二辛酯

DOS	dioctyl sebacate	癸二酸二辛酯
dppf	bis (diphenylphosphino)ferrocene	1,1′-二(二苯膦基)二茂铁
dppp	1,3-bis (diphenylphosphino)propane	1,3-二(二苯膦基)丙烷
DTBP	ditertiary butyl peroxide	二叔丁基过氧化物
DVB	divinylbenzene	二乙烯基苯
EDA	ethylenediamine	乙二胺
EDTA	ethylenediaminetetraacetic acid	乙二胺四乙酸
EE	1-ethoxyethyl	1-乙氧乙基
EG	ethylene glycol	乙二醇
EO	ethylene oxide	环氧乙烷
EPC, EPM	ethylene-propylene copolymer	乙丙共聚物
EPR	ethylene-propylene rubber	二元乙丙橡胶
EPT	ethylene-propylene terpolymer	三元乙丙橡胶
Fc	ferrocenyl	二茂铁基
FcH	ferrocene	二茂铁
GR-A	government rubber-acrylonitrile	丁腈橡胶
GR-I	government rubber-isobutylene	丁基橡胶
GR-P	government rubber-polysulfide	聚硫橡胶
GR-S	government rubber-styrene	丁苯橡胶
1,5-HD	1,5-hexadienyl	1,5-己二烯基
HDPE	high density polyethylene	高密度聚乙烯
HFA	hexafluoroacetone; hexafluoroacetylacetone	六氟丙酮;六氟乙酰丙酮
HMPA	Hexamethylphosphoramide, $[(CH_3)_2N]_3PO$	六甲基磷酰胺
HMPT	hexamethylphosphorus triamide, $P[N(CH_3)_2]_3$	三(二甲氨基)膦
IR	isoprene rubber; infrared	异戊二烯橡胶; 红外
I.R.	India rubber	天然橡胶
LABS	linear alkyl benzene sulfonic acid	直链烷基苯磺酸
LABS	linear alkyl benzene sulfonate	直链烷基苯磺酸盐
LAH	lithium aluminum hydride	氢化铝锂
LDA	lithium diisopropylamide	二异丙基氨基锂,二异丙基锂酰胺
LPE	linear polyethylene	线形聚乙烯
MA	maleic anhydride	马来酸酐,顺丁烯二酸酐
mCPBA	*meta*-cholorperoxybenzoic acid	间氯过氧苯甲酸
MEK	methyl-ethyl-ketone	甲乙酮
MeOH	methyl alcohol	甲醇
MM, MMA	(poly)methylmthacrylate	(聚)甲基丙烯酸甲酯
NBD	norbornadiene	降冰片二烯
NBR	nitrile-butadiene-rubber	丁腈橡胶
NBS	N-Bromosuccinimide	N-溴代丁二酰亚胺
NCS	N-chlorosuccinimide	N-氯代丁二酰亚胺
Ni(R)	Raney Nickel	瑞尼镍,骨架镍
NMO	N-methyl morpholine-N-oxide	N-甲基吗啉-N-氧化物

PA	polyacrylate; polyamide	聚丙烯酸酯;聚酰胺
PAA	polyacrylamide; polyacrylic acid	聚丙烯酰胺;聚丙烯酸
PAM	polyacrylamide	聚丙烯酰胺
PAN	polyacrylonitrile	聚丙烯腈
PB, PBD	polybutadiene	聚丁二烯
PBS	polybutadiene styrene	丁苯橡胶
PEG	polyethylene glycol	聚乙二醇
PET, PETP	polyethylene terephthalate	聚对苯二甲酸乙二醇酯
PIB	polyisobutylene	聚异丁烯
PMA	polymethacrylate; polymethyl acrylate	聚甲基丙烯酸酯;聚丙烯酸甲酯
PS	polystyrene	聚苯乙烯
PTFE	polytetrafluoroethylene	聚四氟乙烯
PVA	polivinyl acetate; polyvinyl alcohol	聚醋酸乙烯酯;聚乙烯醇
PVC	polivinyl chloride	聚氯乙烯
SBR	styrene-butadiene rubbers	苯乙烯-丁二烯橡胶
TBAF	tetrabutylammonium fluoride	四丁基氟化铵
TBDMS, TBS	t-butyldimethylsilyl	叔丁基二甲基硅烷基
TBHP	t-butylhydroperoxide	叔丁基过氧化氢
TCA	trichloroacetic acid	三氯乙酸
TDI	toluene diisocyanate	甲苯二异氰酸酯
TEBA	triethylbenzylammonium	三乙基苄基铵
TEMPO	tetramethylpiperdinyloxy free radical	2,2,6,6-四甲基哌啶-N-氧化物自由基
TEP	triethylphosphate	磷酸三乙酯
TFA	trifluoroacetic acid	三氟乙酸
TFAA	trifluoroacetic anhydride	三氟乙酸酐
Tf, OTf	triflate	三氟甲基磺酸酯(CF_3SO_3R)或盐
THF	tetrahydrofuran	四氢呋喃
THP	etrahydropyrane	四氢吡喃
TMEDA	tetramethylethylenediamine	四甲基乙二胺
TMP	trimethyl phosphate;	磷酸三甲酯;
	2,2,6,6-tetramethylpiperidine	2,2,6,6-四甲基哌啶
TMS	tetramethylsilane; trimethylsilyl	四甲基硅烷;三甲基硅烷基
TMTD	thiurame	秋兰姆,福美双
TNT, T.N.T.	trinitrotoluene	三硝基甲苯
Tol	tolyl	甲苯基
TPA	terephthalic acid	对苯二甲酸
TPP	triphenylphosphate;	磷酸三苯酯
Tr	trityl	三苯甲基
Ts (Tos)	tosyl (p-toluenesulfonyl)	对甲苯磺酰基
VC	vinyl chloride	氯乙烯

APPENDIX Ⅳ

ANALYTIC APPARATUS IN COMMON USE
常用分析测试仪器中英文名对照

仪器中文名称	仪器英文名称	英文缩写
氨基酸组成分析仪	Aminoacid Analyzer	AA
原子吸收光谱仪	Atomic Absorption Spectrometer	AAS
原子发射光谱仪	Atomic Emission Spectrometer	AES
原子荧光光谱仪	Atomic Fluorescence Spectrometer	AFS
老化性能测定仪	Aging Property Tester	APT
自动滴定仪	Automatic Titrator	AT
环境成分分析仪	CHN Analyzer	
燃烧性能测定仪	Combustion Property Tester	CPT
电导仪	Conductivity Meter, Conductometer	
介电常数检测器	Dielectric Constant Detector	DCD
差热分析仪	Differential Thermal Analyzer	DTA
差视扫描量热计(法)	Differential Scanning Calorimeter(Calorimetry)	DSC
直流等离子体发射光谱仪	Direct Current Plasma Emission Spectrometer	DCPES
电性能测定仪	Electrical Property Tester	EPT
电解质分析仪	Electrolytic Analyzer	EA
电子能谱仪	Electron Energy Disperse Spectroscopy	EEDS
电子显微镜	Electron Microscope	E. M.
电子核磁双共振	Electron Nuclear Double Resonance	ENDR
电子顺磁共振波谱仪	Electron Paramagnetic Resonance Spectrometer	EPRS
电子探针微量分析器	Electron Probe Microanalyser	EPMA
电泳仪	Electrophoresis System	ES
能谱仪	Energy Disperse Spectroscopy	EDS
摩擦系数测定仪	Friction Coefficient Tester	FCT
傅里叶变换红外光谱仪	FT-Infrared Spectrometer	FT-IRS
傅里叶变换拉曼光谱仪	FT-Raman Spectrometer	FT-RS
气相色谱仪	Gas Chromatograph	GC
气相色谱-质谱联用仪	Gas Chromatograph- Mass Spectrometer	GC-MS
凝胶渗透色谱	Gel Permeation Chromatograph	GPC
高效液相色谱	High Performance Liquid Chromatography	HPLC
高压液相色谱	High Pressure Liquid Chromatography	HPLC
高效薄层色谱	High Performance Thin Layer Chromatography	HPTLC
电感偶合等离子体发射光谱仪	Inductive Coupled Plasma Emission Spectrometer	ICPES
电感耦合等离子体-质谱联用仪	Inductive Coupled Plasma- Mass Spectrometer	ICP-MS

离子色谱	Ion Chromatograph	IC
同位素X荧光光谱仪	Isotope X-Ray Fluorescence Spectrometer	IXRFS
激光拉曼光谱	Laser Raman Spectra	LRS
液相色谱	Liquid Chromatograph	LC
液相色谱-质谱联用仪	Liquid Chromatograph- Mass Spectrometer	LC-MS
质谱仪	Mass Spectrometer	MS
机械性能测定仪	Mechanical Property Tester	MPT
微波等离子体光谱仪	Microwave Inductive Plasma Emission Spectrometer	MIPES
核磁共振波谱仪	Nuclear Magnetic Resonance Spectrometer	NMRS
光学显微镜	Optical Microscope	OM
光学性能测定仪	Optical Property Tester	OPT
粒度分析仪	Particle Size Analyzer	PSA
pH计	pH Meter	
极谱仪	Polarograph	
流变仪	Rheometer	
扫描探针显微镜	Scanning Probe Microscopy	SPM
表面分析仪	Surface Analyzer SA	
热分析仪	Thermal Analyzer	TA
热重量分析	Thermogravimetry	TG
热重—差视扫描量热联用仪	Thermogravimetry- Differential Scanning Calorimeter	TG-DSC
紫外—可见光分光光度计	Ultraviolet-Visible Spectrophotometer	UV-VS
紫外光谱仪	Ultraviolet Spectrophotometer	UV
粘度计	Viscometer	
X射线荧光光谱仪	X-Ray Fluorescence Spectrometer	XRFS
X射线衍射仪	X-Ray Diffractometer	XRD
水质分析仪	Water Test Kits	WTK

APPENDIX V

ABBREVIATIONS AND ACRONYMS USED IN CAS PUBLICATION
美国《化学文摘》中常用词缩写

英文缩写	英文名称	中文译名
a	atto- (10^{-18})	阿(托)(10^{-18});渺
A	ampere	安(培)
Å	angstrom unit(s)	埃(长度单位,10^{-10}米)
abs.	absolute	绝对的
abstr.	abstract	文摘
Ac	acetyl(CH_3CO, not CH_3COO)	乙酰基
a. c.	alternating current	交流电(流)
addn	addition	加成,添加
addnl	additional	附加的,添加的
alc.	alcohol, alcoholic	醇;醇的
aliph.	aliphatic	脂族的
alk.	alkaline(not alkali)	碱性的
alky	alkalinity	碱度,碱性
a.m.	ante meridiem	上午
amt.	amount	数量
amu	atomic mass unit	原子质量单位
anal.	analysis, analytical(ly)	分析;分析的(地)
anhyd.	anhydrous	无水的
AO	atomic orbital	原子轨(道)函数
app.	apparatus	仪器,装置
approx.	approximate(ly)	近似的,大概的
approxn.	approximation	近似法,概算
aq.	aqueous	水的,含水的
arom.	aromatic	芳香族的
assoc.	associate	缔合
assocd.	associated	缔合的
assocg.	associating	缔合的,缔合作用
assocn.	association	缔合
asym.	asymmetric(al)(ly)	不对称的
at.	atomic(not atom)	原子的
atm	atmosphere(the unit)	大气压=1.01325×10^5帕…
atm.	atmosphere, atmospheric	大气,大气的

av.	average (except as a verb)	平均
b.	(followed by a figure denoting temperature) boils at, boiling at (similarly b13, at 1.3mm, pressure)	沸腾(后面数字毫米汞柱压力)
bbl	Barrel	桶[液体量度单位=163.5升(英国), =119升(美国)]
bcc.	body centered cubic	立方体心
BeV or GeV	billion electron volts	10亿电子伏,10^9电子伏
BOD	biochemical oxygen demand	生化需氧量
μB	Bohr magneton	玻尔磁子[物]
b.p.	boiling point	沸点
Bq	becquerel	贝克勒尔[放射性强度单位,等于1秒$^{-1}$, 1 Bq=27.027pCi]
Btu	British thermal unit	英热单位=1055.06焦
bu.	bushel	蒲式耳=36.368升(英)=35.238升(美)
Bu	butyl (normal)	丁基
Bz	benzoyl (C_6H_5CO, not $C_6H_5CH_2$)	苯甲酰
c-	centi- (10^2)	厘(10^2)
C	coulomb	库仑
℃	degree Celsius (centigrade)	摄氏度
Cal	calorie	千卡,大卡=4186.8焦
calc.	calculate	计算
calcd.	calculated	计算的
calcg.	calculating	计算
calcn.	calculation	计算
CD	circurlar dichroism	圆二色性(物)
c.d.	current density	电流密度
cf.	confer	参见(仅用于图书参考文献)
cfm	cubic feet per minute	立方英尺/分
chem..	chemical(ly), chemistry	化学的,化学
Ci	curie	居里(放射单位)=$3.7×10^{10}$衰变/秒
clin.	clinical(ly)	临床的
CoA	coenzyme A	辅酶A
COD	chemical oxygen demand	化学需氧量
coeff.	coefficient	系数
col.	colour, coloration	颜色
com.	commercial(ly)	工业的,商业的,商品的
compd.	compound	化合物,复合物
compn.	composition	组成,成分
conc.	concentrate (as a verb)	提浓,浓缩
concd.	concentrated	浓的
concg.	concentrating	浓缩(的)
concn.	concentration	浓度

cond	conductivity	导电率,传导性
const.	constant	常数,常量
contg.	containing	包含,含有
cor.	corrected	校正的,改正的,正确的
CP	chemically pure	化学纯的
crit.	critical(ly)	临界的
cryst.	crystalline(not crystallize)	结晶
crystd.	crystallized	使结晶
crystg.	crystallizing	结晶
crystn.	crystallization	结晶,结晶化
cwt	hundredweight	1/20吨[英制重112磅,美制重120磅]
d-	deci-(10^{-1})	分(10^{-1})
d	density (d^{13}, density at 13° referred to water at 4°; d^{20}_{20} at 20° referred to water at the same temperature)	密度,比重
D	debye unit	德拜单位,电偶极矩单位
d.c.	direct current	直流电
decomp.	decompose(s)	分解
decompd.	decomposed	分解的
decompg.	decomposing	分解
decompn.	decomposition	分解
degrdn.	degradation	降解
derive.	derivative	衍生物,导数(数)
det.	determine	测定
detd.	determined	测定的
detg.	determining	测定
detn.	determination	测定
diam.	diameter	直径
dil.	dilute	稀释,冲淡
dild.	diluted	稀释的
diltg.	diluting	稀释
diln.	dilution	稀释
dissoc.	dissociate	离解
dissocd.	dissociated	离解的
dissocn.	dissociation	离解
distd.	distilled	蒸馏的
distg.	distilling	蒸馏
distn.	distillation	蒸馏
d.p.	degree of polymerization	聚合度
dpm	disintegrations per minute	分解量/分钟
ECG	electrocardiogram	心电图
ED	effective dose	有效剂量
EEG	electroencephalogram	脑电图

e.g.	for example 例如	
elec.	electric, electrical(ly)	电的
emf	electromoctive force	电动势
emu	electromagnetic unit	电磁单位
en	ethylenediamine	乙二胺
equil	equilibrium(s)	平衡
equiv.	equivalent	当量,克当量
esp.	especially	特别,格外
est.	estimate	估计
estd.	estimated	估计的
estg.	estimating	估计
estn.	estimation	估计
esu	electrostatic unit	静电单位
Et	ethyl	乙基
et al	and others	等等(人、地方)
etc.	et cetera	等等(不用于人)
eV	electron volt	电子伏[特]
evap.	evaporate	蒸发
evapd.	evaporated	蒸发的
evapg.	evaporating	蒸发
evapn.	evaporation	蒸发
examd.	examined	检验过的,试验过的
examg.	examining	检验,试验
examn.	examination	检验,试验
expt.	experiment	实验
exptl.	experimental	实验的
ext.	extract	提取物,萃,提取
extd.	extracted	提取的
extg.	extracting	提取
extn.	extraction	提取
F	farad	法[拉](电容)F
°F	degree Fahrenheit	华氏度
fcc.	face centered cubic	面心立方体
fermn.	fermentation	发酵
f.p.	freezing point	冰点,凝固点
ft	foot	英尺=0.3048米
ft-lb	foot-pound	英尺磅=0.3048米×0.453592千克
g	gram(s)	克
g	gravitational constant	万有引力常数
(g)	gas, only as in $H_2O(g)$	气态,气体
G	gauss	高斯(磁感应强度单位)
G-	giga-(10^9)	十亿

gal	gallon	加仑=4.546092升(英)=3.78543升(美)
gr	grain(weight unit)	谷(1谷=1/7000磅=0.64799克)
Gy	gray (absorbed radiation dose)	戈瑞(吸附辐射剂量,Gy=100rad)
h	hour	小时
h-	hector- (10^2)	百(10^2)
H	henry	亨利(电感单位)
Ha	hectare	公顷=6.451600×10^{-4}米^2
Hb	hemoglobin	血红蛋白
hcp	hexagonal close-packed	六方密堆积的
Hz	hertz(cycles/sec)	赫[兹],周/秒
ID	inhibitory dose	抑制剂量
i.e.	that is	即,也就是说
Ig	immunoglobulin	免疫球蛋白
i.m.	intramuscular(ly)	肌肉内的
in.	inch	英寸=0.0254米
inorg.	inorganic	无机的
insol.	insoluble	不溶的
i.p.	intraperitoneal(ly)	腹膜内的
IR	infrared	红外线
irradn.	irradiation	照射
IU	International Unit	国际单位
i.v.	intravenous(ly)	静脉内的
J	joule	焦[耳](能量单位)
k.	kilo-(10^3)	千(10^3)
K	kelvin	开[尔文],绝对温度
Km.	Michaelis constant	米氏常数
L	liter	升
(l)	liquid, only as in $NH_3(l)$	液态的,液体
lab.	laboratory	实验室
lb	pound	磅=0.453592千克
LC	lethal concentration	致死浓度
LCAO	linear combination of atomic orbitals	原子轨道的线性组合
LD	lethal dose	致死剂量
LH	luteinizing hormone	促黄体发生激素
liq.	liquid	液体,液态
lm	lumen	流明(光通量单位)
lx	lux	勒[克斯](照度单位)
m	meter	米
m	molal	摩尔的
m-	milli-(10^3)	毫;千分之一
m.	melts at, melting at	熔融
M	molar(as applied to concn.)	摩尔
M.	mega-(10^6)	兆(10^6)

manuf.	manufacture	制造
manufd.	manufactured	制造的
manufg.	manufacturing	制造
math.	mathematical(ly)	数学的
max.	maximum(s)	最大值,最大的
Me	methyl (not metal)	甲基
mech.	mechanical(ly) (not mechanism)	机械的
metab.	metabolism	新陈代谢
mi	mile	英里=1609.344米
min	minute (time)	分钟
min.	minimum (s)	最小值,最小的
misc.	miscellaneous	其它,杂项
mixt.	mixture	混合物
MO	molecular orbital	分子轨道函数
mo	month	月
mol	mole (the unit)	摩尔
mol.	molecule, molecular	分子,分子的
m.p.	melting point	熔点
mph	miles per hour	英里(=1609.344米)/小时
μ-	micro- (10^6)	微米,百万分之一
Mx	maxwell	麦克斯韦(物)(磁量单位)
n	refractive index	折射率,折光率
n-	nano- (10^{-9})	毫微,十亿分之一
N	newton	牛[顿](力的单位)
N	normal (as applied to concentration.)	当量(浓度)
neg.	negative(ly)	阴性的,负的
no.	number	号,数
obsd.	observed	观察,观测
Oe	oersted	奥斯忒(磁场强度单位)
Ω	ohm	欧姆(电阻单位)
org.	organic	有机的
oxidn.	oxidation	氧化
oz	ounce	盎司(常衡=28.349523克)
P-	pico- (10^{-12})	微微
P	poise	泊(粘度单位)
P-	peta- (10^{15})	千万亿
Pa	pascal	帕(斯卡)(压力单位)
p.d.	potential difference	势差,电位差
Ph	phenyl	苯基
phys.	physical(ly)	物理的
p.m.	post meridiem	午后
polymd.	polymerized	聚合的
polymg.	polymerizing	聚合

ploymn.	polymerization	聚合
pos.	positive(ly)	阳性的,正的
powd.	powdered	粉末的,粉状的
ppb	parts per billion	亿万分之(几)
ppm	parts per million	百万分之(几)
ppt.	precipitate	沉淀,沉淀物
pptd.	precipitated	沉淀出的
pptg.	precipitating	沉淀
pptn.	precipitation	沉淀
Pr	propyl (normal)	丙基
prep.	prepare	制备
prepd.	prepared	制备的
prepg.	preparing	制备
prepn.	preparation	制备
psi	pounds per square inch	磅/英寸^2 [=0.453592千克/(6.45100 × 10^{-4}米^2)]
psia	pounds per square inch absolute	磅/英寸^2(绝对压力)
pt	pint	品脱(=0.5682615升)
purifn	purification	精制
py	pyridine(used in Werner complexes only)	吡啶
qt	quality	质量
qual.	qualitative(ly)	定性的
quant.	quantitative(ly)	定量的
R	roentgen	伦琴
redn.	reduction	还原,减小
ref.	reference	参考文献
rem	roentgen equivalent man	人体伦琴当量,雷姆
rep	roentgen equivalent physical	物理伦琴当量
reprodn.	reproduction	再生产,再生
resoln.	resolution	分辨,分解,离析
resp.	respective(ly)	分别地
rpm	revolutions per minute	每分钟转数
RQ	respiratory quotiente	呼吸系数,呼吸商数
s	second (time unit only)	秒
(s)	solid, only as in AgCl(s)	固态,固体
S	siemens	西门子(电)(欧姆的倒数)
sapon.	saponification	皂化
sapond.	saponified	皂化过的
sapong.	saponifying	皂化
sat.	saturate	使饱和
satd.	saturated	饱和的
satg	saturating	饱和的
satn.	saturation	饱和,饱和度

s.c.	subcutaneous(ly)	皮下的
SCE	saturated calomel electrode	饱和甘汞电极
SCF	self-consistent field	自洽场
sec	secondary (with alkyl groups only)	仲,第二的
sep.	separate(ly)	分离
sepd.	separated	分离出的
sepg.	separating	分离的
sepn.	separation	分离
sol.	soluble	可溶的
soln.	solution	溶液
soly.	solubility	可溶性,溶解度
sp.	specific (used only to qualify physical constant)	比的,特殊的
sp.gr.	specific gravity	比重
sp.ht.	specific heat	比热
sr	steradian	球面度,立体弧度
St	stokes	斯托克斯(运动粘度单位)
std.	standard	标准
sym.	symmetric(al)(ly)	对称的
T	tesla	特斯拉(物)(磁通量单位)
T-	tera- (10^{12})	万亿(10^{12})
tbs	tablespoon	大汤匙
tech.	technical(ly)	技术的
temp.	temperature	温度
tert	tertiary (with alkyl groups only)	叔,第三的
thcor.	theoretical(ly)	理论的
thermodn.	thermodynamic(s)	热力学
titrn.	titration	滴定
tsp	teaspoon	茶匙
USP	United States Pharmacopeia	美国药典
UV	ultraviolet	紫外(光)
V	volt	伏[特]
Vmax	maximum velocity	最大速度
vol.	volume (not volatile)	体积
vs.	versus	对
W	watt	瓦[特]
Wb	weber	韦伯(电)(磁通量单位)
wk	week	星期
wt.	weight	重量
yd	yard	码
yr	year	年

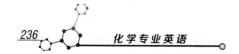

APPENDIX Ⅵ

WORD GROUP
词组

(词组后的数字代表的是该词组出现在第几课课文中)

A

(be)abbreviated from…	1
(be)able to(+*inf.*)	23
accompanied by…	2
act as…	4
added to…	3
affect(*vt.*)…as…	40
a fraction of…	6
a great (good) many of…	6
a host of…	23
(be)akin to…	18
a set of…	25
(be) associate with…	25
a multitude of…	40
analysis of…into…	33
an educated guess…	32
a number of…	4
applicable to…	25
apply to…	10
arise from (out of)…	5
armed with…	21
arrive at a conclusion…	31
arrive at conclusions…	31
as a consequence of…	32
as a result (of…)	4
a series of…	9
a set of…	25
as follows	31
a sine qua non for…	40
as long as…	32
as may be expected (anticipated)	5
as opposed to…	2
assign…to…	17
associated with	35
as to…	26
…as well	1
as with…	4
as well as…	21
as with…	25
at a glance	17
at no time	31
at ordinary condition	2
(be) attached to…	2
attach to…	13
at…temperature	2
at the outset	31
at (three) times the rate of…	33
attributable to…	27
a variety of…	1
aware of…	6
a wide range of…	6

B

(be)based (up) on…	1
be appropriate for…	33
be assigned	5
be attributed to…	5
be available in…	23
be bound up with…	33
be cause for…	23
be characteristic of…	17
be characterized by…	19
be classified as (into)…	19
be cleaved to…	5
become familiar with…	16
be comparable with…	21
be consistent with…	30

be converted into…	17
be crucial to…	30
be derived from…	16
be designated as…	19
be equal to…	17
be eqivalent to (prep.)…	33
be essential for…	30
be followed by…	33
begin with	1
be independent of…	23
be inferior to…	40
be judged by…	21
be known to…	6
be linked to…	16
be of…	21
be of better quality…	23
be of intense concern to…	23
be proportional to…	33
be protected from…	20
be referred to as…	34
be relevant to…	21
be segmented into…	40
be separated from…	20
be steeped in	40
be suitable for…	33
be unsetting to…	23
billions of times	23
bog down	30
bonded to…	2
bounce back off…	34
break down (into)	3
bring about	25
bring into contact with…	34
build up…	4
by contrast (with)	33
by convention	32
by definition	28
by means of…	14
by summing over	39
by virtue of…	37

C

capable of (+ing)	1
carry through…	21
change …for…	21

change from A to B	2
charged with…	37
classify…into…	5
combine with…	2
come about	6
come from…	16
comparison (of…)with…	5
compete (with…)for…	8, 34
complement A with B	5
composed of…	1
concentrate (up) on…	25
concerned in	38
conform to…	21
connect…to…	4
consistent with…	12
consist in…	11
consist of…	1
contrast with…	21
contrast to (with)	8
contribute to…	3
convert A (in)to B…	2
correspond to…	7
cut one's teeth on…	40

D

day to day…	39
deal with…	21
decrease(vi.)in…	2
define as	6
depend (up) on…	8
derive…from…	8
different from…	2
differentiate A from B	7
differ from…	11
differ in… (by…)	10
displacement of A from B by C…	5
distinguish between…	8
distinguished from…	29
divide A by B	28
divide A into B and C	1
do justice to…	34
draw off	37
draw out…	22
(be)due to + n.	9

E

each time…	25
either…or…	17
emerge from…	30
employ…as…	4
enable…to(+ing)	6
end-to-end	22
end (up) with…	27
engage in…	40
enough(+ a.)to (+inf.)	8
enter into…	9
equal to…	1
equivalent to…	35
evolve…into…	40
except for…	1
exist as…	1
exist in…form	2
expand (up) on	31
expose A to B	4
exposed to…	4
express…in (+单位)	10

F

fall off	34
familiar with…	6
feed…into…	37
fit into…	21
focus one's attention to…	5
followed by…	1
follow from…	7
for convenience	1
form from…	2
for purposes of…	26
for the most part	29
for the sake of	37
for this reason	8
free of (from)…	15
free to (+inf.)	31
from…on	7
furnish sb.with sth.	25

G

gain access to…	40
gain familiarity with…	40

give A to B	4
go into…	21
good to (+inf.)	29

H

hammer into…	3
have acquaintance of	32
have both…and	5
have the advantage over (of)…	3
hold together	6

I

identical with…	31
in (a) like manner	29
in a practical sense	32
in a ratio of…	19
inasmuch as…	15
in bulk	31
in common	12
in common use	26
in common with	28
in connection(=connexion) with…	6
in contact with…	31
in contrast	1
in contrast to (with)	2
independently of…	30
in detail	20
in direct propotion to	35
in effect	39
in error	39
in excess of…	40
inferior to…	2
in good agreement with…	30
in modern form	1
in order for… to(inf.)	28
in order of	1
in other words	1
in pairs	17
in particular	23
in place of	26
in principle	13
in (large) quantities	2
in recent years	4
in short	29
in small amounts	4
instead of…	2

in such a manner that	32
insulate from	15
insurance against…	40
in terms of…	21
in the broad sense	23
in the case of	24
in the conventional sense	6
in the final (last) analysis	38
in the form of…	2
in the manner of…	21
in the order of…	16
in the presence of…	8
in the range of…	23
in the search for…	5
in the sense that…	25
(in)this way	23
in turn	12
in view of…	16
irrespective of…	21
it is common practice(+*inf.*)	32
it is no exaggeration to say	23
it remains to be seen…	23
it so happens that	29

J

just as…	27
just as…, so…	6

K

known as…	1

L

learn from…	25
lend itself to…	23
lie in…	9
(as) listed below (above)	4
list in table…	2
look at…	6
look into…	33
lost out	8
low flame	23

M

make possible	25
main group element	1
make over	39

make up	1
(move) away from…	23
more commonly	28
more towards	23
much as…	9
multiplied by…	36

N

neither…nor…	17
no longer	22
no matter what…	22
no more than	11
not…, nor…	32

O

offer…an opportunity for…	40
of… importance	31
of one's (its) own accord	15
of value	27
on…basis	37
on descending (ascending) the group	5
one and the same	13
one…or another	2
on the basis of…	6
on the contrary	30
on the horizon	40
on the occasion of…	10
on the subject of…	27
opposite to…	29
other than…	6
over‐all	32
over a period of…	31
owing to	27
oxidation of…to…	17

P

pass through	26
peer (*vi.*) into…	16
penetrate into…	34
plot A against B…	28
(…)preceded by…	12
preferable to…	15
prefix A to B	10
prepare A from B	18
proportional to…	28
protect A from B	1
provide …with…	25
put demands on…	33

put…into perspective	23

Q

question to consider	24

R

raise to a (third) power	33
range from…to…	2
react by…into…	5
refer to…	1
(be) refer to as…	1
regardless of	27
relate to…	17
relative to…	29
rely on (upon)	25
research into	23
resistant to…	3
responsible for…	3
result from…	16
result in…	2
rich in…	36
run out	24

S

said to be…	29
same as…	28
seem rather strange to do…	5
separate from	26
serve as…	8
set forth…	12
set out…	39
set up…	3
similar to…	1
simply put	25
slow down (up)…	3
so…as to(+*inf.*)	10
so called…	8
speed up	4
spread…out…	22
stimulate…into…	19
subjected to…	2
substitute(…)for…	12
such A as B	2
suitable for (to)…	10
sure to (+*inf.*)	35
sweep through	26

T

take advantage of…	27
take care to (+*inf.*)	15
take into account	23
take off…	37
take on…	28
take the place of…	4
tendency to…	2
tend to (+*inf.*)	36
term by term…	10
that is…	8
the action of A on B	3
the addition of A to B	3
the law of conservation of matter	39
the ratio of…to…	37
the tendency toward…	5
this is not the case	29
thousands and tens of thousands	6
together with…	38
to some extent	1
travel through	35
turn out to be (+*a.*)	35

U

under a given set of conditions	31
under ambient conditions	5
under (certain, present, all…)conditions	21
under consideration	27
under the influence of…	5
use…for (+*ing*)	2
use up	32

V

vary with…	10
vice versa	6

W

wind up	22
with (in) reference	28
with respect to	38
with the aid of…	26
with the exception of…	1
with regard to…	23
what else	8
work with…	25
worthy of	27

APPENDIX VII

PREFIX AND SUFFIX
词头和词尾

（词头或词尾后的数字代表的是该词头或词尾出现在第几课课文中）

A

a-	10
-a	4
-able	2
acet(o)-	38
acetato-	9
aci-nitro-	13
acyl-	19
-ad	4
-adiene	11
-adiyne	11
-al	13
alk-	11
allyloxy-	14
-amide	19
amidino-	13
amino-	13
ammine-	9
-ane	10
anhydr(o)-	6
aquo(a)-	9
-ar	6
arsa-	appx II
-ate	7
-atriene	11
-atriyne	11
aza-	appx II
azido-	13
azo-	19

B

benzeno-	12
benzyloxy-	14
bi-	7
bio-	23
bisma-	appx II
bora-	appx II
bromo-	13
butano-	12
butoxy-	14

C

capr(i,o)-	38
carb(o)-	20
-carbaldehyde	13
carbamoyl-	13
-carbohydrazide	13
carbonato-	9
-carbonitrile	13
carbonyl-	9
-carbonyl	13
-carbonyl halide	13
-carboxamide	13
-carboxamidine	13
carboxy-	13
chem(o)-	6
chloro-	9
chlorosyl-	13
chloryl-	13

chrom(o)-	26			
-cide	2,13	**F**		
cis-	14	ferro-		19
co-	2	fluoro-		13
cyano-	9,13	-fold		19
cyclo-	12	formyl-		13
D		-free		19
de-	11			
deca-	3			
deutero-	28	**G**		
di-	1	-gen		4
dia-	36	germa-		appx Ⅱ
diacetoxyiodo-	13	gluco-		38
diazo-	13	-gram		26
-dienyl	11	-graphy		16
dif-	25	**H**		
dihydroxyiodo-	13			
-dione	14	halo-		19
dis-	8	haloformyl-		13
E		-hedral		2
		-hedron		16
-ecine	appx Ⅱ	hept(a)-		5
electr(o)-	27	hetero-		5
en-	37	hetero-		21
endo-	31	hex(a)-		3
-ene	11	homo-		5
-enyl	11	hydro-		20
ep(i)-	22	hydroxo-		9
-epine	appx Ⅱ	hydroxy-		7
equi-	32	hydroxyl-		20
-er	3	hypo-		7
-ete	appx Ⅱ	**I**		
ethano-	12			
etheno-	12	-ic		1
ethoxy-	14	-ide		1
ethylenedioxy-	14	-ility		23
-etidine	appx Ⅱ	imino-		13
ex-	28	in-		1
exo-	31	-ine		4

inter-	2
iodo-	13
iodosyl-	13
iodyl-	13
-iridine	appx II
-irine	appx II
-ish	3
-ism	17
iso-	4
isopropoxy-	14
-ite	7
-ity	21
-ium	23

K

-ketone	14
kine-	32

L

-less	4
-lysis	18

M

macro-	22
magni-	29
mercapto-	13
mercura-	appx II
meta-(=m-)	12
-meter	26
methano-	12
methoxy-	14
methylamine-	9
-metry	17
micro-	2
mis-	23
mol(e)-	24
mono-	3

N

neo-	10
nitrato-	9
nitro-	13, 19
nitroso-	9, 13
non-	1
nona-	3
nor-	24

O

(metal)···oate	13
-ocene	19
-ocine	appx II
oct(a)-	3
-ode	35
-ohydrazide	13
-oic acid	13
-oid	27
-ol	13
-olate	14
-ole	appx II
-olidine	appx II
olig(o)-	20
-on	1, 4
-one	13, 19
-onine	appx II
-or	6
organo-	18
ortho-(=o-)	12
osm(o)-	25
-ous	7
oxa-	appx II
oxalato-	9
-oxide	19
oxo-	13
oxy-	7
-oyl	13
-oyl halide	13

P

para-(=p-)	12
pent(a)-	3
pentyloxy-	14

per-	7	spectr(o)-	27
perchloryl-	13	stanna-	appx II
perhydro-	12	-stat	20
peri-	12	stereo-	29
phenoxy-	14	stiba-	appx II
-phile	8	sulfo-	13
phospha-	appx II	super-	5
photo-	20	**T**	
plumba-	appx II	tellura-	appx II
poly-	22	telluro-	13
pre-	25	tetra-	3
pro-	36	thermo-	22
propano-	12	thia-	appx II
propoxy-	14	-tion	18
Q		trans-	14
-quinone	14	tri-	3
		trimethylenedioxy-	14
R		**U**	
radio-	23	un-	2
R⋯carboxylate	13	uni-	31
R⋯cate	13	-ure	2
re-	3		
R-oxy-	13	**V**	
R-oxycarbonyl-	13	vapor-	37
R-thio-	13	volt(a)-	35
S			
scopy-	27	**Y**	
selena-	appx II	-yl	10
seleno-	13, appx II	-ylene	19
self-	10	-yne	11
sens-	26	-ynyl	11
sila-	appx II		

VOCABULARY
总词汇表

A

ability[ə'biliti]n.能力
abscissa[æb'sisə]n.横坐标
absolute['æbsəlju:t]a.确实的,绝对的
absorb[əb'sɔ:b]vt.吸收
absorbance[əb'sɔ:bəns]n.吸收率
absorber[əb'sɔ:bə]n.吸收器,吸收剂
absorption[əb'sɔ:pʃən]n.吸收
abstract['æbstrækt]n.摘要
[æb'strækt]vt.提取,转移
abuse[ə'bju:z]vt.n.滥用
academics and researchers 学者
accelerate[æk'seləreit]vt.加速
accept[ək'sept]vt.接受,承认
acceptor[ək'septə]n.接受体
access['ækses]vt.n.存取,接近,通路,访问,入门
accessible[ək'sesəbl]a.能接近的,可进入的
accommodate[ə'kɔmədeit]vt.容纳
accomplish[ə'kɔmpliʃ]vt.达到,完成,实现
account[ə'kaunt]v.计算,衡算
accuracy['ækjurəsi]n.准确度
accurate['ækjurit]a.准确的,精确的
acenaphthene[ˌæsə'næfθi:n]n.苊
acetaldehyde[ˌæsi'tældihaid]n.乙醛
acetamide[ˌæsi'tæmaid]n.乙酰胺
acetate['æsit(e)it]n.醋酸盐(酯)
acetato['æsiteitə](前缀)醋酸合
acetic acid[ə'si:tikæsid]醋酸,乙酸
acetic anhydride[æn'haidraid]醋酸酐
acetone['æsitəun]n.丙酮
acetonide['æsitənaid]n.丙酮化物
acetyl['æsitil]n.乙酰(基)
～ chloride[ˌklɔ:raid]乙酰氯
acetylene[ə'setili:n]n.乙炔
acetylenic[ˌæseti'linik]a.乙炔的
acid['æsid]n.酸 a.酸(性)的,酸式的
acidic[ə'sidik]a.酸的,酸性的

acidification[əˌsidifi'keiʃən]n.酸化
acidify[ə'sidifai]v.酸化
acid ion['aiən]酸根离子
acidity[ə'siditi]n.酸度,酸性
acid radical['rædikəl]酸根
acid salt[sɔ:lt]酸式盐
acid wash 酸洗
acquaintance[ə'kweintəns]n.熟悉,认识(with)
acrilan['ækrilən]n.(商品名)聚丙烯腈纤维,腈纶
acrolein[ə'krəuliin]n.丙烯醛
acrylaldehyde[ˌækrə'lældəhaid]n.丙烯醛
acrylic[ə'krilik]a.丙烯的
～ acid 丙烯酸
acryloyl[ə'kriləwil]n.丙烯酰(基)
actinium[æk'tiniəm]n.Ac 锕
actinide[ˌæktinaid]n.锕系
activate['æktiveit]vt.活化,使活泼
activation[ækti'veiʃən]n.活化
activity[æk'tiviti]n.活动性,活(泼)性,活度
～ series 活动顺序
acyclic[ei'saiklik]a.无环的,非环状的
acyl['æsil]n.酰基
acylation[æsi'leiʃən]n.酰基化
acyldestannylation[ˌæsildistæni'leiʃən]
n.酰化去(脱)锡化
acyl halide['hælaid]酰卤
adamantyl['ædəmentil]n.金刚烷基
adaptable[ə'dæptəbl]a.能适应的
adapter[ə'dæptə]n.接收管
addition[ə'diʃən]n.加成,加合
addition polymerization[ˌpɔlimərai'zeiʃən]
加成聚合,加聚
additive['æditiv]n.添加物(剂)
a.添加的,附加的
additivity[ædi'tiviti]n.加合性
adduct['ædʌkt]n.加成物
adhere[əd'hiə]vi.粘附

adhesive[əd'hi:siv]n. 粘合剂,胶粘剂
adipic[ə'dipik]acid 己二酸
adipoyl[ə'dipəwil]n. 己二酰(基)
adjacent[ə'dʒeisənt]a. 邻近的,相邻的
adjust[ə'dʒʌst]vt. 调整,调节
admit[əd'mit]vi. 通入
adsorb[æd'sɔ:b]vt. 吸附
adsorbate[æd'sɔ:beit]n. 被吸附物,吸附质
adsorbent[æd'sɔ:bənt]n. 吸附剂
advent['ædvent]n. 到来,出现
affect[ə'fekt]vt. 影响
affinity[ə'finiti]n. 亲和性,亲和力
afield[ə'fi:ld]ad. 向野外,离开,离谱
aflatoxin[ˌæflə'tɔksin]n. 黄曲霉素菌
agent['eidʒənt]n. 剂,试剂
aggregate['ægrigit]n.a. 聚集(体),合计
['ægrigeit]v. 聚集,总计
agrochemicals[ˌægrəu'kemikəls] [复]n. 农用化学品
air condenser 空气冷凝器
albeit[ɔ:l'bi:it]conj. 即使,虽然
alcohol['ælkəhɔl]n. 醇,乙醇
alchemist['ælkimist]n. 炼金术士,炼丹家
aldehyde['ældihaid]n. 醛,乙醛
alicyclic[ˌæli'saiklik]a. 脂环的
align[ə'lain]vt. 排列,调整
aliphatic[ˌæli'fætik]a. 脂肪族的
alkadiene[ˌælkə'daii:n]二烯
alkadiyne[ˌælkə'dai'ain]n. 二炔
alkali['ælkəlai]n. ([复]alkali(e)s['ælkəlaiz])碱
~ metal 碱金属
alkaline['ælkəlain]a. 碱的,碱性的
~ earth metal 碱土金属
alkalinity[ˌælkə'liniti]n. 碱度,碱性
alkaloid['ælkəlɔid]n. 生物碱
alkane['ælkein]n. 烷烃
alkatriene[ˌælkə'traii:n]n. 三烯
alkatriyne[ˌælkə'trai'ain]n. 三炔
alkene['ælki:n]n. 烯烃
alkyl['ælkil]n. 烷基,烃基
alkylation[ˌælki'leiʃən]n. 烷基化
alkyne['ælkain]n. 炔烃
alkylpyridinium['ælkil,piri'diniəm] halide 烷基吡啶卤盐
allene['æli:n]n. 丙二烯
allotropic[ˌælə'trɔpik]a. 同素异形体

alloy[ə'lɔi]n.v. 合金,搀合
allyl['ælil]n. 烯丙基
allyl alcohol['ælkəhɔl]烯丙醇
alternative[ɔ:l'tə:nətiv]a. 选择的
n. 可供选择的对象
alumina[ə'lju:minə]n. 氧化铝,矾土
aluminum[ə'lju:minəm]n. Al 铝
~ chloride 三氯化铝
~ hydroxide 氢氧化铝
~ oxide 氧化铝
amalgam[ə'mælgəm]n. 汞齐
ambient['æmbiənt]a. 周围的
~ temperature 室温
americium[ˌæmə'risiəm]n. Am 镅
amic['æmik]acid 酰胺酸
amide['æmaid]n. 酰胺,氨化物
amidine['æmidi:n]n. 脒
amine['æmi:n]n. 胺
aminomethylation['æminəumeθi'leiʃən]
n. 氨甲基化
ammeter['æmitə]n. 安培计,电表
ammonia[ə'məunjə]n. 氨
ammoniacal[ˌæməu'naiəkəl]a. 氨的
ammoniated[ə'məunieitid]
a. 氨合的,与氨化合了的
ammonium[ə'məunjəm]n. 铵
amorphous[ə'mɔ:fəs]a. 无定形的
ampere['æmpɛə]n. 安培
amphoteric[ˌæmfə'terik]a. 两性的
analog['ænəlɔg](=analogue)n. 相似物
analogy[ə'nælədʒi]n. 类似
analysis[ə'næləsis]n. 分析
analytic(al)[ˌænə'litik(əl)]a. 分析的
analytical chemistry 分析化学
analyze['ænəlaiz]vt. 分析
ancestor['ænsistə]n. 先祖,起源
angstrom['æŋstrəm]n. 埃(长度单位)
anhydride[æn'haidraid]n. 酐,酸酐
anhydrite[æn'haidrait]n. 无水石膏,硬石膏
anhydrous[æn'haidrəs]a. 无水的
anion['ænaiən]n. 阴离子,负离子
anioic[ˌænai'ɔnik]a. 阴离子的
anisol['ænisəul]n. 茴香醚
annular['ænjulə]a. 环形的
annulene['ænjuli:n]n. 轮烯
anode['ænəud]n. 阳极

anodic[æ(ə)'nɔdik]a.阳极的
anomalous[ə'nɔmələs]a.不规则的
anthracene['ænθrəsi:n]蒽
anthryl['ænθril]n.蒽基
antibacterial[ˌæntibæk'tiəriəl] agent 抗(细)菌剂
antibonding[ˌænti'bɔndiŋ]a.反键的
anticipate[æn'tisipeit]vt.预期,预想
antimony['æntiməni]n.Sb 锑
antioxidant[ˌænti'ɔksidənt;ˌæntiɑksədənt]
　n.抗氧化剂,防老剂
apart[ə'pɑ:t]ad.拆开,离开
apiezon[ə'pi:zɔn]n.阿匹松,饱和烃
apparatus[æpə'reitəs]n.仪器,装置
apparent[ə'pærent]a.明显的,表现的
appendix[ə'pendiks]n.附录
applied field[ə'plaid'fi:ld]外加(磁)场
appreciate[ə'pri:ʃieit]vt.评价,重视,赏识
apprentice[ə'prentis]n.学徒
approach[ə'prəutʃ]n.方法,步骤,逼近;vt.使接近
approximate[ə'prɔksimeit]v.近似
aqua regia['ækwə'ri:dʒiə]王水
aquatic [ə'kwætik] n.水生动植物;
　　　　　　　　a.水生的
aqueous['eikwiəs]a.水的,含水的
arbitrary['ɑ:bitrəri]a.武断的,随意的
archival[ɑ:'kaivəl]a.档案的
arene['æri:n]n.芳烃
argon['ɑ:gɔn]n.Ar 氩
argue['ɑ:gju:]vt.证明,争辩
arise[ə'raiz]vi.出现,发生
aromatic[ˌærəu'mætik]a.芳香的,芳香族的
aromaticity[ˌærəumə'tisiti]n.芳香性
arrangement[ə'reindʒment]n.排列
arsenane['ɑ:sənein]n.砷杂己环
arsenate['ɑ:sinit]n.砷酸盐(酯)
arsenic['ɑ:snik]n.As 砷
arsenic[ɑ:'senik]acid 砷酸
arsenide[ɑ:'sinaid]n.砷化物
arsenite['ɑ:sinait]n.亚砷酸盐(酯)
artisan[ˌɑ:ti'zæn]n.工匠
aryl['æril]n.芳基
ascend[ə'send]v.上升
ascorbic acid (vitamin C) 抗坏血酸,
　维生素 C
aspect['æspekt]n.方面,观点
assault[ə'sɔ:lt]n.袭击

assay[ə'sei]n.v.化验,鉴定,测试
assess[ə'ses]vt.评价
assemble[ə'sembl]v.组合,聚集
assembly[ə'sembli]n.组合,聚集
assign[ə'sain]vt.指定,指派
associate[ə'səuʃieit]vt.缔合,结合
assume[ə'sju:m]vt.假定,设想
astatine['æstətin]n.At 砹
astronomical[ˌæstrə'nɔmikəl]a.天文学的
asymmetric(al)[ˌæsi'metrik(əl)] a.不对称的
　~ synthetic reaction 不对称合成反应
atmosphere['ætməsfiə]n.大气,空气,气氛,大气压
atmospheric[ˌætməs'ferik]a.大气的
atom['ætəm]n.原子
atomic[ə'tɔmik]a.原子的
　~ number['nʌmbə]原子序数
ATP adenosine[ə'denəsi:n] triphosphate
　三磷酸腺甙
attach[ə'tætʃ]vt.连接,附加
attack[ə'tæk]v.起(化学)反应,侵蚀,进攻
attain[ə'tein]vt.达到,获得,得到
attest[ə'test]vt.证明,证实
attribute[ə'tribju:t]vt.归因于(to)
['ætribju(:)t]n.属性,特质
auric['ɔ:rik]a.金的,三价金的
aurous['ɔ:rəs]a.亚金的,一价金的
automation[ˌɔ:tə'meiʃən] n.自动控制,
　自动操作
auxiliary[ɔ:g'ziljəri]a.辅助的,辅的
awareness[ə'wɛənis]n.意识
axis['æksis]n.轴
azarsetine[æ'zɑ:zəti:n]n.氮杂砷杂环丁烯
azeotrope[ə'zi:ətrəup]n.共沸物,共沸混合
　物
azeotropic[əˌzi:ə'trɔpik]a.共沸的
azeotropic distillation[ˌdisti'leiʃən]共沸蒸馏
azepine['æzəpi:n]n.氮杂环庚烯
azido-['eizidə](前缀)叠氮基
aziridine['æziridi:n]n.氮丙啶,氮杂环丙烷
azocine['æzəusi:n]n.氮杂环辛四烯

B

balance['bæləns]n.天平,平衡
　　　　　　v.使平衡,称
ban[bæn]vt.禁止
bar[bɑ:]n.棒,条,巴(压力单位)
barium['bɛəriəm]n.Ba 钡

barnacle['bɑːnəkl] n. 藤壶
barytes[bə'raitiz] n. 重晶石
base[beis] n. 碱
basic['beisik] a. 碱性的,碱式的
　　~ anhydride 碱酐
　　~ carbonate['kɑːbənit] 碱式碳酸盐
　　~ salt 碱式盐
basicity[bei'sisiti] n. 碱度,碱性
batch process[bætʃ 'prəuses] 分批法,间歇法(过程)
beach['biːtʃ] n. (海,河,湖的)滩头
beaker['biːkə] n. 烧杯
beam[biːm] n. 束
bear[bɛə] vt. 携带
become[bi'kʌm] vi. 变成
　　vt. 适合,与……相称
behaviour[bi'hivjə] n. 行为,性质,特性
bend[bend] vt. 弯曲,变角
benzaldehyde[ben'zældihaid] n. 苯甲醛
benzene[ben'ziːn] n. 苯
benzoate['benzəueit] n. 苯甲酸酯(盐)
benzoic[ben'zəuik] acid 苯甲酸
benzoyl['benzəuil, 'benzəwil] n. 苯甲酰(基)
　　~ iodide['aiədaid] 苯甲酰碘
　　~ peroxide[pə'rəksaild] 过氧化苯甲酰,苯甲酰过氧化物
benzyl['benzil] n. 苄基,苯甲基
　　~ alcohol['ælkəhɔl] 苯甲醇,苄醇
benzylic[ben'zilik] a. 苄基的,苯甲基的
benzyloxy[ˌbenzi'lɔksi] (前缀)苄氧基
berkelium['bəːkliəm] n. Bk 锫
beryllium[be'riljəm] n. Be 铍
bicarbonate[ˌbai'kɑːbənit] n. 酸式碳酸盐
bicyclic[bai'saiklik] a. 双环的
bicyclo [5.2.0] nonane ['nɔnein] n. 双环[5.2.0]壬烷
bicyclo [3.2.1] octane n. 双环 [3.2.1]辛烷
bicyclo [4.3.2] undecane n. 双环[4.3.2]十一碳烷
bidentate[ˌbai'denteit] n.a. 二齿(状)的
　　~ ligand['ligənd] 二齿配体
billion['biljən] num. 十亿
binary['bainəri] a.n. 二元,双
biocatalysis [ˌbaiəukə'tælisis] n. 生物催化(作用)
biochemistry [ˌbaiəu'kemistri] n. 生物化学
biology [bai'ɔlədʒi] n. 生物学,生物
biotechnology[ˌbaiəutek'nɔlədʒi] n. 生物工程,生物工艺学

biphasic [bai'feizik] a. [化]两相的;[植]两阶段的
biphenyl [bai'fenil] n. 联苯
bis-[bis-] (前缀)二,双
bis(2-chloroethyl)ether [bis-tu:-klɔːrə'eθil'iːθə] n. 双(2—氯乙基)醚
bisphenol-A[bis'fiːnɔl-ei] n. 双酚 A
bismuth['bizməθ] n. Bi 铋
bismuth dihydroxynitrate[ˌdaihai'dɔksinaitreit] 二羟基硝酸铋,碱式硝酸铋
bistriflimide[ˌbistrai'flimaid]=bis(trifluoromethane)sulfonimide n. 二三氟甲基磺酰亚胺阴离子[(CF$_3$SO$_2$)$_2$N]−
bis−trifluorosulfonimide[−sʌlfɔ'nimaid]= bis(trifluoromethane)sulfonamide
bisulfate[bai'sʌlfeit] n. 酸式硫酸盐
bisulfite[bai'sʌlfait] n. 亚硫酸氢盐,酸式亚硫酸盐
bitter['bitə] a.n. 苦味
blanket['blæŋkit] n. 毛毯,掩盖,覆盖层
blast[blɑːst] n. 鼓风
bleed[bliːd] vt. 放出(液,浆等)
body['bɔdi] n. 底质,坯体
boiling flask['bɔiliŋ 'flɑːsk] 烧瓶,长颈瓶
boiling flask-3-neck[θriː nek] 三颈烧瓶
boiling point['pɔint] 沸点
bombard[bɔm'bɑːd] vt. 轰击
bond[bɔnd] n. 键,联结;vi. 结合,相接
bond energy['enədʒi] 键能
bonding['bɔndiŋ] n. 键合;a. 成键的,键合的
bonding force['bɔndiŋ 'fɔːs] 价键力
borate['bɔːreit] n. 硼酸盐(酯)
boric['bɔːrik] a. 硼的
　~ acid 硼酸
borin['bɔrin] n. 硼杂环己烯
borinane['bɔrinein] n. 硼杂环己烷
boron['bɔːrɔn] n. B 硼
bottoms['bɔtəms] n. 釜渣,釜(残)液
bounce[bauns] vi. 反弹(off)
bound[baund] n. (pl.) 界限,范围
bpt=b.p.=boilng point 沸点
branch[brɑːntʃ] v. 分支
bridgehead[bridʒ'hed] n. 桥头
brightening agent 光亮剂
brittle['britl] a. 易碎的
brittleness['britlnis] n. 脆性,易脆性
broker['brəukə] n. (买卖)中介人

bromate['brəumeit]n.溴酸盐
bromide['brəumaid]n.溴化物
bromination[brəumi'neiʃən]n.溴化
bromine['brəumi:n]n.Br溴
bromite['brəumait]n.亚溴酸盐
bromobenzene[,brəumə'benzi:n]n.溴苯
bromoglucoside['brəumə'glu:kəsaid]n.溴葡萄糖甙
bubble['bʌbl]n.泡;v.起泡,发泡
Büchner funnel['bu:knə'fʌnl]瓷漏斗,布氏漏斗
bulb[bʌlb]n.烧瓶,球状物,灯泡
Bunsen burner['bunsn'bə:nə]本生灯;煤气灯
burette[bjuə'ret]n.滴定管
~ clamp[klæmp]滴定管夹
~ stand[stænd]滴定管架
burner[bə:nə]n.炉子,燃烧器,灯头
butadieneiron tricarbonyl丁二烯合三羰基合铁
butadienyl[,bju:tə'daii:nil]n.1,3—丁二烯基
butane['bju:tein]n.丁烷
butanoic['bju:tənəuik]acid丁酸
2-butanone['bju:tənəun]n.2—丁酮
butene['bju:ti:n]n.丁烯
2-butenyl['bju:tənil]n.2—丁烯基
butoxy-[bju'tɔksi]-(前缀)丁氧基
butyl['bju:til]n.丁基
butylamine[,bju:tə'læmi:n]n.丁胺
t-butyl chloride['təʃəri-'bju:til-klɔ:raid]
第三氯丁烷,特丁基氯
1-butyl-3-methylimidazolium（bmim）1-丁基-3-甲基咪唑阳离子;
~ nitrate 1-丁基-3-甲基咪唑硝酸盐
butyric[bju:'tirik]acid丁酸
butyryl['bju:tərill]n.丁酰(基)

C

cabinet[kæbinit]n.橱,柜,小室
cadmium['kædmiəm]n.Cd镉
calcium['kælsiəm]n.Ca钙
~ acetate['æsiteit]醋酸钙
calibrate['kælibreit]vt.校准
californium[,kælifɔ:niəm]n.Cf锎
callistephin[kæli'stefin]n.翠菊毒
caproamide[,kæprə'æmaid]n.己酰胺
caprolactam[,kæprəu'læktæm]n.己内酰胺
caproic[kə'prəuik]acid己酸
capture['kæptʃə]vt.n.引起(注意等),俘获,捕获
carbamate['kɑ:bəmeit]n.氨基甲酸酯

carbocyclic['kɑ:bəu'saiklik]a.碳环的
carbohydrate['kɑ:bə'haidreit]
n.碳水化合物,糖类
carbohydroxamic['kɑ:bə,haidrəuk'sæmik]
acid甲酰异羟肟酸
carbon['kɑ:bən]n.C碳
canbonation['kɑ:bə,neiʃən]n.碳酸(盐)化
carbon black 炭黑
carbon disulfide[dai'sʌlfaid]二硫化碳
carbonic acid 碳酸
carbon tetrachloride 四氯化碳
Carbowax[,kɑ:bə'wæks]n.(商品名)聚乙二醇
carboximidic['kɑ:bɔksi'midik]acid甲亚胺(基)酸
carboxylic[,kɑ:bɔk'silik]a.羧基的,羧酸的
~ acid羧酸
carrier['kæriə]n.传递者,载体,担体
~ gas载气
cast[kɑ:st]vt.浇铸
catalysis[kə'tælisis]n.催化作用
catalyst['kætəlist]n.催化剂
catalytic[,kætə'litik]a.催化的
catalyze['kætəlaiz]vt.催化
categorize['kætigəraiz]vt.把……分类
category['kætigəri]n.部门,类型
cathode['kæθəud]n.阴极
cathodic protection[kæ'θɔdik prə'tekʃən]
阴极保护
cation['kætaiən]n.阳离子,正离子
causation[kɔ:'zeiʃən]n.起因
cause[kɔ:z]vt.导致,引起;n.原因,起因
caustic['kɔ:stik]a.苛性的;n.苛性碱
~ wash碱洗
cauterize['kɔ:təraiz]vt.灼,腐蚀
cease['si:s]v.停止,终止
celite[si:'lait]n.硅藻土
cell[sel]n.小室,电池
Celsius['selsjəs]n.摄氏度法,摄氏度;
a.摄氏的
cerium['siəriəm]n.Ce铈
cesium['si:ziəm]n.Cs铯
chain reaction 链锁反应
chalcogen['kælkədʒən]n.硫属,硫族
challenge['tʃælindʒ]n.v.挑战
chamber['tʃeimbə]n.室,房间,箱
change in enthalpy=enthalpy change焓变
channel['tʃænl]black槽法炭黑

characteristic[ˌkæriktəˈristik] n.a. 特性,特点
characterize[ˈkæriktəraiz] vt. 表示……的特征,鉴别
characterization[ˌkæriktəraiˈzeiʃən] n. 鉴别,鉴定
charge[tʃɑːdʒ] n. 电荷,负荷,电量,进料; vt. 使……带电;投料,进料
check[tʃek] v. 核对,检查
chelate[ˈkiːleit] n. 螯合物;a.v. 螯合
chemical[ˈkemikəl] a. 化学的;n. 化学品
~ "building blocks" 化学模块,化学建筑块
~ shift 化学位移
chemisorption[kemiˈsɔːpʃən] n. 化学吸附
chemist[ˈkemist] n. 化学家
chiral[ˈtʃairəl] a. 手性的
chlorate[ˈklɔːrit] n. 氯酸盐
chloride[ˈklɔːraid] n. 氯化物
chlorination[ˌklɔːriˈneiʃən] n. 氯化(作用)
chlorine[ˈklɔːriːn] n.Cl 氯
chlorite[ˈklɔːrait] n. 亚氯酸盐
chloroalkane[ˈklɔːrəˈælkein] n. 氯代烷
chloroauric[ˈklɔːrəˈɔːrik] acid 氯金酸
chlorophyll[ˈklɔːrəfil] n. 叶绿素
chlorosulfonic[ˌklɔːrəsʌlˈfɔnik] a. 氯磺(酸)的
chlorous[ˈklɔːrəs] acid 亚氯酸
choline[ˈkəuliːn] n. 胆碱
chromate[ˈkrəumit] n. 铬酸盐
chromatogram[ˈkrəumətəgræm] n. 色谱图
chromatography[ˌkrəuməˈtɔgrəfi] n. 色层,层析,色谱
chromene[ˈkrəumiːn] n. 苯并吡喃
chromenyl[ˈkrəuminil] n. 苯并吡喃基
chromic[ˈkrəumik] a. 铬的
chromium[ˈkrəumiəm] n.Cr 铬
chromium(III) sulfate 硫酸铬
cinnamaldehyde[ˌsinəˈmældihaid] n. 肉桂醛
cinnoline[ˈsinəliːn] n.1,2—二氮杂萘
cinnolinyl[siˈnɔlinil] n.1,2—二氮杂萘基
circulation[ˌsəːkjuˈleiʃən] n. 运行,循环
circumstance[ˈsəːkəmstəns] n. 情况,环境
citric[ˈsitrik] a. 柠檬的
~ acid 柠檬酸
citrus[ˈsitrəs] n. 柑桔属
claim[kleim] vt. 主张,请求
Claisen distilling head 克莱森蒸馏头
clamp holder[ˈklæmp ˈhəuldə] 持夹器
clay triangle [klei ˈtraiæŋgl] 泥三角
clean[kliːn] a. 彻底的,完全的

cleave[kliːv] vt. 分解,裂解,劈开
cleavage[ˈkliːvidʒ] n. 分解,裂解
closed system 封闭体系
claudiness[ˈklaudinis] n. 阴暗,朦胧
co-associate[kəuəˈsəuʃieit] a. 共缔合的
coat[kəut] vt. 涂,镀;n. 涂层
coating[ˈkəutiŋ] n. 涂料,涂层
cobalt[kəˈbɔːlt] n.Co 钴
coefficient[ˈkəuiˈfiʃənt] n. 系数,率
coinage[ˈkɔinidʒ] n. 造币,货币
coincide[ˌkəuinˈsaid] vi. 一致,符合(with)
cold reflux 冷回流
collate[kɔˈleit] vt. 核对
collect[kəˈlekt] vt. 收集,合并
collision[kəˈliʒən] n. 碰撞
colloidal[kəˈlɔidl] a. 胶质的,胶态的
column[ˈkɔləm] n. 柱,塔,纵列
combination[ˌkɔmbiˈneiʃən] n. 化合,化合物,结合,组合
combinatorial[ˌkɔmbinəˈtɔːriəl] a. 组合的
~ chemistry 组合化学
combine[kəmˈbain] v. 化合,结合
combustible[kəmˈbʌstəbl] a. 易燃的
combustion[kəmˈbʌstʃən] n. 燃烧
compatibility[kəmˌpætəˈbiliti] n. 适合,兼容
compete[kəmˈpiːt] vt. 竞争(with)
competition[ˌkɔmpiˈtiʃən] n. 竞争
compilation[ˌkɔmpiˈleiʃən] n. 编辑,汇编
complementary[ˌkɔmpliˈmentəri] a. 补充的
complete[kəmˈpliːt] vt. 填满,完成; a. 完全的
complex[ˈkɔmpleks] n. 络合物,复杂
component[kəmˈpəunənt] n. 成分,组成,组件; a. 组成的
composition[ˌkɔmpəˈziʃən] n. 构成,成分,组成
compound[kəmˈpaund] n. 化合物
comprehend[ˌkɔmpriˈhend] vt. 包括,领悟
comprehensive[ˌkɔmpriˈhensiv] a. 百科,全面的,综合的
concentrate[ˈkɔnsentreit] v. 提浓,浓缩; n. 浓缩物
concentration[ˈkɔnsenˈtreiʃən] n. 浓度,浓缩
concept[ˈkɔnsept] n. 概念
concerned[kənˈsəːnd] a. 有关的,涉及到的
conclusion[kənˈkluːʒən] n. 结论
concordance[kənˈkɔːdəns] n. 词汇索引

condensate[kən'denseit]n.冷凝物,冷凝液；浓缩物
condensation polymerization 缩聚,缩合聚合
condense[kən'dens]v.冷凝,缩合
condenser[kən'densə]n.冷凝器
conduct['kəndəkt]vt.引导,指导,进行,实施
conductivity[ˌkɔndʌk'tiviti]n.导电率
configuration[kənˌfigju'reiʃən]n.结构,构型
conflict['kɔnflikt]n.冲突,抵触
conform[kən'fɔːm]v.遵照,依照
conformatiom[ˌkɔnfɔː'meiʃən]n.构型
conjugate['kɔndʒugit]a.共轭的
['kɔndʒugeit]v.结合,共轭,配对
conjugation[ˌkɔndʒu'geiʃən]n.共轭
conjugative[kɔdʒu'geitiv]a.共轭的
connect[kə'nekt]vt.接通,连接
connotation[ˌkɔnəu'teiʃən]n.涵义
consequence['kɔnsikwəns]n.结果,后果
conservation[ˌkɔnsə(ː)'veiʃən]n.守恒,不灭
constant['kɔnstənt]n.常数；a.不变的,一定的
~ pressure process 恒压过程
constituent[kən'stitjuənt]a.n.成分,组分
construct[kən'strʌkt]vt.作(图)
consume[kən'sjuːm]vt.消耗,浪费
contact process['kɔntækt'prəuses]接触法
container[kən'teinə]n.容器
contaminant[kən'tæminənt]n.沾污物,污染物
contaminate[kən'tæmineit]vt.污染
content['kɔntent]n.含量；内容,所盛(容)之物(常用复数)
continuous process 连续化过程,连续法
control[kən'trəul]n.对照(物)；控制
controversy['kɔntrəvəːsi]n.争论
convergence[kən'vəːdʒəns]n.融合,集中,收敛
coordinate[kəu'ɔːdinit]a.n.坐标
[kəu'ɔːdineit]v.配位
~ bond 配位键
~ covalent bond 配位共价键
~ valence['veiləns]配价
coordination[kəuˌɔːdi'neiʃən]n.配位
~ compound 配位化合物
copolymer[kəup'ɔlimə]n.共聚物,共聚体
copolymerization[kəu'pɔlimərai'zeiʃən]n.共聚合(作用)
copper['kɔpə]n.Cu 铜

cork borer[kɔːk 'bɔːrə]软木钻孔器
correlation[ˌkɔri'leiʃən]n.相互关系
corrode[kə'rəud]v.腐蚀
corrosion[kə'rəuʒən]n.腐蚀
~ inhibitor 缓蚀剂,腐蚀抑制剂
cost-effective a.低成本的,高效益的
coulomb['kuːlɔm]n.库伦
~ integral['intigrəl]库伦积分
count[kaunt]vt.算为
coupling['kʌpliŋ]n.偶合
~ constant 偶合常数
~ reaction 偶联反应
covalence[kəu'veiləns]n.共价
covalent[kəu'veilənt]a.共价的
co-worker['kəu'wəːkə]n.合作者
cps=cycles per second 周/秒,赫
crack[kræk]v.裂解,断裂
create[kri(ː)'eit]vt.产生,创造,创立,造成
cresol['kriːsɔl]n.甲酚
criterion[krai'tiəriən](复)criteria[kraiˌtiəriə]n.判据,准则,标准
critical['kritikəl]a.临界的,评论的,没有定论的
cross-linked polymer[krɔs'liŋkt'pɔlimə]交联聚合物,
cross-link 交联；交联键
crucial['kruːʃiəl,'kruːʃəl]a.至关紧要的
crucible['kruːsibl]n.坩埚
~ tongs[tɔŋs]坩埚钳
crumbly['krʌmbli]a.易碎的
crush[krʌʃ]vt.压碎,磨细
crystal['kristl]n.结晶,晶体
~ field['fiːld]结晶场,晶体场
~ lattice['lætis]晶格
crystalline['kristəlain]a.结晶状的,透明的
crystallization['kristəlai'zeiʃən]n.结晶
crystallize['kristəlaiz]v.结晶
crystallography[ˌkristə'lɔgrəfi]n.结晶学
cumene['kjuːmiːn]n.异丙基苯
cumenyl['kjuːminil]n.异丙苯基
cumulative['kjuːmjulətiv]a.累积的
cupric['kjuːprik]a.铜的,二价铜的
curium['kjuəriəm]n.Cm 锔
current['kʌrənt]n.(水,气)流,电流
current strenth[steŋθ]电流强度
curriculum[kə'rikjuləm]n.全部课程,必修课
curve[kəːv]n.曲线

customary[ˈkʌstəməri]a.通常的,习惯的
customarily[ˈkʌstəmərili]ad.通常,习惯上
cyanide[ˈsaiənaid]n.氰化物
～ ion 氰离子
cyanohydrin[ˈsaiənəuˈhaidrin]n.氰醇
cycle[ˈsaikl]n.周波,周期
cyclic[ˈsaiklik]a.环的,环状的
cycloalkane[ˌsaikləˈælkein]n.环烷烃
cycloalkyl[ˌsaikləˈælkil]n.环烷基
1,3-cyclohexadiene[ˈsaikləuˈheksədaiin]
　n.1,3—环己二烯
cyclohexane[ˌsaikləuˈheksein]n.环己烷
cyclohexanecarboxylic acid 环己烷(基)甲酸
cyclohexyl[ˌsaikləˈheksil]n.环己烷基
cyclopentadiene[ˌsaikləˌpentəˈdaiiːn]n.环戊二烯
cyclopentadienide[ˌsaikləˌpentəˈdaii(ə)naid]n.环戊二烯化(合)物
cyclopentadienyl[ˌsaikləˈpentəˈdaiənil]
　n.环戊二烯基(合)
cyclopentadienylium - η³-cycloheptatrienylium molybdenum dicarbonyl 环戊二烯基合-η³-环庚三烯基合二羰基合钼
2-cyclopenten-1-yl[ˌsaikləˈpentin-wʌn-il]
　n.环戊—2—烯—1—基
cyclopropane[ˌsaikləˈprəupein]n.环丙烷
cyclopropyl[ˌsaikləˈprəupil]n.环丙(烷)基

D

cytosine[ˈsaitəsiːn]n.胞核嘧啶,氧氨嘧啶
dacron[ˈdeikrɔn]n.(商品名)的确凉,涤纶,聚对苯二甲酸乙二醇酯
damage[ˈdæmidʒ]n.损害,损伤
darken[ˈdɑːkən]vi.变黑
data base 数据库
datum[ˈdeitəm](复 data [ˈdeitə])n.资料,论据
DDT 滴滴涕,二氯二苯三氯乙烷
deactivate[diˈæktiveit]vt.使失活
deal[diːl]vi.论及,涉及(with)
debate[diˈbeit]n.争论
decane[ˈdekein]n.癸烷
decelerate[diːˈseləreit]v.减速,减慢
decompose[diːkəmˈpəuz]v.分解
decomposition[ˌdiːkɔmpəˈziʃən]n.分解
deconvolution[ˌdiːkɔnvəˈluːʃən]n.去卷积,反褶积
deduce[diˈdjuːs]vt.推断,推论

deduction[diˈdʌkʃən]n.推论,推断
deficiency[diˈfiʃensi]n.缺乏
define[diˈfain]vt.阐明,下定义
definition[ˌdefiˈniʃən]n.确定,定义;精确度
degrade[diˈgreid]v.降解,降低
degree[diˈgriː]n.程度,度
dehydration[ˈdiːhaiˈdreiʃən]n.脱水
deliberately[diˈlibəritli]ad.故意地,审慎地
delicate[ˈdelikit]a.精密的;严谨的,周密的
delineate[diˈlinieit]vt.叙述,描绘
deliver[diˈlivə]vt.供应,放出,出产
delocalize[diːˈləukəlaiz]vt.离域,非定域
delrin[ˈdelrin]n.(商品名)聚甲醛(塑料)
demetallation[diːmetəˈleiʃən]n.脱金属化
density[ˈdensiti]n.密度,比重
depict[diˈpikt]v.描述
deplete[diˈpliːt]vt.用尽,枯竭
deposit[diˈpɔzit]vt.n.沉淀,附着(物),析出
derivation[ˌderiˈveiʃən]n.衍生,由来
derivative[diˈrivətiv]n.衍生物;a.衍生的
derive[diˈraiv]v.衍生,得到
descend[diˈsend]vi.下降,递降
deshielded[diːˈʃiːldid]a.去屏蔽的
desiccator[ˈdesikeitə]n.干燥器
destroy[disˈtrɔi]vt.消灭,破坏
detect[diˈtekt]vt.检测,发现,检定
detector[diˈtektə]n.检测器
～ signal[signl]检测器信号
determination[diˌtəːmiˈneiʃən]n.测定
determine[diˈtəːmin]v.测定
deuteriochloroform[djuːˈtiəriəˈklɔrəfɔːm]
　n.重氢(化)氯仿
deviation[ˌdiːviˈeiʃən]n.偏离,偏差
device[diˈvais]n.装置,设备,器
devise[diˈvaiz]vt.设想,计划
diacid[daiˈæsid]n.二元酸
diagram[ˈdaiəgræm]n.图,图表
1,3-dialkylimidazolium[daiˈælkilˌimidæˈzəuliəm]
halide 1,3-二烷基咪唑卤盐,1,3-二烷基咪唑卤化盐
dialkyl phthalate[daiˈælkil fˈθæleit]
　邻苯二甲酸二烷基酯
diamagnetic[ˌdaiəmægˈnetik]a.抗磁(性)的
diamine[daiˈæmiːn]n.二胺
diamminediaquacarbonatocobalt(Ⅲ) nitrate
　[daiˈæmiːndaiˈækwəˈkɑːbənitəkəˈbɔ:

lt'naitreit]碳酸合二氨二水合硝酸钴(Ⅲ)
diamminesilver(Ⅰ)[dai'æmi:n'silvə']chloride 氯化二氨合银(Ⅰ)
diamond['daiəmənd]n.金刚石
diatomic[daiə'tɔmik]a.双原子的
diazodehalogenation
　　[dai'æzəudihæləd3e'neiʃən]n.重氮化脱卤(化)
diazodesulphonation
　　[dai'æzəudi'sʌlfə'neiʃən]n.重氮化脱磺基
diazonium[ˌdaiə'zəuniəm]n.重氮化
dibenzenechromium[daiben'zi:n'krəumiəm]
　　n.二苯铬
dichromate[dai'krəumit]n.重铬酸盐
di-η5-cyclopentadienyliron(=ferrocene)
　　二茂铁,二环戊二烯基铁
diet['daiət]n.饮食,食物
diethyl ketone['ki:təun]二乙酮
dietyl malonate[mə'lɔneit]丙二酸二乙酯
diethyl sulfate['sʌlfeit]硫酸二乙酯
diffraction[difrækʃən]n.衍射
diffusion[di'fju:3ən]n.扩散
dihedral[dai'hedrəl]a.二面的
1,4-dihydronaphthalene[ˌdaihaidrɔ'næfθəli:n]n.1,4—二氢化萘
dilute[dai'lju:t]a.稀的
dimethylsulfoxide[-'sʌlfəuksaid]n.二甲基亚砜
dimethyl terephthalate[teref'θæleit]对苯二甲酸二甲酯
diminish[di'miniʃ]v.减少,缩小
dioxane[dai'ɔksein]n.二噁烷,二氧杂环己烷
dioxide[dai'ɔksaid]n.二氧化物
dip[dip]v.浸渍
dipole['daipəul]n.偶极
dipole moment　偶极矩
direct[di'rekt]vt.指向,指引,指示方向
discipline['disiplin]n.学科
disappearance[ˌdisə'piərens]n.消失,失踪
discard[dis'ka:d]v.排放,丢弃,抛弃
discernible[di'sə:nəbl]a.可辨别的
discrete[dis'kri:t]a.不连续的,离散的;
　　~ molecule 离散分子
dismutation[ˌdismju'teiʃən]n.歧化
disperse[dis'pə:s]v.分散
dispersion[dis'pə:ʃən]n.分散
displace[dis'pleis]vt.代替,置换
displacement(reaction)取代(反应),置换(反应)

display[dis'plei]vt.n.显示,表现
disproportionate[disprə'pɔ:ʃənit]a.不均衡的
　　　　　　　　[disprə'pɔ:ʃəneit]vi 歧化
disprove[dis'pru:v]vt.证明……不成立,反驳
dispute[dis'pju:t]vt.争论
dissociate[di'səuʃieit]vt.离解
dissociation[di'səusi'eiʃən]n.解离
dissolve[di'zɔlv]vt.溶解
dist.col.=distillation column　蒸馏塔
distil[dis'til]v.蒸馏
distillation[ˌdisti'leiʃən]n.蒸馏
distilling[dis'tiliŋ]head 蒸馏头
distilling tube[tju:b]蒸馏管
distinguish[dis'tiŋgwiʃ]v.区别,识别
distinguishable[dis'tiŋgwiʃəbl]
　　a.可区别的,可辨别的
distort[dis'tɔ:t]vt.使扭曲
distribute[dis'tribju(:)t]vt.分配,分布
dithiazine[daiθai'æzi:n]n.二硫杂氮杂环己烯
diversity[dai'və:siti] n.差异,多样性
division[di'vei3ən]n.区别,划分
DMSO = dimethyl sulfoxide　二甲基亚砜
DNA　deoxyribonucleic acid
　　[di:'ɔksiˌraibəu'nju:kli:ik'æsid]脱氧核糖核酸
D2O=heavy water=deuterium[dju'(:)tiəriəm] oxide　重水
docosane['dɔ:kəsein]n.二十二(碳)烷
document['dɔkjumənt]vt.证明
dodecane[dəu'dekein]n.十二(碳)烷
donate[dəu'neit]v.给出,捐赠
donor['dəunə]n.给予体
dot[dɔt]n.圆点;v.打点
dotriacontahectane['dɔtraiə'kɔntə'hektein]
　　n.一百三十二(碳)烷
dotriacontane['dɔtraiə'kɔntein]n.三十二(碳)烷
double['dʌbl]n.v.a.ad.加倍,二倍,双重
　　~ bond['bɔnd]双键
　　~ salt['sɔ:lt]复盐
doublet['dʌblit]n.二重峰,双重峰;电子对
downfield[ˌdaun'fi:ld]n.ad.低磁场
drain[drein]v.徐徐流出,渐次排出;
　　　　　　n.排水,排放口
dramatic[drə'mætik]a.引人注目的
dry ice　干冰
dual['dju(:)əl]a.二重的
ductail['dʌktail]a.延性的

dummy[dʌmi]a.假的,虚的
dust[dʌst]n.粉
dye[dai]n.染料
dyestuff['daistʌf]n.染料
dynamic[dai'næmiks]a.动力学的
dynamics[dai'næmiks]n.动力学
dysprosium[dis'prəusiəm]n.Dy镝

E

effect[i'fekt]vt.实现;n.效应,影响
effectiveness[i'fektivnis]n.效力
efficiency[i'fiʃənsi]n.效率
effluent['efluənt]a.流出的,n.流出物
einsteinium[ain'stainiəm]n.Es锿
elaborate[i'læbərit]a.精心的,复杂的
elastic[i'læstik]a.有弹性的,弹性的
electric couple[i'lektrik'kʌpl]电对,电偶
eletricity[ilek'trisiti]n.电,电学,电流
electrochemical[i'lektrəu'kemikəl]a.电化学的
electrochemical cell[sel]化学电池
electrode[i'lektrəud]n.电极
electrolysis[slek'trɔlisis]n.电解
electrolyte[i'lektrəulait]n.电解质
electrolytic[i,lektrəu'litik]a.电解的
electromagnetic[i'lektrəumæg'netik]a.电磁的
electromagnetic radiation[reidi'eiʃən]电磁辐射
electromotive force[i'lektrəuməutivfɔːs]
　　=EMT电动势
electron[i'lektrɔn]n.电子
electronegativity[i'lektrəunegə'tiviti]n.电负性
electron cloud　电子云
electronic[ilek'trɔnik]a.电子的
electron pair['pɛə]电子对
electron paired　电子(成)对的
electron-supplying a.给电子的
electron-withdrawing a.吸电子的
electrophile[i'lectrəfail]n.亲电试剂
electrophilic[i'lektrəu'filik]a.亲电(子)的
electroplate[i'lektrəupleit]vt.电镀
electrostatic[i'lektrəu'stætik]a.静电的
electrovalent[i'lektrəuveilənt]a.电价的
element[eliment]n.元素,要素,成分
elementray[eli'mentəri]a.初步的,基本的
elevate[eliveit]vt.提高,抬高
elide[i'laid]vt.取消,省略,删去
eliminate[i'limineit]vt.消除,消去

elimination[i'limineiʃən]n.消除,消去
elision[iːliʒən]n.省略
elucidation[i,luːsi'deiʃən]n.剖析,阐明,说明
emerge[i'məːdʒ]vi.出现,形成
EMF series　电动序
emission[i'miʃən]n.(光、热等的)散发,发射
emphasis['emfəsis]n.强调,着重于(on)
empirical formula['fɔːmjulə]实验式
empty['empti]a.空的,未占用的
enable[i'neibl]vt.使……能够
encase[in'keis]vt.把……放入套内
enclose[in'kləuz]vt.封入,包入
encode[in'kəud]vt.编码,译码
endeavor[in'devə]n.努力,尽力
endothermic[,endəu'θəːmik]a.吸热的
endpoint['end'pɔint]n.终点
energetics[,enəːdʒətiks]n.动能学
energy content　内能
energy level[levl]能级
en masse[aŋ'mæs]ad.整体,全体地,一同地
enolic[iː'nɔlik]a.烯醇的
enrich[in'ritʃ]vt.富集(in)
enthalpy[en'θælpi]n.热焓
entire[in'taiə]a.全部的,整个的
entity['entəti]n.实体
entrainer[in'treinə]n.携带剂,夹带剂
entraining agent 携带剂
entropy['entrəpi]n.熵
environment[in'vaiərənmənt]n.环境,环绕
epichlorohydrin[,epə,klɔːrə'haidrin]n.3-氯-1,2-环
　　氧丙烷,环氧氯丙烷
epoxy[e'pɔksi]a.环氧的;n.环氧
　～resin环氧树脂
equate[i'kweit]vt.使相等
equation[i'kweiʃən]n.方程式
equilibrium[,iːkwi'libriəm]
　　(复数equilibria[,iːkwi'libriə])n.平衡
　～state平衡(状)态
equimolar[,iːkwi'məulə]a.等摩尔的
equipment[i'kwipmənt]n.设备,装备
equivalent[i'kwivələnt]a.相同的,等价的,当量的;
　　　　　　　　　　　　n.当量
　～weight当量
erbium['əːbiəm]n.Er铒
Erlenmeyer['erlənmaiə]flask　三角烧瓶,艾伦迈耶
　　烧瓶

error['erə] n. 错误,误差
e.s.r=electron-spin resononce 电子自旋共振
ester['estə] n. 酯
estimate['estimeit] n. 评价,判断
ethanal['eθənəl] n. 乙醛
ethane['eθein] n. 乙烷
1,2-ethanediol['eθein'daiəul] n. 乙二醇
ethanoic[eθə'nəuik] acid 乙酸
ethanol['eθənɔl] n. 乙醇
ethenyl['eθənil] n. 乙烯基
ether['i:θə] n. 醚,乙醚
etherate['i:θərit] n. 醚化物
ethyl['eθil] n. 乙基
　~ acetate['æsitit] 醋酸乙酯,乙酸乙酯
　~ alcohol['ælkəhɔl] 乙醇
ethylammonium['eθil ə'məunjəm] nitrate 硝酸乙基铵
ethylating['eθi'leitiŋ] n.a. 乙基化
ethyl chloroformate['eθilklɔ:rə'fɔ:mit] 氯甲酸乙酯
ethylene['eθili:n] n. 乙烯;1,2-亚乙基,乙撑,
　—CH₂CH₂—
ethylene (di)chloride 1,2-二氯乙烷,氯化乙烯
ethylenediamine['eθili:ndai'æmi:n]
　n. 乙二胺,乙二胺合
ethylene glycol['glaikɔl] 乙二醇
ethylene oxide['ɔksaid] 环氧乙烷
ethyleneplatinum trichloride 乙烯合三氯化铂
ethylenic[,eθi'lenik] a. 烯的,乙烯的
ethyl halide['eθil'hælaid] 卤乙烷
ethyl methyl ether['eθil'meθil'i:θə] 甲乙醚,甲氧基乙烷
ethyl methyl ketone['kitəun] 甲乙酮
ethylsulfuric[sʌl'fjuərik] acid 乙基硫酸,硫酸单乙酯
ethyl vinyl ether['eθil'vainil'i:θə] 乙基乙烯基醚
ethynyl[e'θainil] n. 乙炔基
europium[juə'rəupiəm] n. Eu 铕
evacuate[i'vækjueit] vt. 抽空
evaluate[i'væljueit] vt. 评价,估价
evaluation[i'vælju'eiʃən] n. 评价,估价
evaporate[i'væpəreit] vt. 蒸发
evaporating dish[i'væpəreitiŋ diʃ] 蒸发皿
evaporation[i,væpə'reiʃən] n. 蒸发
evolve[i'vɔlv] vt. 离析,放出;发展,进化
exaggeration[ig'zædʒə'reiʃən] n. 夸张,夸大之词
execute['eksikju:t] vt. 执行
exert[ig'zə:t] vt. 产生,采用

exhaust[i'gzɔ:st] vt. 耗尽
exhibit[ig'zibit] vt.n. 表示,显示,呈现
exothermic[,eksəu'θə:mik] a. 放热的
expansion[iks'pænʃən] n. 扩大,伸展,膨胀
experienced[iks'piəriənst] a. 已有经验的,熟练的
explicit[iks'plisit] a. 明确的,清楚的
exploit[iks'plɔit] vt. 利用,开拓,开发,剥削
explosion[iks'pləuʒən] n. 爆炸
explosively[iks'pləusivli] ad. 爆炸式地
exponent[eks'pəunənt] n. 指数
exponential[ekspəu'nenʃəl] a. 指数的
exposure[iks'pəuʒə] n. 曝露,曝光,揭示
extension clamp 延长夹,铁夹子
extensive property['prɔpəti] 量度性质
extent[iks'tent] n. 程度,限度
external[eks'tə:nl] a. 外部的,外界的,表面的
external reference['refrəns] 外标(物),外部标准
extract[iks'trækt] vt. 抽提,萃取
　['ekstrækt] n. 萃出物
extractive[iks'træktiv] a. 抽提的,萃取的
extractive distillation 萃取蒸馏
extraction[iks'trækʃən] n. 抽提,萃取
extractor[iks'træktə] n. 抽提器,萃取器
extrude[eiks'tru:d] v. 挤压成,使……射出

F

fabric['fræbrik] n. 织物,纤维品
facet['fæsit] n. 方面
facilitate[fə'siliteit] v. 促进,使容易
factor['fæktə] n. 因素
fade[feid] v. 褪色
fallacy['fæləsi] n. 谬误
family['fæmili] n. (周期表的)族
faraday['færədi] n. 法拉第(电量单位)
fashion['fæʃən] n. 方式,样子
fat[fæt] n. 脂肪
fatal['feitl] a. 致命的
fatality[fə'tæliti] n. 死亡
fatty['fæti] acid 脂肪酸
feasible['fi:zəbl] a. 易实现的,可实行的
feed[fi:d] v.n. 进料,供料,原料
feed-stock[fi:d stɔ:k] n. 原料
fermium['fɛəmiəm] n. Fm 镄
ferric['ferik] a. 铁的,三价铁的
ferrocene['ferəusi:n] n. 二茂铁
ferrous['ferəs] a. 亚铁的,二价铁的

~ bromide['brəumaid]溴化亚铁
ferrocyanide['ferəu'saiənaid]n.亚铁氰化物
fertilizer['fə:tilaizə]n.肥料
fiber['faibə]n.纤维
figure['figə]n.图,插图
filament['filəmənt]n.灯丝
file[fail]n.锉刀,文件
filler['filə]n.填料,填充剂
film[film]n.膜,胶片
filter['filtə]n.过滤器;v.过滤
filter flask[flɑ:sk]吸滤瓶
filter-paper thimbl['θimbl]滤纸(套)筒
filtrate['filtreit]v.过滤;n.滤液
final[fainl] state终止状态,终态
finely['fainli]ad.细小地
firebrick[faiə'brik]n.耐火砖
first-order spin-spin splitting 一级自旋—自旋裂分
fission['fiʃən]n.分裂,裂开
fit[fit]v.a.适合,符合,安装
flask[flɑ:sk]n.烧瓶,瓶
flame spreader [fleim spredə] 火焰扩张器,鱼尾灯头
flat[flæt]a.平坦的,平的
flexible['fleksəbl]a.易操作,易适应的
flip[flip]v.n.跃迁,猝然跳动
Florence['florəns]flask 平底烧瓶
flow sheet[fləu ʃi:t]流程图
fluoride['fluəraid]n.氟化物
~ ion 氟离子
fluorine['flu(:)ərin]n.F氟
foam[fəum]n.泡沫;v.起泡沫,发泡沫
forceps['fɔ:seps]n.镊子
foreign material 异质物料,外来杂质
formaldehyde[fɔ:mældihaid]n.甲醛
formalism['fɔ:məli:zəm]n.形式主义,拘泥形式
formanilide[fɔ:m'ænilaid]n.N—甲酰苯胺
format['fɔ:mæt]n.格式,形式
former['fɔ:mə]n.前者
formic['fɔ:mik]acid 甲酸,蚁酸
formula['fɔ:mjulə]n.分子式,公式
formulate['fɔ:mjuleit]vt.用公式表示
formulated['fɔ:mjuleitid]a.公式化了的,一般(化)了的;配制的,配方的
formylate['fɔ:mileit]v.甲酰化
forward['fɔ:wəd]a.向前的,正向的
~ reaction 正反应

fraction['frækʃən]n.馏分,部分,分数
fractional['frækʃnl]a.分馏的,分级的
fractional crystallization 分级结晶
fractional distillation 精馏,精密分馏
fractionate['frækʃəneit]vt.分馏,精馏
fractionating column['frækʃə'neitiŋ 'kɔləm]精馏柱
fractionating tower['tauə]精馏塔
fractometer['fræktəu'mi:tə]n.色层分离仪
fragment['frægmənt]n.碎片,断片
francium['frænsiəm]n.Fr 钫
free energy[fri:'enədʒi]自由能,吉氏函数
free radical['rædikəl]自由基,游离基
freeze dry 冷冻干燥,冷冻脱水
freezing point depression['fri:ziŋ pɔint di'preʃən] 冰点降低(法)
freon['fri:ɔn]n.氟里昂
frequency['fri:kwənsi]n.频率
frontier['frʌntjə]n.前沿,前线
frost[frɔst]n.霜;vt.霜化
fructose['frʌktəuz;'frʌktəus]n.果糖
fuel[fjuəl]n.燃料
fulvene['fulvi:n]n.亚甲基环戊二烯
function['fʌŋkʃən]n.官能团,函数;vi.起作用
functional['fʌŋkʃənl]group 官能团
functionality['fʌŋkʃənliti]n.官能度
fundamental[ˌfʌndə'mentl]a.基本的,根本的
funding['fʌndiŋ]n.(提供)基金
fungal['fʌŋgəl]a.真菌的,由真菌引起的
fungicide['fʌndʒisaid]n.杀(霉)菌剂
funnel['fʌnl]n.漏斗
~ support   漏斗架
2-furaldehyde[fju'rældihaid]n.呋喃醛,糠醛
furan['fjuərən]n.呋喃
2-furanamine[ˌfjuərə'næmi:n]n.2-氨基呋喃
furfural['fə:fərəl]n.呋喃醛
furyl['fjuəril]n.呋喃基
fusion['fju:ʒən]n.熔化,熔合,合成

G

gadolinium[ˌgædə'liniəm]n.Gd 钆
galena[gə'li:nə]n.方铅矿
gallium['gæliəm]n.Ga 镓
galvanic cell[gæl'vænik sel]原电池,自发电池
galvanize['gælvənaiz]vt.电镀,镀锌
galvanized iron   镀锌铁,白铁
gas chromatography[gæs ˌkrəumə'tɔgrəfi]

气相色谱(法)
gaseous['geizjəs]a.气体的,气态的
gas-liquid partition chromatography 气—液分配色谱法
gas measuring tube 气体量管
gasoline['gæsəli:n]n.汽油
gauss[gaus]n.高斯
Geiser['gaisə]burette 活塞滴定管,酸滴定管
generality['dʒenə'ræliti]n.概论,概念,普遍性,通性
generalization['dʒenərəlai'zeiʃən]n.概括,一般法则
generate['dʒenəreit]vt.产生,造成
gentiobiose[ˌdʒentiə'baiəz]n.龙胆二糖
gentle['dʒentl]a.温和的,轻轻的
germanium[dʒə:'meiniəm]n.Ge 锗
gilt[gilt]n.炫目的外表,涂层
glass filter crucible 玻璃过滤坩埚
glucose['glu:kəus]n.葡萄糖
glycerol['glisərol]n.甘油,丙三醇
glycinate['glaisineit]n.甘氨酸盐(酯)
glycol['glaikɔl]n.1,2,—乙二醇,二元醇
gold['gəuld]n.Au 金
graduate['grædjueit]a.研究院的,已取得学士学位的
~ student 研究生
graduated cylinder['silində]量筒
gram molecular weight 克分子量
granule['grænju:l]n.颗粒
graph[græf]n.曲线图,图
graphic['græfik]a.图表的
graphite['græfait]n.石墨
graphitic[græ'fitik]a.石墨的
gray-black 黑灰色
grease[gri:s]n.润滑脂,脂肪
green chemistry 绿色化学
greenish['gri:niʃ]a.微带绿色的
Griffin beaker['grifin'bi:kə]烧杯,格里芬烧杯
Grignard reagent[ˌgri'njɑ:ri:'ei-dʒənt]格氏试剂
ground joint[graund dʒɔint]磨口接头
group[gru:p]n.族,基,团;vt.把……分组
~ theory 群论
guesswork['geswə:k]n.臆测,猜测
gypsum['dʒipsəm]n.石膏
gyromagnetic ratio[ˌdʒaiərəumæg'netik 'reiʃiəu]磁旋比,回转磁化率

H

hafnia['hæfniə]n.氧化铪
hafnium['hæfniəm]n.Hf 铪
half-wave potential 半波电位
halide['hælaid]n.卤化物
haloethoxy[ˌhæləui:'θɔksi]n.卤乙氧基
haloformyl[ˌhælə'fɔmil]n.卤甲酰基
halogen['hælədʒən]n.卤素
halogenate['hælədʒəneit]v.卤化
halogenation[ˌhælədʒə'neiʃən]n.卤化(作用)
handle['hændl]vt.处理,掌握,操作
hard-to-separate 难分离的
harmful['hɑ:mful]a.有害的
hazardous['hæzədəs]a.危险的
heat capacity[hi:t kə'pæsiti]热容
heat content[kən'tent]热焓,热含量
heat of combustion[kəm'bʌstʃən]燃烧热
hectane['hektein]n.一百(碳)烷
helium['hi:ljəm]n.He 氦
helix['hi:liks](复-es['helisi:z])n.螺旋线
heme[hi:m]n.原血红素
hemoglobin[hi:məu'gləubin]n.血红蛋白
henicosane[he'naikəsein]n.二十一(碳)烷
hentriacontane[ˌhentraiə'kɔntein]n.三十一(碳)烷
heptane['heptein]n.庚烷
heptanedioic[ˌhepteindai'əuik]acid 庚二酸
heptanoic[ˌheptə'nəuik]acid 庚酸
heteroatom[ˌhetərə'ætəm]n.杂原子
heterocycle[ˌhetərə'saikl]n.杂环
heterocyclic[ˌhetərə'saiklik]a.杂环的
heterodiatomic[ˌhetərədaiə'tɔmik]a.杂双原子的
heterogeneous[ˌhentərəu'dʒi:niəs]a.多相的,异相的,非均一相的
heterolytic[ˌhetərə'litik]a.异(性)的
heteropolar[ˌhetərə'pəulə]a.异极的
hexacontane[ˌheksə'kɔntein]n.六十(碳)烷
hexadiene[ˌheksə'daii:n]n.己二烯
hexafluorophosphate n.六氟磷酸盐
hexamethylenediamine 己二胺
hexanal['heksənəl]n.己醛
hexane['heksein]n.己烷
hexanenitril[ˌheksein'naitril]n.己腈
2-hexene[tu:'heksi:n]n.2-己烯
hiding['haidiŋ]n.掩蔽,躲藏
hindrance['hindrəns]n.障碍
Hirsch funnel[həʃ'fʌnl]赫尔什漏斗
hold[həuld]vi.有效,适用
holmium['hɔlmiəm]n.Ho 钬

homocubyl[ˌhɔmauˈkju:bil]n.高立方烷基
homogeneous[ˌhɔmeˈdʒi:niəs]a.均相的,均一的
homolog[ˈhɔmələg]n.同系物
homologous[hɔˈmɔləgəs]a.同系列的
homolysis[hɔˈmɔlisis]n.均裂
homolytic[ˌhɔməˈlitik]a.均(裂)的,同性的
hood[hud]n.通风橱
horizontal[hɔriˈzɔntl]a.水平的,卧式的,横式的
hybridize[ˈhaibridaiz]v.杂化
hydrate[ˈhaidreit]n.水合物;v.水合,水化
hydration[haiˈdreiʃən]n.水合,水化
hydrazide[ˈhaidrəzaid]n.酰肼
hydrazine[ˈhaidrəzi:n]n.肼,联氨
hydride[ˈhaidraid]n.氢化物
hydroboration[ˌhaidrəbɔ:ˈreiʃən]n.硼氢化(作用)
hydrobromic[ˈhaidrəˈbroumik]acid 氢溴酸
hydrocarbon[ˈhaidrəuˈkɑ:bən]n.烃,碳氢化合物
hydrocarbonylation[ˌhaidrəˌkɑ:bəniˈleiʃən]
  n.氢甲酰化
hydrochloric[ˈhaidrəˈklɔrik]acid 氢氯酸,盐酸
hydrochloride[ˈhaidrəˈklɔ:raid]n.盐酸化物,盐酸盐
hydrocyanic[ˈhaidrəusaiˈænik]acid 氢氰酸,氰化氢
hydrogen[ˈhaidrədʒən]n.H 氢
hydrogenate[ˈhaidrədʒəneit]vt.氢化,加氢
hydrogenation[ˈhaidrədʒəˈneiʃən]n.加氢,氢化
hydrogen bonding 氢键
hydrogen bromide[ˈbroumaid]溴化氢
hydrogen carbonate=bicarbonate 酸式碳酸盐
hydrogen cyanide[ˈsaiənaid]氰化氢
hydrogenolysis[ˌhaidrədʒəˈnɔlisis]n.氢解
hydrogen sulfate[ˈsʌlfeit]酸式硫酸盐
hydrogen sulfite[ˈsʌlfait]酸式亚硫酸盐
hydrolysis[haiˈdrɔlisis]n.水解(作用)
hydrolytic[ˌhaidrəˈlitik]a.水解的
hydrolyze[ˈhaidrəlaiz]v.水解
hydronium[haiˈdrouniəm]ion 水合氢离子
hydroquinone[ˈhaidrəu(ə)kwiˈnəun]
  n.对苯二酚,氢醌
hydroxide[haiˈdrɔksaid]n.氢氧化物
hydroxydethalliation n.羟化脱铊化(作用)
7-hydroxy-2-heptanone[ˈheptənoun]
  n.7—羟基—2—庚酮
hydroxy group 羟基团
hyperconjugation[ˌhaipəkɔndʒuˈgeiʃən]
  n.超共轭作用
hyperfine[ˌhaipəˈfain]a.超精细的

hypobromite[ˌhaipəˈbroumait]n.次溴酸盐
hypochlorite[ˌhaipəˈklɔ:rait]n.次氯酸盐
hypoiodite[ˌhaipəˈaiədait]n.次碘酸盐
hypophosphite[ˌhaipəuˈfɔsfait]n.次磷酸盐
hypothetical[ˌhaipəu(ə)ˈθetikəl]a.假定的,臆测的

I

iatrochemist[aiˌætrəˈkemist]n.医疗化学家
icosane[ˈaikəsein]n.二十(碳)烷
idea[aiˈdiə]n.概念,计划,幻想
  ~ solution[aiˈdiəl səˈlju:ʃən]n.理想溶液
identification[aiˌdentifiˈkeiʃən]n.鉴定,鉴别
identity[aiˈdentiti]n.本身,相同,同一性质,特性,
  身份
i.e.[ˈaiˈi:]=id est[idˈest]也就是,即
ignore[igˈnɔ:]vt.忽视
illustrate[ˈiləstreit]vt.说明,证明,阐明,举例
imagine[iˈmædʒin]vt.vi.想象
imidazole[imiˈdæzəul]n.咪唑
imidazolium halogenoaluminate salt 咪唑卤代铝酸盐
imidazolyl[imiˈdæzəulil]n.咪唑基
imide[ˈimaidˌimid]n.酰亚胺
imidic[iˈmidik]acid 亚胺酸
imine[iˈmi:n]n.亚胺
immediate[iˈmi:djət]a.最接近的
immerse[iˈmə:s]vt.浸入
immiscible[iˈmisəbl]a.不混溶的,难溶的
implication[ˌimpliˈkeiʃən]n.含意,牵涉
impose[imˈpəuz]vt.强使,加于
impregnate[ˈimpregneit]vt.浸渍
impurity[imˈpjuəriti]n.杂质
inaccessible[ˌinækˈsesəbl]a.不能进入的,不能到达的
inactivity[ˌinækˈtiviti]n.不活泼性
inadequate[iˈnædikwit]a.不适当的
increment[ˈinkrimənt]a.增量,增加
indazole[ˈindəzəul]n.吲唑
indazolyl[ˈindəzəulil]n.吲唑基
indefinite[inˈdefinit]a.无限的,不确定的
indelible[inˈdelibl]a.不能拭去的
indene[ˈindi:n]n.茚
2-indenyl[ˈindi:nil]n.2—茚基
index of refraction[ˈindeks əv riˈfrækʃən]折光指数
indicator[ˈindikeitə]n.指示剂
indifferent[inˈdifrənt]a.惰性的,无关紧要的
indigo[ˈindigəu]n.靛蓝,靛青;靛蓝类染料
indium[ˈindiəm]n.In 铟

indole['indəu(ɔ)l]n.吲哚
indolizine[in'dɔlizi:n]n.中氮茚
indolizinyl[in'dɔlizinil]n.中氮茚基
indolyl['indɔlil]n.吲哚基
induce[in'dju:s]vt.感应,引诱
inductive[in'dʌktiv]a.引入的
inert[i'nə:t]a.惰性的,不活泼的
inertia[i'nəʃiə]n.惰性
influence['influəuns]n.影响
infrared['infrəred]a.红外线的
infrequently[in'fri:kwəntli]ad.偶尔
infusible[in'fju:zəbl]a.不熔的
ingest[in'dʒest]vt.咽下
ingredient[in'gri:di:ənt]n.配料,成分
inhalation[,inhə'leiʃən]n.吸入
inherent[in'hiərənt]a.固有的,内在的
inhibit[in'hibit]vt.防止,阻止
initial state 起始状态,始态
inject[in'dʒekt]vt.注射(into)
injurious[in'dʒuəriəs]a.有害的
innocuous[i'nɔkjuəs]a.无害的,无毒的
inorganic[,inɔ:'gænik]a.无机的
input['in-put]n.进料量,输入
insecticide[in'sektisaid]n.杀虫剂
insight['insait]n.见识
insoluble[in'sɔljubl]a.不溶的
instantaneous[,instən'teinjəs]a.立即的,瞬时的
institutional[,insti'tju:ʃənl]a.规定的,学校的
instructor[in'strʌktə]n.教师,导师
instrumental[,instru'mentl]a.仪器的,工具的
instrumentation[,instrumen'teiʃən]n.测试设备
intake['inteik]n.吸入,摄入
integral['intigrəl]n.整体,积分
integrator['intigreitə]n.积分仪,求积仪
intensely[in'tənsli]ad.强烈地
intensity[in'tensiti]n.强度
intensive property[in'tensiv 'prɔpəti]强度性质
interact[,intə'rækt]vt.相互作用
interaction[,intə'rækʃən]n.相互作用
interchangeably[,intə'tʃeindʒəbli]
    ad.可交替地,可互相交换地
interconversion[,intəkən'və:ʃən]n.互变,相互转化
interface['intə'feis]n.界面,分界面
interfacial['intə'feiʃəl]polymerization 界面聚合
interhalogen[,intə(:)'hælədʒen]n.杂卤素
intermidiary[,intə(:)'mi:dʒəri]n.媒介

intermediate[,intə(:)'midʒət]n.中间体;a.中间的
internal combustion engine 内燃机
internal energy 内能
internal reference[,refrəns]内标(物),内部标准
interpret[in'tə:prit]vt.解释
intersection[,intə'sekʃən]n.交插,相交
intervene[,intə'vi:n]vi.插入,介于其间
intuition[,intju(:)'iʃən]n.直观,直觉
invariably[in'vɛəriəbli]ad.不变地,总是
investigator[in'vestigeitə]n.研究者
invisible[in'vizəbl]a.肉眼看不到的,不可见的
involve[in'vɔlv]vt.包含,包括,涉及
iodide['aiədaid]n.碘化物
iodate['aiədeit]n.碘酸盐
iodinate['aiədineit]vt.碘化
iodine['aiədi:n]n.I 碘
iodite['aiədait]n.亚碘酸盐
ion['aiən]n.离子
    ～ exchange resin['resin]离子交换树脂
ionic[ai'ɔnik]a.离子的
ionic liquid (IL) 离子液
ionization[,aiənai'zeiʃən]n.离子化
ionize['aiənaiz]vt.离子化,电离
ir.=infrared spectra['spektrə]红外光谱
iridium[ai'ridiəm]n.Ir 铱
iron['aiən]n.Fe 铁
iron(Ⅱ) bromide['brəumaid]二溴化铁,溴化亚铁
iron pyrite[pai'rait]黄铁矿
irradiate[i'reidieit]vt.照射
isobenzofuran[aisə,benzəu'fjuərən]异苯并呋喃
isobenzofuranyl n.异苯并呋喃基
isobutane[,aisə'bju:tein]n.异丁烷
isobutylene[,aisə'bju:tili:n]n.异丁烯
isocyanide[,aisə'saiənaid]n.异氰
isoelectronic[,aisauilek'trɔnik]a.等电子的
isoindole[aisə'indəul]n.异吲哚
isoindolyl[aisə'indɔlil]n.异吲哚基
isolate['aisəleit]vt.分离,隔离,使脱离
isomer['aisəumə]n.异构体,同分异构体
isomeriation[ai'sɔmərai'zeiʃən]n.异构化
isophthalic[,aisəf'θælik]acid 间苯二甲酸
isophthaloyl[,aisəf'θæləwil 'aisəf'θæləuil]
    n.间苯二甲酰(基)
isoprene['aisəpri:n]n.异戊二烯
isopropenyl[aisə'prəupənil]n.异丙烯基,
    1—甲基乙烯基

isopropoxy-[ˌaisəprə'pɔki](前缀)异丙氧基

1-isopropoxypropane[ˌaisəprə'pɔksi'prəupein] n.1—异丙氧基丙烷

isopropyl['aisəu'prəupil]n.异丙基

isopropylbenzene n. 异丙苯

isoquinoline[ˌaisə'kwinəli:n]n. 异喹啉

isoquinolinyl[ˌaisə'kwinəlinil]n. 异喹啉基

isothermal[ˌaisəu'θə:məl]a. 等温(线)的

isothiazole[ˌaisə'θaiəzəul]n. 异噻唑

isothiazolyl[ˌaisə'θaiəzəulil]n. 异噻唑基

isotope['aisətəup]n. 同位素

isotopic[ˌaisəu'tɔpik]a.同位素的

item['aitəm]n. 条,项目

J

journal['dʒə:nl]n. 杂志

justifiable['dʒʌstifaiəbl]a.情有可原的

K

Kelvin['kelvin] n.绝对温度(标)

ketone['ki:təun]n. 酮

kieselgu(h)r['ki:zəlguə]n.硅藻土

kilocalorie['kiləˌkæləri]n.大卡,千卡

kinetic[kai'netik]a.动力学的

kinetics[kai'netiks]n.动力学

krypton['kriptɔn]n.Kr氪

L

labile['lei'bail]a.不稳定态的,易变的

laboratory[lə'bɔrətəri]n.实验室

laborious[lə'bɔ:riəs]a.费劲的,吃力的

lack[læk]vt.没有,缺少,需要

lacquer['lækə]n.漆,真漆,清漆,中国漆

lactam['læktæm] n.内酰胺

lactic['læktik]a.乳的

~ acid 乳酸

lactone['læktəun]n. 内酯

laminating resin['læmineitiŋ'rezin]层压树脂

lampblack['læmp'blæk]n.灯黑

lanthanide['lænθənaid]n.镧系

lanthanum['lænθənəm]n.La镧

latent['leitənt]a.潜在的

lattice['lætis]n.格子,点阵

lauch[lɔ:tʃ]vt.发起

laurate['lɔreit]n.月桂酸盐(酯),十二烷酸盐(酯)

lauric['lɔ:rik]acid月桂酸,十二烷酸

lauroyl['lɔ:rəuil]n.月桂酰(基)

lawrencium[lɔ:'rensiəm]n.Lr铹

layer['leiə]n.层

lead[led]n.Pb铅

lead[li:d]n.先导(化合)物,领导,领先

leather['leðə]n.皮革

legal['li:gəl]a.法律上的

lemon['lemən]n.柠檬

lengthen['leŋθən]v.加长,延长,变长

leveling effect['levəliŋ i'fekt]均衡效应

library['laibrəri]n.图书馆,(数据)库

ligand['laigənd]n.配位体,配位基

ligand field['fi:ld]配位场

limitation[ˌlimi'teiʃən]n.限制

limited['limitid]a.有限的,狭窄的

linear combination of atomic orbitals=LCAO
    原子轨道的线性组合

linear polymer['liniə'pɔlimə]线形聚合物

linkage['liŋkidʒ]n.键(合)

linoleum[li'nəuljəm]n.漆布,油毡

lipophilic[ˌlipə'filik] n. 亲油的,亲脂的

liquid['likwid]n.a.液体

literature['litəritʃə]n.文献

liquor['likə]n.液,液体

liquid mirror['mirə] 液体镜

lithium['liθiəm]n.Li锂

litmus['litməs]n. 石蕊

~ paper石蕊试纸

load[ləud]n.v.负荷,负载,装填

locant['ləukənt]n.位次

locate[ləu'keit] v.定位,位于,确定…的位置

location[lau'keiʃən]n.定域

logical['lɔdʒikəl]a.逻辑的,合理的

long-stem funnel['fʌnl]长颈(柄)漏斗

loose[lu:s]a.松散的

loss[lɔ:s]n.损失

lucite['lu:sait]n.(商品名)有机玻璃,聚甲基丙烯酸甲酯

lunar caustic['lu:nə'kɔ:stik](医用)硝酸银

lustron['lʌstrɔn]n.(商品名)聚苯乙烯塑料

lutecium[lju:'ti:siəm]n.Lu镥

luxury['lʌkʃəri]n.奢华,享受,奢侈品

M

macromolecule[ˌmækrə'mɔlikju:l]n.大分子

macroscopic[ˌmækrə'skɔpik]
　　a. 宏观的, 肉眼可见的
magnesium[mæg'ni:ziəm]n.Mg 镁
magnet['mægnit]n. 磁铁
magnetic[mæg'netik]a. 磁性的, 磁的
　　~ field 磁场
　　~ moment['məumənt]磁矩
magnetically[mæg'netikəli]ad. 磁性地, 磁性上
magnitude['mægnitju:d]n. 大小, 数量, 长度
maintain['men'tein]vt. 保持, 维持
maintenance['meintinəns]n. 保持
major['meidʒə]a. 主要的
makeup['meikʌp]n. 化妆品;
　　compositional ~ 组合化妆品, 结构化妆品
maleate[mə'li:it]n. 顺丁烯二酸盐(酯)
maleic[mə'li:ik]acid 顺丁烯二酸, 马来酸
maleoyl[mə'li:əwil,mə'li:əuil]n. 顺丁烯二酰(基)
malleable[mæliəbl]a. 展性的
malonic[mə'lɔnik]acid 丙二酸
malonyl['mælənil]n. 丙二酰(基)
manganate['mæŋgəneit]n. 锰酸盐
manganic[mæŋ'gænik]a. 锰的, 三价锰的
manganous['mæŋgənəs]a. 亚锰的, 二价锰的
manganese['mæŋgəni:z]n.Mn 锰
manipulate[mə'nipjuleit]vt. 操作, 处理, 改造
manipulation[mə,nipju'leiʃən]n. 操作, 操纵
manner['mænə]n. 方式, 方法
marketable['mɑ:kitəbl]a. 适销的, 可销售的
marking['mɑ:kiŋ]n. 识别标志, 刻度, 记号
marking ink['iŋk]不褪色墨水
mask[mɑ:sk]vt. 掩饰, 伪装;n. 面具; 面罩
　　~ ed a. 隐形的, 隐蔽的
mass['mæs]n. 质量
　　~ balance=wight balance 物料平衡
　　~ number 质量数
　　~ spectrometry[spek'trɔmitri]质谱测定法
match[mætʃ]v. 相等, 适合, 一致
material[mə'tiəriəl]n. 物料, 材料
　　~ accounting[ə'kauntiŋ]物料换算
　　~ balance 物料平衡, 物料换算
matrix['meitriks]n. 摇篮, 基质, 矩阵
maze[meiz]n. 迷宫, 迷津
Mcs=megacycles['megə'saikls]per second
　　兆周/秒, 兆赫
meaningful['mi:niŋful]a. 意味深长的, 重要的
measure['meʒə]n. 量度, 标准, 尺度;v. 度量

measurement['meʒəmənt]n. 测量
mechanism['mekənizəm]n. 机理, 历程
medicine droper['drɔpə]医用滴管
medieval[medi'i:vəl]a. 中世纪的
medium['mi:diəm]n. 介质, 中间物
MEK=methyl ethyl ketone 甲乙酮
melt[melt]v. 熔化
melting point['meltiŋ'pɔint]熔点
mendelevium[ˌmendə'liviəm]n.Md 钔
meniscus[mə'niskəs]n. 液面, 弯月面
mercuric[mə'kjuərik]a. 汞的, 二价汞的
mercurous['məkjurəs]a. 亚汞的, 一价汞的
mercury['mə:kjuri]n.Hg 汞
mesityl['mezitil]n.1,3,5－三甲苯基
mesitylene[mi'sitəli:n]n.1,3,5-三甲苯, 对称三甲苯
metabolism[me'tæbəlizəm]n. 新陈代谢
metabolite[mi'tæbəlait]n. 代谢物
metalation[metə'leiʃən]n. 金属化
metallocene['metələsi:n]n. 二茂金属化物
metaloid['metəlɔid]n. 准金属
methacrylic[ˌmeθə'krilik]acid 甲基丙烯酸
methacryloyl[ˌmeθə'krilɔwil]n. 甲基丙烯酰(基)
methane['meθein]n. 甲烷
methanoic[meθə'nəuik]acid 甲酸, 蚁酸
methanol['meθənɔl]n. 甲醇
methine['meθain]n. 甲川
methodology[meθə'dɔlədʒi]n. 方法学, 方法论
methyl['meθil]n. 甲基
methylamine['meθi'læmi:n]n. 甲胺
methylbutylamine['meθil'bju:tə'læmi:n]
　　n.N—甲基丁胺
methyl ester['estə]甲(基)酯
methylol['meθəlɔl]n. 羟甲基, 甲醇基
methyl orange['ɔrindʒ]甲基橙
N-methyl pyrrolidone[pai'rəulədəun]
N-甲基吡咯烷酮
methyl red 甲基红
microfiche['maikrəufi:ʃ]n. 缩微胶片
micron[maikrɔn]n. 微米(μ), 百万分之一米
microscope['maikrəskəup]n. 显微镜
microscopic(al)[maikrəs'kɔpik(əl)]a. 微细的, 微观的
microwave['maikrəweiv]n. 微波
migrate[mai'greit]vi. 移动, 迁移
millisecond['milisekənd]n. 毫秒
mimic['mimik]vt.a. 模拟, 模仿
mineral['minərəl]a. 矿物的, 无机的;n. 矿物

~ water 矿泉水,矿物水
miniaturize['miniətʃə,raiz] vt. 使小(微)型化
　　-ation n.
minimal['miniməl] a. 最小的,最低的
minor['mainə] a. 较小的,较次要的
minute[mai'nju:t] a. 微小的,细小的
miscellaneous[misileinjəs] a. 各种各样的
miscible ['misibl] a. 易混合的,可混溶的
miscibility[,misi'biliti] n. 可混溶性,易混合性
misuse['mis'ju:s] n. 误用
mixture['mikstʃə] n. 混合物
moderatly['mɔdəritli] ad. 中等地,适度地
modification[mɔdifi'keiʃən] n. 变体
modify['mɔdifai] v. 修饰,修改,变更
modification [,mɔdifi'keiʃən] n.
Mohr [mɔə]burette 莫尔滴定管,碱滴定管
Mohr measuring pipette[mɔə'meʒəriŋ pi'pet]
　　莫尔吸量管
moist[mɔist] a. 潮湿的
moisture['mɔistʃə] n. 湿气
molar heat capacity['məulə hi:t kə'pæsiti] 摩尔热容
molarity[məu'læriti] n. 摩尔浓度
mold[məuld] vt. 铸塑(into)
molded article 模制品
mole[məul] n. 克分子,摩尔
molecular[məu'lekjulə] a. 分子的
　　~ -orbital theory-MOT 分子轨道理论
　　~ structure 分子结构
　　~ weight 分子量
molecule ['mɔlikju:l] n. 分子,摩尔
　　~ spectra['spektrə] 分子光谱
mole fraction 摩尔分数
mole weight (克)分子量
molybdenum[mɔ'libdi:nəm] n. Mo 钼
momentum[məu'mentəm] n. 动量
monazite['mɔnəzait] n. 独居石
monatomic[mɔnə'tɔmik] a. 单原子的
monitor['mɔnitə] n. 监视器;v. 监控,检验
monobasic acid 一元酸
monoclinic['mɔnə'klinik] a. 单斜的
monocyclic[,mɔnə'saiklik] a. 单环的
monomer['mɔnəmə] n. 单体
mortar['mɔ:tə] n. 研钵
mortality [mɔ:'tæliti] n. 死亡率
mortar pestle['mɔ:tə pestl] (捣研用的)杵,碾槌
mother liquor['likə] 母液

motion['məuʃən] n. 运动
multiple['mʌltipl] a. 多重的,重复的,多次的
multiplet['mʌltiplit] n. 多重峰
multitude['mʌltitju:d] n. 许多,大量
mutual ['mju:tiuəl] a. 相互的,共同的
mylar['mailə] n. (商品名)涤纶,聚酯薄膜,聚对苯二甲酸乙二醇酯
mythical ['miθikəl] a. 虚构的

N

name[neim] vt. 命名,给……取名
nanoparticle [,nænɔ'pɑ:tikl] n. 纳米粒子
naphthalene['næfθəli:n] n. 萘
naphthol['næfθɔl] n. 萘酚
naphthyl['næfθil] n. 萘基
naphthyridine[næfθə'ridi:n] n, 1,5—二氮杂萘
naphthyridinyl[næfθə'ridinil] n.1,5—二氮杂萘基
natural product 天然产物
needle['ni:dl] n. 针,指针
negative['negətiv] a. 负的,阴的;n. 负片,底片
　　~ ion 负离子,阴离子
neighboring['neibəriŋ] a. 邻近的
neodymium[,ni(:)ə'dimiəm] n. Nd 钕
neopentane[,ni:ə'pentein] n. 新戊烷
neon['ni:ɔn] n. Ne 氖
neptunium[nep'tjuniəm] n. Np 镎
net[net] a. 净的,纯粹的
neutral['nju:trəl] a. 中性的
neutralize['nju:trəlaiz] vt. 中和
neutralization['njutrəlai'zeiʃən] n. 中和
neutron['nju:trɔn] n. 中子
nickel['nikəl] n. Ni 镍
nicotinamide (vitamin B3) [,nikə'tinəmaid] n. 烟碱
niobium[nai'əubiəm] n. Nb 铌
nitrate['naitreit] n. 硝酸盐(酯);v. 硝化
nitration[nai'treiʃən] n. 硝化(作用)
nitric['naitrik] acid 硝酸
nitride['naitraid] n. 氮化物
nitrile['naitrail] (=nitril['naitril]) n. 腈
nitrite['naitrait] n. 亚硝酸盐(酯)
nitrogen['naitridʒən] n. N 氮
p-nitrophenol[naitrə'fi:nɔl] n. 对硝基酚
nitrophenyl['naitrəfenəl] n. 硝基苯基
nitrophenylsulfenyl['sʌlfenil] n. 硝基苯基亚磺酰基
nitrosodethalliation n. 亚硝化脱铊化
nitrosyl ['naitrəsil] n. 亚硝酰(基)

~ chloride 亚硝酰氯
nitrosyl sulphuric[sʌl'fjurik]acid 亚硝酰硫酸
n.m.r=nuclear magnetic resonance 核磁共振
nobelium[nəu'beliəm]n.No 锘
noble['nəubl]a.惰性的
~ gas 惰性气体
nomenclature[nəu'menklətʃə]n.命名
nonacontane[nɔnə'kɔntein]n.九十(碳)烷
nonane['nɔnein]n.壬烷
non-electrolyte[i'lektrəlait]a.非电解质
nonflammable[nɔn'flæməbl]a.不燃的
nonionic[nɔnai'ɔnik]a.非离子的
nonpolar[nɔn'pəulə]a.非极性的
nontoxic[nɔn'tɔksik]a.无毒的
normal['nɔ:məl]a.正常的,当量的
normal ester 等当量酯,中性酯
normality[nɔ:'mæliti]n.当量浓度
normal salt 中性盐,正盐
noticeable['nəutisəbl]a.明显的,显著的
nuclear['nju:kliə]a.核的
nuclear megnetic resonance spectroscopy
['spek'trɔskəpi]核磁共振波谱
nucleic[nju:'kliik]acid 核酸
nucleophile['nju:kliəfail]n.亲核试剂
nucleophilic[,nju:kliə'filik]a.亲核的,亲质子的
nucleophilic agent['eidʒənt]亲核试剂
nucleotide['nju:kliətaid]n.核苷酸
nucleus['nju:kləs](复 nuclei['nju:kliai])
n.核,晶核,环
number['nʌmbə]vt.给……编号
numerator['nju:məreitə]n.(分数中的)分子
numerical[nju:'merikəl]a.数字的,以数字表示的
nylon['nailɔn]n.尼龙,锦纶,聚酰胺

O

obscure[əb'skjuə]a.不清楚的
observable[əb'zɛ:vəbl]a.观察得出的,引人注目的
observation[,ɔbzə(:)'veiʃən]n.观察
obsolesence[,ɔbsə'lesns]n.萎缩,过时
occurrence[ə'kʌrens]n.埋藏量
octacontane[,ɔktə'kɔntein]n.八十(碳)烷
octane['ɔktein]n.辛烷
octet[ɔk'tet]n.八偶
oil-forming['ɔil'fɔ:miŋ]生成石油,成油
oil-resistant rubber['rʌbə]耐油橡胶
oily['ɔili]a.油的,油状的,含油的

olefin['əuləfin]n.烯烃
oleum['əuliəm]n.发烟硫酸
oligonucleotide[,ɔligəu'njukli:ətaid]
n.低(聚)核苷酸
online['ɔnlain]n.联机,在线
open system 敞开体系
opposing[ə'pəuziŋ]n.相反,相对,反抗
optical['ɔptikəl]a.光学的
~ purity 光学纯度
option['ɔpʃən]n.选择
orange['ɔrindʒ]n.a.柑,桔;橙色
orbit['ɔ:bit]n.轨道
order['ɔ:də]n.级数
ordinate['ɔ:dnit,'ɔ:dinit]n.纵坐标
organic[ɔ:'gænik]a.有机的
organism['ɔ:gənizəm]n.有机体,生物体
organolithium[,ɔ:gənə'liθiəm]a.有机锂化合物
organometallics[,ɔ:gənəumi'tæliks]
n.金属有机化学,金属有机化合物
orient['ɔ:riənt]vt.定向,取向
orientation[ɔ:rien'teiʃən]n.方向,定向,取向
origin['ɔ:rədʒin]n.根源,起源;起因
orlon['ɔ:lən]n.(商品名)腈纶,奥纶,聚丙烯腈纤维
ortho-fused[ɔ:θəu-fju:zd]邻位稠(并),单边绸(并)
osmium['ɔzmiəm]n.Os 锇
osmometry[ɔz'mɔmitri]n.渗透压测定(法)
outlet['autlet]n.出口,排水口
outpour[aut'pɔ:]v.泻出,流出
output['autput]n.产量,出料
oven['ʌvn]n.炉,烘箱
overall yield['əuvərɔ:l ji:ld]总产率
overhead['əuvə'hed]n.塔顶馏出物
overlap[əuvə'læp]n.v.重叠
~ integral 重叠积分
oxalate['ɔksəleit]n.草酸盐(酯)
oxalic[ɔk'sælik]acid 草酸
oxalyl[ɔk'sælil,'ɑ:ksəlil]n.草酰(基)
oxathiolane[,ɔksə'θaiəlein]n.噁噻烷,氧杂硫杂环戊烷
oxazole['ɔksəzəul]n.噁唑
oxidation[ɔksi'deiʃən]氧化
~ state 氧化态
~ level 氧化水平,氧化级(数)
~ number 氧化值
~ potential[pə'tenʃəl]氧化电位
oxidative['ɔksideitiv]a.氧化的

oxide['ɔksaid]n.氧化物
oxidize['ɔksidaiz]v.氧化
oxime['ɔksi:m]n.肟
oxirane['ɔksirein]n.环氧乙烷
oxo reaction 羰基合成,羰化反应
oxyanion[,ɔksi'ænaiən]n.含氧阴离子
oxychloride[,ɔksi'klɔraid]n.氧氯化物
oxygen['ɔksidʒən]n.氧
oxygenate[ɔk'sidʒineit]vt.氧化
ozonolysis[auzə'nɔlisis]n.臭氧(分)解(作用)

P

pack[pæk]vt.填充,塞满
paint[paint]n.油漆,涂料
pale blue[peil blu:]淡蓝色
pale yellow['jeləu]浅黄色
palladiation[pə,leidi'eiʃən]n.钯化(作用)
palladium[pə'leidiəm]n.Pd 钯
palmitate['pælmiteit]n.软脂酸盐(酯),十六烷酸盐(酯)
palmitic[pæl'mitik]acid 软脂酸,十六烷酸
palmitoyl[pæl'mitəuil]n.十六碳烷酰
pamphlet['pæmflit]n.小册子
paradoxical[,pærə'dɔksikəl]a.反常的,反论的
paraffin['pærəfin]n.石蜡,链烷烃
parallel['pærəlel]a.平行的,同一方向的
paramagnetic[,pærəmæg'netik]a.顺磁的
parent['pærənt]n.a.母体
~ compound 母体化合物
parenthesis[pə'renθisis](复 parentheses[pə'renθisi:z]n.圆括号
partial pressure['pɑ:ʃəl'preʃə]分压
partition[pɑ:'tiʃən]n.vt.分开,分离,分配
patent['pæ(ei)tent]n.专利
pattern['pætən]n.模型,型式;vt.仿制(on)
peel[pi:l]vi.脱皮,剥落(off)
peer[piə]vi.凝视
penetrate['penitreit]vt.渗入
penicillin [,peni'silin] n.青霉素,盘尼西林
pentacontane[,pentə'kɔntein]n.五十(碳)烷
pentane['pentein]n.戊烷
2,4-pentanedione[,pentein'daiəun]2,4-戊二酮
1,3,5-pentanetricarbaldehyde 1,3,5-戊烷三甲醛
1,3,5-pentanetricarbonitrile 1,3,5-三氰基戊烷
1,3,5-pentanetricarboxylic acid 1,3,5-三羧基戊烷,戊烷-1,3,5-三甲酸

4-penten-1-ol['fɔ:-'penti:n-wʌn-ɔl]4-戊烯-1-醇
peptide['peptaid]n.肽,缩氨酸
perbromate[pə'brəumeit]n.过溴酸钾
perchlorate[pə'klɔ:reit]n.过(高)氯酸盐
perception[pə'sepʃən]n.感性认识,观念
perchloric[pə'klɔ:rik]acid 高氯酸
perfluoroalkyl[pə'fluərə'ælkil]n.全氟代烷基
peri-fused[,peri'fju:zd]迫位稠(产);萘环的[1,8]或[4,5]位稠(并)
perinaphthene n.1,8-(或 4,5)取代萘
periodate[pə'raiədeit]n.高碘酸盐
period['piəriəd]n.周期
periodic[piəri'ɔdik]a.周期的
~ law 周期律
~ table 周期表
periodicity[,piəriə'disiti]n.周期性
permanganate[pə'mæŋgənit]n.高锰酸盐
permanganic[,pə:mæŋ'gænik]acid 高锰酸
permeate['pə:mieit]vt.渗入
peroxide[pə'rɔksaid]n.过氧化物
pervasive[pə:'veisiv]a.扩大的,渗透的
pestle['pesl,'pestl]n.(捣研用的)杵,碾槌
petrochemical[,petrəu'kemikəl]a.石油化学的 n.石油化学品
petroleum[pi'trəuljəm]n.石油
~ ether['i:θə]石油醚
pharmaceuticals[,fɑ:mə'sju:tikəls][复]n.医药品
pharmaceutisch 药物(德文)
phase[feiz]n.相(态)
phenacyl[fi'næsil]n.苯甲酰甲基
phenanthrene[fə'nænθri:n]n.菲
phenanthryl[fə'nænθril]n.菲基
phenol['fi:nɔl]n.酚,苯酚,石炭酸
phenolic[fi'nɔ:lik]a.酚的,苯酚的
phenolphthalein[,fi:nɔl'fθæliin]n.酚酞
phenomenon[fi'nɔminən]n.现象
phenyl['fenil,fenəl]n.苯基
phenylene['fenilin]n.1,4-亚苯基,苯撑
phenyl ethyl alcohol 苯乙醇
phenylmagnesium bromide 溴化苯基镁
phloroglucinol[,flɔ:rə'glu:sinɔl]n.间苯三酚,均苯三酚
pH meter[pi:eitʃ'mi:tə]pH 计(仪)
phosphate['fɔsfeit]n.磷酸盐(酯)
phosphide['fɔsfaid]n.磷化物

phosphite['fɔsfait]n.亚磷酸盐(酯)
phosphoric[fɔs'fɔrik]acid 磷酸
phosphorus['fɔsfərəs]n.P 磷
photochemical[ˌfəutəu'kemikəl]a.光化学的
photolysis[fəu'tɔlisis]n.光解
photolytic[fətɔ'litik]a.光解的
phthalate['fθæleit]n.邻苯二甲酸盐(酯)
phthalazine[f'θælǝziːn]n.2,3—二氮杂萘
phthalazinyl[f'θælǝzinil]n.2,3—二氮杂萘基
phthalic[f'θælik]acid(邻)苯二甲酸
phthalic anhydride[æn'haidraid]邻苯二甲酸酐
phthaloyl[f'θælǝuil,f'θælǝwil]邻苯二甲酰(基)
physical state 物态
physiological[fiziə'lɔdʒikəl]a.生理的
physisorption[fizi'sɔːpʃən]n.物理吸附
pigment['pigmənt]n.颜料,色素
pilot plant['pailət plɑːnt]中间工厂,试验工厂
pinacol['pinəkɔl]n.口片呐醇,2,3—二甲基—2,3—丁二醇
pinch clamp[pintʃ klæmp]弹簧夹
pinchcock['pintʃkɔk]n.节流夹
pipe[paip]n.管道,输送管
pipeline['paip'lain]n.管线
pipette[pi'pet]n.移液管
piston['pistən]n.活塞
~ -like a.活塞的
planimeter[plæ'nimitə]n.面积仪
plastic['plæstik,plɑːstik]a.可塑的,塑性的
n.塑料
~ squeeze bottle[skwiːz'bɔtl]塑料挤压瓶
plate[pleit]n.板,塔板
platinum['plætinəm]n.Pt 铂
plausible['plɔːzəbl]a.似乎合理的,似有道理的
player ['pleiə] n.从业者,演员,表演者
plexiglass['pleksiglɑːs]n.(商品名)有机玻璃,聚甲基丙烯酸甲酯
plot[plɔt]vt.绘图,画出与……关系曲线(against); n.图表,地点
plumba-['plʌmbə-](前缀)铅杂
plutonium[pluː'təunjəm]n.Pu 钚
poisonous['pɔiznəs]a.有毒的
polar['pəulə]a.极性的
polarimetry[ˌpəulə'rimitri]n.旋光测定法
polarisability['pəuləˌraizə'biliti]n.极化率
polarity[pəu'læriti]n.极性
polarizable['pəuləraizəbl]a.可极化的

pollution[pə'luːʃən]n.污染
polonium[pə'ləuniəm]n.Po 钋
polyacrylonitrile[ˌpɔli'ækrilǝu'naitril]n.聚丙烯腈
polyamide[pɔliæmaid]n.聚酰胺
polyatomic[pɔliə'tɔmik]a.多原子的,多元的
polybasic[bɔli'beisik]a.多元的
~ acid 多元酸
polycaprolactam[ˌpɔli'kæprǝu'læktæm]n.聚己内酰胺
polychlorinate[ˌpɔlik'lɔ(ː)rineit]vt.多氯化
polycyclic[ˌpɔli'saiklik]a.稠环的,多环的
n.多环(或稠环)化合物
polyene['pɔliiːn]n.聚烯,多烯
polyethylene[ˌpɔli'eθiliːn]n.聚乙烯
polyethylene glycol['glaikɔl]聚乙二醇
poly(hexamethylene)adipamide
[pɔli'heksə'meθiliːnædi'pæmaid]n.聚-1,6—亚己基己二酰胺,聚己二撑己二胺
polyhexamethylenesebacamide
[pɔliheksə'meθiliːnsibə'kæmaid]n.聚-1,6-亚己基癸二酰胺,聚己二撑癸二酰胺
polymer['pɔlimə]n.聚合体,聚合物
polymer-forming 形成聚合物
polymeric[ˌpɔli'mɛərik]a.聚合的,聚合体的
polymerization[ˌpɔliməraiˈzeiʃən]n.聚合(作用)
polymethyl methacrylate[pɔli'meθil meθə'krileit]聚甲基丙烯酸甲酯
polyolefinic['pɔli'ǝulǝ'finik]a.聚烯(烃)的
polystyrene[ˌpɔli'staiəriːn]n.聚苯乙烯
polyvinyl acetate[pɔli'vainil'æsiteit]聚醋酸乙烯酯
polyvinyl chloride 聚氯乙烯
popular['pɔpjulə]a.普遍的,流行的
population[ˌpɔpju'leiʃən]n.群体;族,组
porcelain['pɔːslin]n.瓷;a.瓷的
porous['pɔːrəs]a.多孔的
porphyrin['pɔːfərin]n.卟啉
pose[pəuz]vt.提出(问题),拿出(要求)
positive['pɔzətiv]a.正的,阳的;n.正片
~ ion 正离子
possession[pə'zeʃən]n.持有,财产
postulate['pɔstjuleit]vt.假定;a.假说
potassium[pə'tæsjəm]n.K 钾
~ acetate['æsitit]醋酸钾
~ bisulfate[bai'sʌlfeit]酸式硫酸钾,硫酸氢钾
~ carbonate['kɑːbənit]碳酸钾
~ ferrocyanide[ferǝu'saiənaid]亚铁氰化钾

~ hexacyanoferrate[heksə-'saiənəu'fereit]六氰合铁(Ⅱ)酸钾

~ tetrafluoroborate[ˌtetrə'flu(ː)ərə'bɔːreit]四氟合硼(Ⅲ)酸钾

potential[pə'tenʃəl]n. 电位,电势

~ energy[enədʒi]位能,势能

potentiostat[pə'tenʃiəˌsteit]n. 恒电位器

potting['pɔtiŋ]a. 陶器制造的

pour[pɔː]v. 注,流出

power[pauə]n. 功率

ppm=parts per million 百万分之……

practice['præktis]n. 实践,操作

praseodymium[ˌpreiziəu'dimiəm]n. Pr 错

preheat['priː'hiːt]vt. 预先加热,预热

precede[priˈsiːd]vt. 先于……,在前,在先

preceding[pri(ː)'siːdiŋ]a. 前面的

precipitate[pri'sipiteit]v. 沉淀

precise[pri'sais]a. 精确的

preclude[pri'kluːd]vt. 排除,消除

preconcentrate[ˌpri'kɔnsentreit]n. 预(先)浓缩物

precursor[pri(ː)'kəːsə]n. 前体

predecessor[ˈpriːdisesə]n. 前辈

preliminary[pri'liminəri]a.n. 初步,开端,开始

prerequisite[ˌpriː'rekwizit]a.n. 先决条件

prescribe[pris'kraib]v. 规定,指定

press[pres]vt. 压,挤压

pressure['preʃə]n. 压力

presumably[pri'zjuːməbli]ad. 推测,可能

pretreatment[pri'triːtmənt]n. 预处理

primary['praiməri]a. 第一的,伯的

principal quantum number 主量子数

prior-art ['praiəɑːt]优先技巧,优先策略

priority[prai'ɔriti]n. 优先(顺序),优先权

probability[ˌprɔbə'biliti]n. 或然率,概率

procedure[prə'siːdʒə]n. 步骤,方法,过程

proceed[prə'siːd]vi. 进行,继续进行,开始

process unit['prəusesˈjuːnit]工艺设备

product[prɔ'dəkt-'dʌkt]n. 产物,生成物;积,乘积

profound[prə'faund]a. 意义深远的

prolific[prə'lifik]a. 富有成果的,丰富的

promethium[prə'miːθiəm]n. Pm 钷

promote[prə'məut]vt. 引发,助长,促进

prone[prəun]a. 有……之倾向,易于……的,习惯于

propagate['prɔpəgeit]vt. 传播,增长,传达

propagation[ˌprɔpə'geiʃən]传播,传递

propane['prəupein]a. 丙烷

propanoic[ˌprəupə'nəuik]acid 丙酸

2-propanol['prəupənɔl]a. 2—丙酸

1-propenyl['prəupənil]n. 1—丙烯基

properly['prɔpəli]ad. 正当地,精确地

property['prɔpəti]n. 性质,特性

propionaldehyde[ˌprəupiə'nældəhaid]n. 丙醛

propionic[ˌprəupi'ɔnik]acid 丙酸

propionyl['prəupiənəl]n. 丙酰(基)

proposition[ˌprɔpə'zeiʃən]n. 意见,见解

propylene['prəupiliːn]n. 丙烯

propylene (di)chloride 1,2—二氯丙烷,二氯化丙烯

propyl['prəupəl, 'prəupil]n. 丙基

propynyl['prəupainil, 'prəupainl]丙炔基

prospective[prə'spektiv]a. 预期的

protactinium[ˌprəutæk'tiniəm]n. Pa 镤

protective[prə'tektiv]a. 保护的

~ group 保护基

proton['prəutɔn]n. 质子

~ magnetic resonance 质子磁共振

prove[pruːv]vt. 证明,证实

provide[prə'vaid]vt. 供给,提供

provision[prə'viʒən]n. 规定,供应

p.s.i.g=pounds per square inch gauge['paundspəːskwɛə intʃ geidʒ]磅/平方英寸

pucker['pʌkə]v. 折叠,缩拢

pump[pʌmp]n. 泵;v. 用泵打,用泵抽

purchase['pəːtʃəs]n.vt. 购买

pure[pjuə]a. 纯的

purification[ˌpjuərifi'keiʃən]n. 提纯

purify['pjuərifai]vt. 纯化,净化,清净

purine['pjuəriːn]n. 嘌呤

purinyl['pjuərinil]n. 嘌呤基

pursue[pə'sjuː]vt. 追踪,跟踪

pursuit[pə'sjuːt]n. 工作

push[puʃ]vt. 推,斥

pyran['pairən]n. 吡喃

pyranyl['pairənil]n. 吡喃基

pyrazine['pirəziːn]n. 吡嗪

pyrazinyl['pirəzinil]n. 吡嗪基

pyrazole['pairəzəul]n. 吡唑

pyrazolyl['pairəzəulil]n. 吡唑基

pyridazine[pairi'dæziːn]n. 哒嗪

pyridazinyl[pairi'dæzinil]n. 哒嗪基

pyridine['piridiːn]n. 吡啶

pyridoxol (vitamin B6)[ˌpiri'dɔksɔl]n. 吡哆醇

pyridyl['piridil]n. 吡啶基

pyrimidine[pai'rimidi:n]n.嘧啶
pyrimidinyl[pai'rimidinil]n.嘧啶基
pyrocatechol[,pairə'kætəkəul]邻苯二酚,儿茶酚
pyrogallic[,pairəu'gælik]acid 焦性没食子酸
pyrogallol[pairə'gælɔl]n.连苯二酚,1,2,3—苯三酚
pyrolysis[pai'rɔlisis]n.热解,高温分解
pyrolytic[paiərə'litik]a.热解的,高温分解的
pyrrole[pi'rəul]n.吡咯
pyrrolyl[pi'rəulil]n.吡咯基

Q

quadruple['kwɔdrupl]v.n.a.ad.四倍,四重
qualitative['kwɔlitətiv]a.定性的
quantify['kwɔntifai]vt.定理,定……之量
quantitative['kwɔntitətiv]a.定量的
quantity['kwɔntiti]n.量,数量,定量
quantum['kwɔntəm]n.量子,量
　　~ mechanics 量子力学
　　~ theory 量子论
quench[kwentʃ]vt.骤冷,淬火
quinazoline[kwinə'zəuli:n]n.1,3—二氮杂萘
quinazolinyl[kwinə'zəulinil]n.1,3—二氮杂萘基
quinoline['kwinəli:n]n.喹啉
quinolyl['kwinəlil]n.喹啉基
quinolizine[kwi'nɔlizi:n]n.喹嗪
quinolizinyl[kwi'nɔlizinil]n.喹嗪基
quinone[kwi'nəun]n.苯醌;(词尾)醌
quinoxaline[kwi'nɔksəli:n]n.1,4—二氮杂萘
quinoxaliny1[kwinɔk'sælinil]n.1,4—二氮杂萘基

R

radar['reidə]n.雷达
radiation[,reidi'eiʃən]n.辐射,照射
radical['rædikəl]a.n.基,根,原子团
radioactive[reidiəu'æktiv]a.放射性的
radioactivity['reidiəuæk'tiviti]n.放射性
radio-frequency['fri:kwənsi]射频,无线电频率
radium['reidiəm]n.Ra 镭
radius['reidiəs](复 radii[reidiai])n.半径
radon['reidɔn]n.Rn 氡
random['rændəm]n.a.随意,任意
Raoult's law 拉乌尔定律
rare-earth[rɛə ə:θ]稀土
rare earth element['elimənt]稀土元素
rare-rare earth　稀有稀土

rarity['rɛəriti]n.罕有,稀有
rate law　速度定律
raw material 原料
react[ri(:)'ækt]vi.反应(with,on)
reactant[ri'æktənt]n.反应物
reaction[ri(:)'ækʃən]n.反应
reactive[ri'æktiv]a.活泼的,反应的
reactivity[,ri(:)æk'tiviti]n.反应性
reactor[ri'æktə]n.反应堆,反应器
reading['ri:diŋ]n.(仪器的)读数
reagent[ri(:)'eidʒənt]n.试剂,反应物
reality[ri(:)'æliti]n.真实
realm[relm]n.领域
rearrange[ri:ə'reindʒ]vt.重排,重新排列
reassure[,ri:ə'ʃuə]v.使安心
recall[ri'kɔ:l]vt.想起,忆起,恢复
record[ri'kɔ:d]v.记录
['rekɔ:d]n.记录,报告
recorder[ri'dɔ:də]n.记录器(仪)
recover[ri'kʌvə]v.回收
recovery n.回收,(回)收率
recrystallization['ri:kristəlai'zeiʃən]n.再结晶,重结晶
recrystallize[ri:'kristəlaiz]v.再结晶,重结晶
recyclability[ri:,saiklə'biliti:]n.可再循环性
recycle[ri:'saik]vt.n.再循环
red-brown　红棕色
reddish-brown　浅红棕色
red-shift['red ʃift]n.红移,红向移动
reduce[ri'dju:s]vt.n.还原
reducing agent[ri'dju:siŋ'eidʒənt]还原剂
reducing bushing[buʃiŋ]减压套管,减压衬圈
reduction[ri'dʌkʃən]n.还原
　　~ potential[pə'tenʃəl]还原电位
reference['refrəns]n.参考,参考书目,参考文献
refinement[ri'fainmənt]n.改良,精制
reflux['ri:flʌks]n.回流
　　~ ratio['reifiəu]回流比
refractory[ri'fræktəri]n.耐火材料
refrigerant[ri'fridʒərənt]n.致冷剂
regenerate[ri'dʒənəreit]v.再生
region['ri:dʒən]n.区域,范围
register['redʒistə]vt.指示
regular['regjulə]a.正规的,正式的
regulation[,regju'leiʃən]n.规则,规程,条例
reinforce[,ri:in'fɔ:s]vt.增强,加强
relationship[ri:'leiʃənʃip]n.关系,联系

relative['relətiv]a.有关的,相对的
relay['riːlei]vt.传递
release[ri'liːs]vt.n.释放
relevant['relivənt]a.有关的,中肯的
reliable[ri'laiəbl]a.可靠的,可信赖的
reliability[riˌlaiə'biliti]n.可靠性
relief[ri'liːf]n.减轻,解除,释放
reluctance[ri'lʌktəns]n.(to,at)抵抗
remainder[ri'meində]n.残留物,剩余物
removal[ri'muːvəl]n.除掉,排除
remove[ri'muːv]vt.除去,移去
render['rendə]vt.致使,使成,使变成
replace[ri(ː)'pleis]vt.取代,以……代替(with)
representation[ˌreprizen'teiʃən]n.表示,描述,表象,代表,象征
reproducible[ˌriːprə'djuːsəbl]a.能再现的,可再生的,可复制的
repulsion[ri'pʌlʃən]n.推斥
requisite['rekwizit]a.需要的,必要的
research and development(产品等的)研究与开发
resemble[ri'zembl]vt.类似,像
reservoir['rezəvwaː]n.贮存器,水库,蓄水池
residue['rezidjuː]n.釜渣,残留物,残液
resin['rezin]n.树脂
resolution[ˌrezə'ljuːʃən]n.溶解;分辨率
resolving power[ri'zɔlviŋ pauə]分辨率
resonance['rezənəns]n.共振,中介现象
resorcinol[re'zɔːsinɔl rez'ɔːsinɔl]n.间苯二酚
resort[ri'zɔːt]vi.求助
response[ris'pɔns]n.感应,反应,回答
restriction[ris'trikʃən]n.限制
resuliting[ri'zʌltiŋ]a.生成的,得到的
retain[riː'tein]vt.保持,保留
retention[ri'tenʃən]n.保留,保持
  ~ time保留时间
retrieval[ri'triːvəl]n.(可)修补
retrospective[ˌretrəus'pektiv]a.追溯的
reveal[ri'viːl]vt.呈现,展现,揭示
reverse[ri'vəːs]v.n.a.逆转,颠倒,可逆
  ~ reaction逆反应
review[ri'vjuː]n.综述,述评
reward[ri'wɔːd]v.值得做
rhenium[ri:niəm]n.Re铼
rhodium['rəudiəm]n.Rh铑
rhombic['rɔmːbik]a.斜方(形)的
riboflavin (vitamin B2)[ˌraibəu'fleivin]n.核黄素

rigid foam['ridʒid'fəum]硬泡沫(塑料)
rigorous['rigərəs]a.严格的,精确的,激烈的
ring[riŋ]n.环,(铁)圈
  ~ clamp[klæmp]铁环夹
  ~ stand铁架台,铁环架
rinse[rins]vt.清洗,冲洗,漂洗
RNA = ribonucleic acid 核糖核酸
roast[rəust]vt.焙烧
robotics[rəu'bɔtiks]n.机器人技术,计算机技术
rod[rɔd]n.细长棒
roll[rəul]vt.滚,转,辗,滚压平
  n.卷状物
room-temperature 室温,常温
  ~ ionic liquid (RTIL)室温离子液
rot[rɔt]vt.完全破坏,腐烂
rotational[rəu'teiʃənəl]a.转动的,旋光的
round bottomed flask[raund'bɔtəmd flɑːsk]圆底烧瓶
route[ruːt]n.路线,途径
  vt.定路线;输送(through)
routine[ruː'tiːn]n.常规,例行手续
row[rəu]n.排,横列
rubber['rʌbə]n.橡胶
  ~ rubber pipette bulb[bʌlb] 橡皮吸球
rubbery['rʌbəri]a.橡胶状的
rubidium[ruː'bidiəm]n.Rb铷
ruin[ruin]vt.破坏,毁灭
rule[ruːl]n.规律
rupture['rʌptʃə]n.裂解,裂开,断裂
rust[rʌst]v.生锈;n.铁锈
ruthenium[ruː'θiːniəm]n.Ru钌
ruthenocene[ruː'θiːnəsiːn]n.二茂钌
rutile['ruːtail]n.金红石,氧化钛

S

saliant['seiliənt]a.显著的,突出的
salt bridge 盐桥
saltlike['sɔːlt'laik]a.类似盐的,盐状的
samarium[sə'mɛəriəm]n.Sm钐
sample['sɑːmpl]n.试样
sandpaper['sændˌpeipə]n.砂纸
sandwich-type['sænwidʒ-taip]n.夹心式的,三明治型的
saturate['sætʃəreit]vt.a.饱和,浸透
scaffold ['skæfəld] n.脚手架
scale[skeil]n.大小,规模,尺度,阶
scan[skæn]vt.n.扫描

scandium['skændiəm]n.Sc 钪
scarce[skɛəs]a.稀有的,不足的
　~ element 稀有元素
scater['skætə]vt.散射;n.散布
schematic[ski'mætik]a.图解的,示意性的
　~ diagram['daiəgræm]示意图
scheme[ski:m]n.图表,图式,方式
school of thought 学派
scrape[skreip]vt.刮落,擦,掏
screen[skri:n]vt.筛选
　　　　　n.筛子,掩蔽物,屏风
screw[skru:] clamp 螺旋夹
seal[si:l]vt.密封
sea level 海平面,海拔
sebacoyl[si'bæsəuil]n.癸二酰
sec.=secondary['sekəndəri]a.第二的,仲的
second-order spin-spin splitting 二级自旋—自旋裂分
section['sekʃən]n.断面,部分
secular['sekjulə]equation 久期方程
seed[si:d]n.晶种,种子
　~ crystal 晶种
segment['segmənt]vt.分割
selectivity[silek'tiviti]n.选择性
selenium[si'li:niəm]n.Se 硒
sense[sens]vt.(自动)检测
　　　　　n.感觉,器官
sensitive['sensitiv]a.敏感的,灵感的
　~ film 感光胶片
　~ paper 感光纸
sensitivity[ˌsensi'tiviti]n.灵敏度
seperate['sepəreit]v.分离
seperator['sepəreitə]n.分离器
seperatory funnel['sepərətəri'fʌnl]分液漏斗
sequence['si:kwəns]n.连续,继续,顺序,结果
series['siəri:z]n.系列,组
sever['sevə]vi.断裂,裂开
shapeless['ʃeiplis]a.不定形的,无形状的
share [ʃɛə]v.共享,分享
sheet[ʃi:t]n.纸张,薄板,板材
　　　　　v.覆盖,铺开
shell [ʃel]n.壳,外壳
shield[ʃi:ld]v.屏蔽,防御
shift[ʃift]n.vt.移动,位移,变动
shock [ʃɔk]n.冲击
side chain 侧链,支链
side reaction 副反应

signal['signəl]n.信号
significantly[sig'nifikəntli]ad.值得注目地
silane['silein]n.硅烷
silica['silikə]n.硅石,二氧化硅
silicate['silikit]n.硅酸盐(酯)
silicic[si'lisik]a.(含)硅的
　~ acid 硅酸
silicon['silikən]n.Si 硅
silicone(s)['silikəun]n.(聚)硅氧烷,(聚)硅酮
silicone rubber 硅橡胶
silolene['siləuli:n]n.硅杂啉,硅杂环戊-3-烯
silver['silvə]n.Ag 银
silylethoxy['silil'iθɔksi]n.甲硅烷基乙氧基
simple molecular orbital theory 简单分子轨道理论
simultaneously[ˌsiməl'teinjəsli]ad.同时地
sine qua non['saini kwei'nɔn]必要条件
single kind　同一类
sink[siŋk]n.污水槽
situation[ˌsitju'eiʃən]n.位置,情况
size[saiz]n.大小,尺寸
skill[skil]n.技能,技巧;技术人员
skillfully['skilfəli]ad.熟练地,巧妙地
slippery['slipəri]a.滑的
smokestack['sməuk'stæk]n.烟囱
soak[səuk]v.n.浸泡,浸渍
soapy['səupi]a.肥皂般的,滑腻的
sodium['səudjəm]n.Na 钠
　~ carbonate 碳酸钠
　~ chloroacetate[ˌklɔ:rə'æsitit]氯乙酸钠
　~ chloride 氯化钠
　~ heptanoate[heptə'noueit]庚酸钠
　~ hexafluoroaluminateⅢ
　　['heksə'flu(:)ərə ə'lju:mineit]六氟合铝(Ⅲ)酸钠
　~ hydroxide[hai'drɔksaid]氢氧化钠
　~ methanolate[meθə'nɔleit]甲醇钠
　~ methoxide[me'θɔksaid]甲醇钠
　~ phenolate['fi:nəleit]酚钠
　~ polysulfide[pɔli'sʌlfaid]多硫化钠
　~ potassium sulfate 硫酸钾钠
　~ thiosulphate[ˌθaiə'sʌlfeit]硫代硫酸钠
soften ['sɔfn]v.变软,软化
solid['sɔlid]n.a.固体
solubility[ˌsɔlju'biliti]n.溶解度
soluble['sɔljubl]a.可溶的
solute['sɔlju:t]n.溶质;a.溶解的
solution[sə'lju:ʃən]n.溶液;溶解,解释

solvat[ˈsɔlveit]n.溶剂化物;vi.溶剂化
solvation[sɔlˈveiʃən]n.溶剂化(作用)
solve[sɔlv]vt.解释,解答;溶解
solvent[ˈsɔlvənt]n.溶剂;a.溶解的
solvent-free 无溶剂的
sophisticate[səˈfistikeit]vt.使复杂
sorcerer[ˈsɔːsərə]n.魔术师
sort[ˈsɔːt]vt.分级,拣选
sound[saund]a.健全的,充分的
sour[ˈsauə]a.酸的,酸味的;v.变酸
space[speis]vt.隔开,分隔
spatial[ˈspeiʃəl]a.空间的
spatula[ˈspætjulə]n.刮刀,小铲
species[ˈspːʃiːz]n.种,种类,形式
specific[spiˈsifik]heat 比热
specific heat capacity[kəˈpæsiti]比热容
specificity[ˌspesiˈfisiti]n.特异性,专一性
specific radiofrequency energy 比射频能
specific value 比值
spectrometer[spekˈtrɔmitə]n.波谱仪,分光计
spectroscopy[spekˈtrɔskəpi]n.光谱学,光谱研究
spectrum[ˈspktrəm]n.谱,波谱,光谱
Sp.Gr=specific gravity[ˈɡræviti]比重
sphere[sfiə]n.范围,领域,界
spill [spil] n.v.溢出,溅出,流出
spin[spin]n.v.自旋,自转
spin-spin splitting 自旋—自旋裂分
spinning [ˈspiniŋ]a.旋转的;自旋的
split[split]v.裂分,裂开
spontaneous[spɔnˈteinjəs]a.自发的,自然的
spot[spɔt]n.斑点,地点
sprinkle[ˈspriŋkl]vt.喷雾,洒,散布
spur[spəː]vi.驱赶(on)
stable[ˈsteibl]a.稳定的,安定的
stability[stəˈbiliti]n.稳定性,安定性
stainless[ˈsteinlis]a.不锈的
~ steel 不锈钢
standard solution[ˈstændəd səˈljuːʃən]标准溶液
standard taper[ˈteipə]equipment 标准铥度仪器,标准接口仪器
standpipe[ˈstændpaip]n.竖管
stanna-[ˈstænə-](前缀)锡杂
stannic[ˈstænik]a.锡的,四价锡的
stannous[sˈtænəs]a.亚锡的,二价锡的
starch[ˈstɑːtʃ]n.淀粉
starch-iodide paper 碘化物—淀粉试纸

state[steit]v.说明,陈述,认为;
　　　　n.状态
~ function 状态函数
statement[ˈsteitmənt]n.陈述,记载
statistical[stəˈtistikəl]a.统计的,统计学的
~ mechanics[miˈkæniks]统计力学
steadily[ˈstedili]ad.有规则地,均匀地
steady[stedi]state 稳态,守恒状态
steam[stiːm]n.蒸汽
~ bath 蒸汽浴
stearate[ˈstiːəreit]n.硬脂酸盐(酯),十八烷酸盐(酯)
stearic[stiˈærik,ˈstirik]acid 硬脂酸,十八烷酸
stearone[ˈstiərəun]n.十八烷酮
stearoyl[ˈstiərəuil]n.十八碳烷酰
stemless[sˈtemlis]funnel 无茎(柄)漏斗
stepwise[ˈstepwaiz]a.逐渐的,逐步的
stereochemical[ˌsteriəˈkemikəl]a.立体化学的
stereochemistry[ˌsteriəˈkemistri]n.立体化学
stereoisomer[ˌsteriəuˈaisəmə]n.立体异构体,空间异构体
stereoseletivity[ˌsteriəsilekˈtiviti]n.立体选择性
steric[ˈsterik,ˈstiərik]a.立体的,空间的
steroid[ˈsterɔid]n.a.甾族化合物,类固醇
stiff[stif]a.浓粘的,硬的
stimulate[ˈstimjuleit]vt.刺激
stir[stəː]vt.搅拌
stirring rod[ˈstəːriŋ rɔd] 搅拌棒
stock[stɔk]n.原料,材料
stoichiometric[stɔikiːəˈmetrik]a.化学计量的
stoichiometry[stɔikiˈɔmitri]n.化学计量学(法)
stopcock[ˈstɔpkɔk]n.活栓,活塞,龙头
straight-run[streit-rʌn]n.直馏馏分,直馏
straightforward[streitˈfɔːwəd] a.直接的,简单的,易懂的
strain[strein]vt.拉紧,使变形
strategy[ˈstrætidʒi]n.战略,策略,方案
straw[strɔː]yellow 稻草黄色
stream[striːm]n.流体,流,流动
stretch[stretʃ]v.伸长,拉长,伸展
strike[straik]vt.碰撞,攻击
strip[strip]n.长条
stripper[ˈstripə]n.汽提塔
strong acid 强酸
strong alkali[ˈstrɔŋˈælkəlai]强碱
strong base[beis] 强碱
strontium[ˈstrɔnʃiəm]n.Sr 锶

structure['strʌktʃə] n. 结构
styrene['staiəri:n] n. 苯乙烯
styron[stairɔn] n. (商品名)斯蒂龙,肉桂塑料,
  聚苯乙烯塑料
styryl['staiəril] n. 苯乙烯基
subatomic['sʌbə'tɔmik] a. 逊原子的,亚原子的
subject['sʌbdʒekt] vt. 遭受,蒙受;n. 主题,科目
sublethal[sʌb'li:θəl] a. 亚致死(量)的
subscript['sʌbskript] n. 下标(符)
substance['sʌbstəns] n. 物质
substantial[səbs'tænʃəl] a. 实质的,相当多的
substitute['sʌbstitju:t] n. 取代物;v. 取代,代替
substituent[sʌb'stitjuənt] n. 取代基
substitution[sʌbsti'tju:ʃən] n. 取代
substitutive['sʌbstitju:tiv] a. 取代的,取代物的
substrate['sʌbstreit] n. 底物,基质
subtle['sʌtl] a. 巧妙的,精细的
succinic[sək'sinik] acid 丁二酸,琥珀酸
succinyl['sʌksinəl] n. 丁二酰(基)
sucrose['sju:krəus] n. 蔗糖
sulfate['sʌlfeit] n. 硫酸盐
sulfide['sʌlfaid] n. 硫化物
sulifinic[sʌl'finik] acid 亚磺酸
sulfite['sʌlfait] n. 亚硫酸盐
sulfonate['sʌlfəneit] n. 磺酸盐(酯)
sulfonic[sʌl'fɔnik] acid 磺酸
sulfur['sʌlfə] n. S 硫
sulfuric[sʌl'fjuərik] a. 硫的,含硫的
  ~ acid 硫酸
sulfuryl['sʌlfəril] chloride 磺酰氯
sulphonation['sʌlfə'neiʃən] n. 磺化(作用)
superficial['sju:pə'fiʃəl] 表面的; a. 表面的
superposable['sju:pə'pəuzəbl] a. 能重叠的
supersaturate['sju:pə'sætʃəreit] vt. 过饱和
superscript['sju:pəskript] n. 上标符
supersede['sju:pə'si:d] vt. 代替
support[sə'pɔ:t] vt. 支持,附载;n. 载体,担体
surface['sə:fis] n. 表面; a. 表面的
surface-active agent 表面活性剂
surroundings[sə'raundiŋz] n. 环绕
survey[sə:'vei] n. 概况,综览
survive[sə'vaiv] vi. 残存,留存下来
susceptibility[sə,septə'biliti] n. 敏感度,感受性
suspect[səs'pekt] v. 猜想,怀疑
suspend[səs'pend] v. 悬浮
suspicion[səs'piʃən] n. 猜想,怀疑

syllable['siləbl] n. 音节
symbol['simbəl] n. 符号
symmetrical[si'metrikəl] a. 对称的
symmetry['simitri] n. 对称(性)
synthesis['sinθisis](复-ses[si:z]) n. 合成,综合
synthesize['sinθisaiz] vt. 合成
synthetic[sin'θetik] a. 合成的,综合的
systematic[,sisti'mætik] a. 系统的

T

tabular['tæbjulə] a. 表的
tabulate['tæbjuleit] v. 制成表格
tag[tæg] n. 标签,标记符,名称
tailor['teilə] vt. 裁制,裁剪;n. 裁缝
tall form beaker[tɔ:l fɔ:m'bi:kə] 高形烧杯
tantalum['tæntələm] n. Ta 钽
tap[tæp] water 自来水
target['tɑ:git] n. 靶,目标
tarry['tɑ:ri] a. 焦油状的,柏油状的
taste[teist] n. 味道,气味,味觉
technetium[tek'ni:ʃiəm] n. Tc 锝
technique[tek'ni:k] n. 技术,方法,工艺
Teflon['teflɔn] n. (商品名)聚四氟乙烯
tellurate['teljuəreit] n. 碲酸盐(酯)
telluration[,teljuə'reiʃən] a. 碲酸化
telluric[te'ljuərik] a. 碲的
  ~ acid 碲酸
tellurium[te'ljuəriəm] n. Te 碲
temperature['tempritʃə] n. 温度
temporarily['tempərərili] ad. 临时地
tempt[tempt] vt. 诱,引诱,引起
tendency['tendənsi] n. 倾向,趋势
tensile strength['tensail streŋθ] 抗张强度
tentative['tentətiv] a. 不明确的,无把握的
terbium['tə:biəm] n. Tb 铽
terephthalic[,teref'θælik] acid 对苯二甲酸
terephthaloyl[,teref'θæləuil] n. 对苯二甲酰(基)
term[tə:m] n. 术语,名称;项(比例)
  vt. 把……叫做(称做)
terminal['tə:minl] n. 词尾; a. 末端的
terminology[tə:mi'nɔlədʒ] n. 术语
tert-butyl alcohol['bju:til'ælkɔhɔl] 叔丁醇,第三丁醇
terylene['terili:n] n. (商品名)特丽纶,涤纶,聚对苯
  二甲酸乙二醇酯
test[test] vt. 检验,验证,试验
  ~ tube[tju:b] 试管
  ~ tube brush 试管刷

~ tube holder 试管夹

~ tube rack 试管架

tetraammine copper(Ⅱ) sulfate 硫酸四氨合铜(Ⅱ)

tetraaminedinitrocobalt(Ⅲ)nitrate 硝酸二硝基四氨合钴(Ⅲ)

tetrachloride[ˌtetrəˈklɔːraid]n. 四氯化物

tetracontane[ˌtetrəˈkɔntein]n. 四十(碳)烷

tetradecahydroanthracene=perhydroanthracene 十四氢化蒽 全氢化蒽

tetrafluoroborate[ˌtetrəˈfluərəˈbɔːreit]n. 四氟硼酸盐

tetrahedral[ˌtetrəˈhedrəl]a. 四面体的

tetrahedron[ˌtetrəˈhedrən]n. 四面体

tertahydrofuran[ˈtetrəˈhaidrəˈfjuræn]n. 四氢呋喃

tetrahydropyranyl n. 四氢吡喃基

tetramethylsilane n.=TMS 四甲基硅

tetrazole[ˈtetrəzɔ(əu)l]n. 四氮唑

thallium[ˈθæliəm]n. Tl 铊

theoretical plate 理论塔盘

therapeutic[ˌθerəˈpjuːtik] a. 治疗(学)的; n. 治疗剂,治疗学家

thermo-[ˌθəːməu](前缀)热

thermodynamic[ˈθəːməudaiˈnæmik]a. 热力学的

thermodynamics[ˌθəːməudaiˈnæmiks]n. 热力学

thermoplastic[ˌθəːməuˈplæstik]a. 热塑(性的)

thermosetting[ˌθəməuˈsetiŋ]a. 热固(性)

thermostat[ˈθəːməstæt]n. 恒温器

thiadiazine[θziədaiˈæziːn]n. 噻二吖嗪

thiamin (vitamin B1)[ˈθaiəmiːn]n. 硫胺

Thiele melting point tube 梯勒熔点管

thenyl[ˈθaiənil]n. 噻吩基

thin[θin]a. 薄的,稀薄的;n. 薄层

thiolation[θaiəˈleiʃən]n. 硫醇化

thiophene[ˈθaiəfiːn]n. 噻吩

thiokol[ˈθaiəkɔl]=polysulfide rubber 聚硫橡胶

thiol[θaiəul]n. 硫醇

Thistle[ˈθisl] tube 蓟头漏斗,长梗漏斗

thrive[θraiv]vi. 兴旺,繁荣

thorium[ˈθɔːriəm]n. Th 钍

three-dimensional[diˈmenʃənəl]a. 立体的,体型的,三度的

thulium[ˈθjuːliəm]n. Tm 铥

thymine[ˈθaimiːn] n. 胸腺嘧啶

time-consuming[kənˈsjuːmiŋ]费时,浪费时间

times[taims]n. 乘,倍

tin[tin]n. 锡

tiny[ˈtaini]a. 微小的,极小的

titanium[tiˈteiniəm]a. Ti 钛

titrate[ˈtitreit]v. 滴定

titration[tiˈtreiʃən]n. 滴定

tolerance[ˈtɔlərəns]n. 耐药量,容忍

toluene[ˈtɔljuːiːn]n. 甲苯

~ diisocyanate[-daiˈaisəˈsaiəneit]甲苯二异氰酸酯

toluic[təˈluːik, ˈtɔljuik]acid 甲基苯甲酸

p-toluidine[təˈluːidin]n. 对甲苯胺

tolyl[ˈtɔlil]n. 甲苯基

tosylate[ˈtɔsileit] n. 对甲苯磺酸盐(酯)

total pressure[ˈtəutlˈpreʃə]总压(力)

tough[tʌf]不易磨损的,耐磨的

tower[ˈtauə]n. 塔

toxicity[tɔkˈsisiti]n. 毒性

trace[treis]vt. 探索,考察;n. 微量,痕量

transfer[ˈtrænsfə]n. 转移,转化,转换
[trænsˈfəː]v. 转移,转化,转换

transfer pipette[piˈpet]移液管

transform[trænsˈfɔːm]vt. 使转变

transformation[ˌtrænsfəˈmeiʃən]a. 转移,变化

transistor[trænˈzistə]n. 晶体管

transition[trænˈziʃən]n. 过渡,转移

transmit[trænzmit]vt. 传送,传导

transmittance[trænzˈmitəns]n. 透过率,透射比

transparent[trænsˈpɛərənt]a. 透明的,透过的

tray[trei]塔,盘

~ tower 盘式塔

treat[triːt]vt. 处理

treatise[ˈtriːtiz]n. 论文

tremendous[triˈmendəs]a. 非常重要的,极大的

triacontane[ˌtraiəˈkɔntein]n. 三十(碳)烷

triad[ˈtraiəd]n. 三素组,三个一组,三价元素

trial[ˈtraiəl]n. 试验,考验,努力

triamminetrinitrocobalt(Ⅲ)n. 三硝基三氨合钴(Ⅲ)

triazine[traiˈæziːn]三(吖)嗪,三氮己因

triazole[traiˈæzəul]n. 三唑

tricosane[ˈtraikəsein]n. 二十三(碳)烷

tridecane[traiˈdekein]n. 十三(碳)烷

tridentate[traiˈdenteit] a. 三齿的,三叉的

~ ligand 三齿配体

triethylamine[traieθəlˈæmiːn]n. 三乙胺

triflate[ˈtraifleit]n. 三氟甲基磺酸盐(酯)

trifluoroacetoxylation n. 三氟乙酸化

trigonal[ˈtrigənəl]a. 三角形的

trihalogenoaluminate[traiˈhælədʒənouəˈljuːmineit] n. 三卤代(化)铝酸盐

trimethylacetic acid 三甲基乙酸
trimethylacetonitrile n. 三甲基乙腈
triphenylmethanol n. 三苯甲醇
tripenylphosphine n. 三苯基膦
triple['tripl] a. 三重的
  ~ bond 三键
triplet['triplit] n. 三重峰
tripod['traipɔd] n. 三脚架,三脚支撑物
tritriacontane[trai'traiə'kɔntein] n. 三十三(碳)烷
trityl['traitl] n. 三苯甲基(游离基)
tropylium[trɔ'piliəm] n. 卓鎓离子,环庚三烯芳香型阳离子
tuneable ['tju:nəbl](= tunable) a. 可调整的,可协调的
tungsten['tʌŋsten] n. W 钨
  ~ hexacarbonyl 六羰基(合)钨
turning['tə:niŋ] n. 削屑

U

ultrasound['ʌltrə,saund] n. 超声波
ultra-violet['ʌltrə'vaiəlit] a. 紫外的
unambiguous['ʌnæm'bigjuəs] a. 明确的,不含糊的,清楚的
unbonded ['ʌn'bɔndid] a. 未键合的,自由的
undecane[ʌn'dekein] n. 十一(碳)烷
undecanedial[,ʌndekein'daiəl] n. 十一(碳)烷二醛
undecyl[ʌn'desil] n. 十一(碳)烷基
undergo[,ʌndə'gəu] vt. 遭受,经过
unidentate[,ju:ni'denteit] n.a. 一齿的,一价配位基
uniform['ju:nifɔ:m] a. 单一的,均匀的,一致的
uniquely[ju:'ni:kli] ad. 唯一地,独特地
unit['ju:nit] n. 装置,单元
univalent[ju:ni'veilənt] a. 一价的
universe['ju:nivə:s] n. 宇宙,整体
unknown['ʌn'nəun] a. 未知的;n. 未知物,未知数
  ~ test sample 未知试样
unpaired[ʌn'pɛəd] a. 未配对的
unpredictable['ʌnpri'diktəbl] a. 不可预测的
unshared[ʌn'ʃɛəd] a. 独有的,未共享的
unsupported['ʌnsə'pɔ:tid] a. 未附载的
update[ʌp'deit] vt.n. 现代化
upfield['ʌp'fi:ld] n.ad. 高磁场
uracil['juərəsil] n. 尿嘧啶
uranium[juə'reinjəm] n. U 铀
urethans(es)['juərəθənz] n. 聚氨酯(橡胶)
urine['juərin] n. 尿

usable['ju:zəbl] a. 可用的,有用的
u.v.(=ultraviolet)['ʌltrə'vaiəlit] a. 紫外线的,紫外光谱

V

vacate[və'keit] vt. 空出,腾出
vacuum['vækjuəm] n. 真空
vade mecum ['veidi 'mi:kəm] n. 手册,袖珍指南
valence['væləns,'veiləns] n.(原子)价
  ~ shell 价电子层
valence-bond theory(=VBT) 价键理论
valent['vælənt,'veilənt] a.(原子)价的
valeric[və'lɛərik] acid 戊酸
valerone['vælərəun] n. 戊酮
valeryl[və'lɛəril] n. 戊酰(基)
valve[vælv] n. 阀,真空管
vanadium[və'neidiəm] n. V 钒
Van der Waals force 范德华力
vapor['veipə] n. 蒸汽,汽
vaporisation(=vaporization)[veipərai'zeiʃən] n. 蒸发
vaporize['veipəraiz] v. 汽化
vapo(u)r phase [veipə'feiz] 气相
vapor-phase chromatography[krəumə'tɔgrəfi] 气相色谱(法)
vapor pressure['preʃə] 蒸气压
variable['vɛəriəbl] a. 易变的;n. 变量,参数
  ~ power['pauə] 可变电源,可变动力
variation[,vɛəri'eiʃən] n. 变化
varnishe['vɑ:niʃ] n. 漆清,罩光漆
vat[væt] n. 大盆,桶
vat dye [væt dai] 瓮染料,还原染料
versatile['və:sə tail] a. 多方面的,万能的,易变的
versus['və:səs] …对…,比较
vertex['və:teks] n. 顶点
vertical['və:tikəl] a. 竖的,垂直的
vessel['vesl] n. 器皿,(容)器
veterinary['vetərinəri] a. 兽医的
  ~ product 兽医用产品
viable['vaiəbl] a. 实际的,可实现的
vibration[vai'breiʃən] n. 震动
vibrational[vai'briʃnl] a. 震动的,摆动的
vice versa['vaisi'və:sə] 反之亦然
vigorous['vigərəs] a. 有力的,活泼的,激烈的
vinegar['vinigə] n. 醋
vinyl['vainil] n. 乙烯基
  ~ bromide 溴代乙烯

vinylidene[vaiˈnilidiːn]n.乙烯叉,亚乙烯基
violent[ˈvaiələnt]a.激烈的,猛烈的
virtually[ˈvəːtjuəli]ad.几乎,差不多
virtue[ˈvəːtjuː]n.效力,功效,长处
viscous[ˈviskəs]a.粘(稠)的,胶粘的
viscosity[visˈkɔsiti]n.粘度
visible[ˈvizəbl]a.可见的,显而易见的
vital[ˈvaitl]a.极重要的,不可缺的
viz.(=namely,that is)也就是说,即
VOC (= volatile organic compound)易挥发有机化合物
vol.(=volume)n.册,卷;体积,容积
volatile[ˈvɔlətail]a.易挥发的,挥发性的
volatility[ˌvɔləˈtiliti]n.挥发性,挥发度
volcanic[vɔlˈkænik]a.火山的
volt[vəult,vɔlt]n.伏特
voltage[ˈvəultidʒ,ˈvɔltidʒ]n.电压
voltaic[vɔlˈteiik]a.伏特的
voltameter[vɔlˈtæmitə]n.伏特计,电压表
volume[ˈvɔljum]n.卷,册,体积
volumetric[ˌvɔljuˈmetrik]a.体积的,容量的
～ flask 容量瓶
v.s.(=versus)[ˈvɔːsəs]与…相对,… 对 …
VSEPR(=valence-shell electron pair repulsion theory)价层电子对互斥理论

W

wand[wɔnd]n.棒,棍,杖
warrant[ˈwɔrənt]vt.证明……为正当
wash tank[wɔʃ tæŋk]洗(涤)槽
watch glass 表(面)玻璃
watchword[ˈwɔtʃwəːd]n.口号,格言
waterway [ˈwɔːtəwei] n.水路,排水沟
water-sorting 水(力)拣选,水选
wavelength[ˈweivleŋθ]n.波长
wave function 波函数
wave number　波数
wax[wæks]n.蜡
weak[wiːk]a.弱的,差的

wedge[wedʒ]n.楔形
weigh[wei]vt.称(重)
well-being n.健康
welfare[ˈwelfɛə]n.a.福利
white lead 铅白,碱式碳酸锌
wide-mouth bottle　广口瓶
widespread[ˈwaidspred]a.广泛的,广布的
wind[ˈwaind]wound v.缠绕,盘绕,卷
windup[ˈwaindʌp]a.装有发条的
wing top[wiŋ tɔp]鱼尾灯头
wire-gauze(=wire mesh)(石棉)铁丝网
wise[waiz]a.明智的,考虑周到的,慎重的
withdraw[wiðˈdrɔː]vt.吸引,收回
withdrawal[wiðˈdrɔːəl]n.收回
work energy　静压能
world[wəːld]n.领域
worthwhile[ˈwəːθˈhwail]a.值得的

X

xenon[ˈzenɔn]n.Xe 氙
X-ray diffraction[ˈeksˈrei difˈrækʃən] X-射线衍射
xylene[ˈzailiːn]n.二甲苯
xylyl[ˈzailil]n.二甲苯基

Y

yellow-green 黄绿色
yield[jiːld]v.产生,生成;n.产率,产量
ytterbium[iˈtəːbjəm]n.Yb 镱
yttrium[ˈitriəm]n.Y 钇

Z

zinc[ˈziŋk]n.Zn 锌
zincate[ˈziŋkeit]n.锌酸盐
zinc blend[ˈziŋk ˈblend]闪锌矿
zinc-coated[ˈkəutid]iron 镀锌铁
zirconia[zəːˈkəuniə]n.氧化锆
zirconium[zəːˈkəuniəm]n.Zr 锆